Textile Science and Clothing Technology

Series Editor

Subramanian Senthilkannan Muthu, SgT Group & API, Hong Kong, Kowloon, Hong Kong

This series aims to broadly cover all the aspects related to textiles science and technology and clothing science and technology. Below are the areas fall under the aims and scope of this series, but not limited to: Production and properties of various natural and synthetic fibres; Production and properties of different yarns, fabrics and apparels; Manufacturing aspects of textiles and clothing; Modelling and Simulation aspects related to textiles and clothing; Production and properties of Nonwovens; Evaluation/testing of various properties of textiles and clothing products; Supply chain management of textiles and clothing; Aspects related to Clothing Science such as comfort; Functional aspects and evaluation of textiles; Textile biomaterials and bioengineering; Nano, micro, smart, sport and intelligent textiles; Various aspects of industrial and technical applications of textiles and clothing; Apparel manufacturing and engineering; New developments and applications pertaining to textiles and clothing materials and their manufacturing methods; Textile design aspects; Sustainable fashion and textiles; Green Textiles and Eco-Fashion; Sustainability aspects of textiles and clothing; Environmental assessments of textiles and clothing supply chain; Green Composites; Sustainable Luxury and Sustainable Consumption; Waste Management in Textiles; Sustainability Standards and Green labels; Social and Economic Sustainability of Textiles and Clothing.

More information about this series at http://www.springer.com/series/13111

Miguel Ángel Gardetti ·
Rosa Patricia Larios-Francia
Editors

Sustainable Fashion and Textiles in Latin America

Editors
Miguel Ángel Gardetti
Center for Studies on Sustainable Luxury
Buenos Aires, Argentina

Rosa Patricia Larios-Francia
University of Lima
Lima, Peru

ISSN 2197-9863 ISSN 2197-9871 (electronic)
Textile Science and Clothing Technology
ISBN 978-981-16-1852-9 ISBN 978-981-16-1850-5 (eBook)
https://doi.org/10.1007/978-981-16-1850-5

This Springer imprint is published by the registered company Springer Nature Singapore Pte Ltd.
The registered company address is: 152 Beach Road, #21-01/04 Gateway East, Singapore 189721, Singapore

Preface

Sustainability is one of the issues that has been strengthened after the Covid-19 pandemic, consumers are now more sensitive to the negative actions of different industries on climate change. Thus, the textile and fashion industry, considered one of the most polluting requires designers, entrepreneurs and researchers to develop strategies that lead companies to generate processes and products more environmentally, socially and economically sustainable. The fashion industry in Latin America and the world is growing exponentially. Latin America has great potential because of the wealth of raw materials such as cotton and camelid hair that are recognized worldwide for their high quality, for the talent of its highly qualified labor and for its artisans. All this represents an opportunity to generate a value chain around a sustainable industry.

This book *Sustainable Fashion and Textiles in Latin America* aims to present case studies of initiatives launched in different Latin American countries, scientific articles and literature reviews that are a guide to good practice in sustainability in the textile and fashion industry.

In the first chapter, titled Socioeconomic and Environmental aspects of the production of silk cocoons in the Brazilian sericulture, the authors, Silvia Botoloto, Rodrigo Salvador, Graça Guedes, Eliane Pinheiro, Cassiano Moro Piekarski and Antonio Carlos de Francisco, present a case study of the sericulture value chain in Brazil, and the impact of the production process on the socioeconomic and environmental conditions of Brazilian communities.

The next chapter, Adapting sustainable design assessment tools for local development. Some insights over Argentina's clothing industry by Carola Ruppel; Celina Monacchi and Rocío Belén Canetti, through the development of a case study, show us the development of a self-declaration tool in sustainable design management implemented in Mar del Plata, Argentina by The Sustainable Design Research Group (GIDSu), and that due to its effectiveness could be scalable to other sectors in the Argentine industry.

Next, Elizabeth Davelaar and Marsha A. Dickson in the chapter Nongovernmental Organization Support for Sustainable Artisan Business, through an analysis of case studies of NGO intervention in groups or communities of artisans, show that the relationship with the NGO results in the acquisition of new skills through training, and knowledge of the use of new materials, which will have a positive impact on the existence of demand and the accompaniment of the NGO.

Following this, Sanjoy Debnath, in the chapter titled Flax Fibre Extraction to Textiles and Sustainability: A Holistic Approach, discusses general aspects of the flax value chain worldwide and highlights the opportunity it has to develop in the midst of sustainable processes that make its use possible in different industries.

In the next chapter, Strategic Design for Social Innovation in the Fashion System: the Sustainable Fashion Ecosystem Case, Karine Freire shows how through strategic design processes open collaborative ecosystems can be developed in search of a common good that promotes values towards sustainability.

Then, Sebastian Garcia and Anthony Halog in their chapter titled Pursuing a circular and sustainable textile industry in Latin America, develop an analysis of the textile sector in Latin American countries, showing that a common behavior is the decline of local textile markets by the introduction of Asian products. They also observe a trend of development of artisanal textiles and sustainability programs with the transformation of recycled textile materials.

Later on, Diana Patricia Gómez García, in the Sustainable Latin America Aesthetic, presents a review of the literature on the context in Latin America and the formulation of the concepts of aesthetics, fashion and minimalism and how they are related to give birth to a strategy to design a sustainable aesthetic.

The next chapter, Value embroidery. Design-oriented actions to support Paraguayan crafts for local female self-determination, by Giovanni Maria Conti, presents the results of a collaborative project of Italian institutions with specialized artisan and design institutions in Paraguay, which had as an objective the training of instructors and best practices of crafts in communities, with a vision of sustainable design.

Then, in the chapter entitled Public Policy and Legislation in Sustainable Fashion, Analia Pastran and Evangelina Colli, emphasize the importance of policies and standards that will be the guidelines for sustainable fashion, highlighting that the same industry must take responsibility for generating change.

We continue with Circular business model alternative from pineapple waste for textile fiber rope production: Case Costa Rica, by the authors Roberto Quirós, Esteban Valverde, Luis Torres, Rodrigo Chamorro, Paola Gamboa, Carolina Va'squez, Diego Camacho and Anthony Hallogc, who present an evaluation of the alternative circular business model, from the use of pineapple waste, generating alternative sources of income for agricultural communities.

Subsequently, María Eugenia Correa in the chapter entitled Fashion, design and sustainability. New horizons in the ways of conceiving production processes; presents a new production model in a sustainable logic with principles of ecological and ethical awareness.

In the next chapter, Incorporating consumer perspective into the value creation process in the fashion industry: A path to circularity; Eliane Pinheiro, Rodrigo Salvador, Antonio Carlos de Francisco, Cassiano Moro Piekarski and Anthony

HalogRodrigo, suggest circularity strategies for incorporating the consumer perspective in the creation of value in the fashion industry, developing a data collection work with surveys to 20 companies in Brazil.

In the present chapter, Cultural perspectives for fashion sustainability: Learning from Brazil, Julia Valle-Noronha, Namkyu Chun , through case studies demonstrate the development of sustainable fashion from a cultural perspective of fashion design, from a symbolic, social and material dimension.

The following chapter, Brazilian Organic Cotton Network: sustainable driver for the textile and clothing sector, presented by Larissa Oliveira Duarte, Marenilson Batista da Silva, Maria Amalia da Silva Marques, Barbara Contin, Homero Fonseca Filho and Julia Baruque-Ramos, describes the value chain of organic cotton and highlights the opportunity to generate an industry that is aware of its effect on environmental costs with organic crops of raw materials such as cotton.

In the chapter Sustainable Fashion, Luis Chaves and Shirley Villalobos analyze the relationship between the concepts of fashion and sustainability through a literature review and comparison of different textile fibers that the industry uses, highlighting the importance of establishing indicators that help companies to measure and take appropriate decisions based on their environmental impacts.

Finally, in the chapter entitled, Upcycling as a tool for participatory critical reflection, Lucia Lopez Rodriguez presents success cases of Latin American designers who implemented upcycling, through the intersection of design, art, social projects and education, promoting synergies between actors to generate a transformation in the current fashion system that contribute to awareness and reduction of environmental impact.

Buenos Aires, Argentina
Lima, Peru

Miguel Ángel Gardetti
Rosa Patricia Larios- Francia

Contents

About the Editors

Miguel Ángel Gardetti, Ph.D. founded the Centre for Study of Sustainable Luxury, first initiative of its kind in the world with an academic/research profile. He is also the founder and director of the "Award for Sustainable Luxury in Latin America". For his contributions in this field, he was granted the "Sustainable Leadership Award (academic category)," in February, 2015 in Mumbai (India). He is an active member of the Global Compact in Argentina—which is a United Nations initiative-, and was a member of its governance body—the Board of The Global Compact, Argentine Chapter—for two terms. He was also part of the task force that developed the "Management Responsible Education Principles" of the United Nations Global Compact. This task force was made up of over 55 renowned academics worldwide pertaining to top Business Schools.

Rosa Patricia Larios-Francia, Ph.D. Candidate in Strategic Management with mention in Business Management and Sustainability, by the Consortium of Universities of Peru. Master's Degree in Industrial Engineering, Industrial Engineer, with a specialization diploma in Innovation Management.Research directed to the management of the MSME Sector, Sustainability, textile sector, Cluster and Innovation, Humanitarian Logistics. Founder and director of the Center for Textile Studies and Innovation at the University of Lima. Researcher and University Professor in the career of Industrial Engineering, more than 20 years of experience in the textile and apparel sector; member of the Technical Committees of the National Quality Institute (INACAL): Management and Quality Assurance; Management of research, technological development and Innovation, Management of MSMEs and Textile and Apparel. Member of the Sustainable Fashion Cluster.

Socioeconomic and Environmental Aspects of the Production of Silk Cocoons in the Brazilian Sericulture

Silvia Mara Bortoloto Damasceno Barcelos, Rodrigo Salvador, Graça Guedes, Eliane Pinheiro, Cassiano Moro Piekarski, and Antonio Carlos de Francisco

Abstract Sericulture is extremely important in the socioeconomic context in Brazil. The country is the fifth largest silk producer in the world, and the silk produced in Brazil is internationally recognized for its quality. Nonetheless, sericultural activities have direct impacts on the silkworm rearers (sericulturists) and are also responsible for environmental impacts, caused by the exchanges of energy and material between the natural environment and the technical/technological environment. Based on that, this exploratory and qualitative study aimed to identify the aspects that affect the (i) socioeconomic and (ii) environmental conditions of silk cocoon production in Brazilian sericulture. Two surveys were designed to collect information on the socioeconomic and environmental aspects of sericulture. The data sets collected through surveys were treated using the NVivo software tool. The results present the socioeconomic and environmental aspects of the production of silk cocoons, and what measures could improve these conditions in order to provide a better quality of life for sericulturists and to improve the relationship of sericulture with the environment in order to guarantee continuity of the activity. In general, the results show that sericulture proved to be ahead of many other cultures with regard to socioeconomic aspects, as it is an alternative to the diversification of small rural properties, generating monthly income for families in rural areas. In addition, there is evidence of efforts to make sericultural activities less impacting on the environment.

Keywords Sericulture · Cocoon production · Silk cocoon · Silk · Socioeconomic aspects · Environmental aspects · Bioeconomy

---

S. M. B. D. Barcelos (✉) · E. Pinheiro
Universidade Estadual de Maringá (UEM), Cianorte, Brazil
e-mail: smbdamasceno@uem.br

R. Salvador · C. M. Piekarski · A. C. de Francisco
Universidade Tecnológica Federal do Paraná (UTFPR), Ponta Grossa, Brazil

G. Guedes
Universidade do Minho, Guimarães, Portugal

M. Á. Gardetti and R. P. Larios-Francia (eds.), *Sustainable Fashion and Textiles in Latin America*, Textile Science and Clothing Technology,
https://doi.org/10.1007/978-981-16-1850-5_1

1 Introduction

The fashion and textile industry is extremely important for the socioeconomic context in Brazil, accounting for an average production of 1.2 million tons of textile products, absorbing 1.5 million direct employees, and being the second largest employer in the manufacturing industry [1]. The socioeconomic importance of the textile and fashion chain in Brazil can be seen in the turnover of US$ 48.3 billion, with just over US$ 1.2 billion being the results of the export of almost 665 thousand tons of textile products, only in the first quarter of 2020. In the specific case of the export of textile yarns and fibers, focusing on silk, Brazil exported 144 tons, in the same period, obtaining a turnover of US$ 7,586 million [2].

Focusing on the sericulture (cultivation of silkworms to produce silk), Brazil is the world's fifth largest silk producer [24], where the state of Paraná is responsible for 84% of the country's production, representing 2,533 tons of green cocoons. The silk produced in Brazil is internationally recognized in terms of the quality of fibers and for its sustainable economy [23], making the Brazilian silk greatly demanded worldwide by fashion designers [3]. However, such volume and importance of the production of silk also raise environmental and socioeconomic awareness over the activity.

In the current context of the textile/fashion value chain, where there is a need for a paradigm shift toward less environmental and socioeconomic impacts, silk is a product of increasing importance. Silk is a renewable raw material with low environmental impacts, in addition to its exceptional characteristics related to thermophysiological comfort and touch [7, 19].

The cultivation of mulberry trees and the rearing of silkworms comprise the two main activities performed by the sericulturists (who are the silkworm rearers), mainly done in small rural properties, characterized by family labor, providing them with a monthly income throughout the nine-to-ten months a year, using a small piece of land [30]. The socioeconomic aspects and the relatively low environmental impacts of the activity contribute to Brazil's sustainable development [10]. Sericulture is an agroindustrial activity and requires relatively small investments and simple technology, thus being considered a booster of family businesses in rural areas.

Sericulture is essentially important in improving the quality of life of the sericulturists and their families, in the Brazilian economy, and especially in the economy of Paraná, the state that concentrates the largest production of silk cocoons in the country [11]. Moreover, sericulture can help keep the rural population employed, preventing migration to large cities and guaranteeing paid employment. It requires small investments and provides the raw material for the textile industries. In this sense, Brazilian silk is an agent of economic and social transformation, strengthening small municipalities, cities, and sustainable communities in Brazil [3]. Nevertheless, although seemingly simple, both the cultivation of mulberry trees and the rearing of silkworms require the use of chemicals, which might bring negative impacts to both human health and the environment, on top of requiring from sericulturists long labor-intensive working hours.

Most of the studies addressing sericulture and silk production in the world that can be found in the existing literature address the context of Asian countries. Moreover, existing studies are usually focused on production processes, production costs, and productivity. Therefore, there is a lack of studies on the socioeconomic and environmental aspects of sericulture in the context of the textile chain, especially in Brazil, as many studies that refer to silk and its production processes do not reflect the production reality of Brazilian silk [11].

In light of the aforementioned, this chapter sought to answer the following research question: what aspects of silk cocoon production affect the socioeconomic and environmental conditions of sericulture in Brazil? Therefore, the objective of this chapter is twofold: identify the aspects that affect the (i) socioeconomic and (ii) environmental conditions of the silk cocoon production in Brazilian sericulture.

2 Background

2.1 Global Scenario of Silk Production

As a natural fiber, silk is a fibroin secreted by the silkworm in the form of a continuous filament, which forms the cocoon. This continuous filament is of great commercial importance. Its natural properties make silk unique, besides having natural luster and exceptional characteristics related to thermophysiological comfort and touch [25].

Many types of silkworms have been reported to be used in silk production. There are silkworms that are reared outdoor such as *Antheraea assamensis* (fed on *Persea bombycina* or *Litsea monopetala*, which are non-mulberry trees), and some that are reared indoor, such as *Samia cynthia* (fed on *Ricinus communis*), *Bombyx textor*, and *Bombyx mori* (fed on *Morus* trees) [28, 31, 44]. Moreover, there are five types of silk of commercial importance obtained from different species of silkworms, with the main one being a mulberry-silkworm [9].

The various types of silk have diverse applications in different sectors, such as luxury clothing, haute couture, furniture, upholstery, carpets, sewing, knitting, and embroidery [25]. Silk has excellent mechanical properties when tested as a biomaterial, and has shown to be biodegradable and compatible with cell interaction, in addition to being used in technical applications in filters (of different types), membranes, paper, textiles, leather, in the field of biosensors, and in the medical and biomedical sectors [7]. The use of silk in those and various other areas such as nutrition, cosmetics, biomaterials, bioengineering, pharmaceuticals, automobile manufacturing, home building, crafts, and arts contributes to the increase in the global demand for silk [8].

Silk production is widespread in 60 countries worldwide, 90% of which is concentrated in Asia and, more recently, in Brazil, whose production is 100% of mulberry silk [24]. The greatest silk producers in the world, considering production volume, are China, India, Uzbekistan, Thailand, Vietnam, and Brazil; Brazil being the only producer of silk yarn on a commercial scale in the West.

In the global textile scenario, silk represents only 0.2% of the textile market. However, through sericulture, China employs around 1 million workers, India employs 7.9 million people, Thailand employs 20,000 families, and Brazil generates jobs for 2,310 families. Still in this scenario, the main world consumers of silk are the USA, Italy, Japan, India, France, China, United Kingdom, Switzerland, Germany, United Arab Emirates, Korea, Vietnam, among others [20, 23, 24].

Sericulture is considered extremely important in the fight against poverty since it prevents the migration of families from rural to urban areas, and it is also an employment opportunity for the rural population in several developing countries, Brazil being a good example [36]. Therefore, it is observed that silkworm rearing can be a lucrative activity for rural livelihoods [39].

Silk production also raises political and economic interests around the globe (see [32, 42]), greatly because silk production is concentrated in a few countries. International cooperation efforts have been launched in order to achieve better production and distribution of silk, mainly in Eurasia. Inter and intranational efforts contribute not only to political and economic agreements but also to technology and innovation exchange [42, 49].

Different levels of technicality can be found in different countries and different areas, which contribute to the varied quality of silk produced around the world. This permeates the feed production (silkworm feed, silkworm rearing and cocoon production, and yarn manufacturing). In Brazil, for instance, the level of technicality in the cultivation of mulberry trees (silkworm feed) and cocoon production is majorly supervised by the yarn manufacturer [12]. In other countries, though, this might be left at the sericulturists' discretion, which influences standardization, mainly in outdoor rearing.

Silk production also results in byproducts. The main product of all this process is the silk thread; however, portions of silk that are unreelable or tainted can be used to produce exquisite silk products [12], and the pupae, which is considered edible in a few markets [35].

2.2 *Silk Production in Brazil*

Silk production, as a sericultural activity, was introduced in Brazil in the nineteenth century, specifically in Rio de Janeiro. The first Brazilian silk industry was installed in the municipality of Itaguaí, the Imperial Companhia Seropédica Fluminense [38]. In the twentieth century, sericulture was established in the state of São Paulo, and later in Paraná [15, 38]. However, it was from the 1970s that sericulture began to spread in Paraná through the installation of factories and other investments [13, 15].

Until 1980, the national silk production was concentrated in São Paulo [14]. A few years later, in 1985, Paraná surpassed the production of other states, becoming the largest national producer of green cocoons [34]. The local climate was a factor that favored the rearing of silkworms in the region of Paraná, in addition to incentives from the State government, making resources available to small producers [14].

Paraná faced a 46% drop in 2007, followed by an additional fall of 8% in 2010 in the production of green cocoons [14]. After this period of recession, it recovered due to the resumption of world demand for silk yarn, since most of Paraná's silk production was exported to countries such as Vietnam, France, Italy, and Japan [20].

The region known as the Silk Valley is located in the northwest region, with the highest cocoon production in Paraná, formed by 29 producing municipalities [20]. In addition, projections indicate that Paraná has the potential to increase its production by 50% in the coming years, where the intention is innovating to produce more on smaller properties [4, 17].

Silk production has been thriving in Brazil through its profitability, as a way to settle families in the countryside [22]. Sericulture is an activity that requires human, technical, and material resources and can optimize sustainable local, regional, and even the country's development [16], considering primarily its social aspects, followed by low environmental impacts [33].

Sericulture in Paraná is essentially developed by small and medium-sized farmers with family labor; and with the increasing mechanization of the activity, many producers continue investing in sericulture, due to the reduction of manual labor. The new technologies introduced in the sector aid the development of the activity [22], however, the technologies still need to reach the majority of silkworm rearers.

Brazilian silk production, in the 2016/2017 harvest, reached the figure of 2,968,849.61 kg of cocoons, where Paraná alone represented 83.28% of such production with 2,471,959.16 kg, followed by São Paulo with 349,312.95 kg and Mato Grosso do Sul with 147,577.50 kg, which were sold to the only spinning company in Brazil, Bratac [18].

2.3 Socioeconomic Impacts of the Silk Cocoon Production

Sericulture comprises two main activities, the cultivation of mulberry trees and the production of silk cocoons [30]. In this activity, sericulturists are susceptible to the dynamic international market, where the price of their products is subject to fluctuation and uncertainty, considering that 90% of the production of Brazilian silk is destined for exports [37].

The production of silk cocoons is important in the socioeconomic development of a given location, improving the quality of life of families in small rural properties, in addition to contributing to the generation of jobs, income, and taxes [30]. However, the majority of sericultural activities still comprise mostly manual labor, requiring exhaustive physical effort, in addition to some activities causing harm to workers' health. Among the possible problems, one can mention physical tiredness, body aches, health problems, reduced hours of sleep, psychological pressure, stress, contact with toxic chemicals, long working hours, night shifts, risk of accidents involving fire and snakes, and environmental degradation caused by pesticides and other chemicals [37].

The aspects addressed are related to the cultivation of mulberry trees (the only food for silkworms) and the rearing of silkworms. Sericulturists come into contact with herbicides, pesticides, and fertilizers, in the cultivation of mulberry trees, and chemicals such as formaldehyde, chlorine, and lime, in the rearing of silkworms, for the disinfection and control of pests (sheds and grids), as well as contact with fire in the scorching of the grids.

Socioeconomic impacts that directly affect rearers can be observed in the production of silk cocoons, such as freedom of association and collective bargaining, child labor, working hours, forced labor, equal opportunity/discrimination, health and safety (working environment), social benefits/social security, education, and psychological working conditions.

A few aspects of the socioeconomic impacts in the silk cocoon production have been reported in research on silk production in India (see [29, 40, 41, 43]). In Brazil, existing research is generally focused on genetic improvement, the cost of producing silkworms, and cocoon quality and productivity. However, there are several aspects that contribute to the socioeconomic impacts related to the production of silk cocoons in Brazil, which are later addressed in this study.

2.4 Environmental Impacts of the Silk Cocoon Production

On top of socioeconomic impacts, silk production can have significant environmental impacts, especially for the indoor rearing of silkworms, which requires specific conditions such as the cultivation of mulberry trees, in order to feed the silkworms, as well as materials and workforce for feeding them [12], cleaning and disinfecting facilities, on top of specific climate conditions [48] for the development of silkworms.

Impacts can be observed on the environment, such as land use, toxicity (human and ecological—both to water and land), eutrophication (both to water and land), climate change, and human health. Environmental impacts can be considered as the effects on the environment caused by energy and material exchanges between the natural environment (biosphere) and the technical/technological environment (technosphere). These inputs and outputs, from and to the environment, respectively, are environmental aspects. Inputs encompass (e.g.) water, raw materials, and energy. Outputs can be finished products, byproducts, and wastes, including emissions to air, water, and soil.

Inputs and outputs related to environmental aspects have been reported in existing research on the production of silk cocoons, in India [5, 6, 48] and Brazil [12, 22]. Aspects reported in existing research include general statements on their contribution to the impacts of cocoon production. The most complete research on the impacts of silk cocoon production in Brazil thus far seems to be the research of Barcelos et al. [12], who conducted a life cycle assessment (LCA) of the mulberry and cocoon production under Brazilian conditions.

The aspects commonly addressed in existing research are related to the use of mulberry leaves (including, e.g., fertilizers, pesticides, etc.) and the use of chemicals for disinfecting the rearing beds and the sheds. However, there are many more aspects that contribute to the impacts inherent to the production of silk cocoons. Therefore, a mapping of such aspects will be provided later in this chapter.

3 Methods

This study is exploratory and qualitative. Two surveys were designed to collect information on the socioeconomic and environmental aspects of the sericulture in the Silk Valley (a region comprising 29 municipalities) in Brazil.

The information on the socioeconomic and environmental aspects were collected between February and August 2017, and between November 2017 and April 2018, in 15 municipalities within the Silk Valley, being them: Terra Boa, Cianorte, Santa Isabel do Ivaí, Cidade Gaúcha, Tapira, Tuneiras do Oeste, Indianópolis, Rondon, Doutor Camargo, São Manoel do Paraná, Ivaté, Aparecida d'Oeste, Nova Olímpia, Nova Esperança, and Alto Paraná.

The survey on the socioeconomic aspects was answered by 69 sericulturists (silk cocoon producers), and the data collection was given via a structured questionnaire. To build the socioeconomic survey, the aspects in the guidance for assessing and managing the social impacts of projects [47], EVALSED: The resource for the evaluation of Socio-Economic Development [21], guidelines for social life cycle assessment of products [45], and the methodological sheets for subcategories in social life cycle assessment (S-LCA) were considered [46]. Taking into consideration that some sericulturists were illiterate, in order to follow the same pattern all questionnaires were filled by the researchers during face-to-face meetings with the sericulturists in each property. Besides this survey, interviews (based on a structured guide) were conducted with seven focus groups. All interviews were recorded and transcribed, and later transferred into the NVivo 12 Plus software tool for data treatment.

The survey on the environmental aspects was answered by 43 sericulturists. The data collection was given via a structured questionnaire. To build the environmental survey, the researchers used as guides the standards on life cycle assessment (LCA) ISO 14040 [26] and ISO 14044 [27]. Once again, taking into consideration that some sericulturists were illiterate, in order to follow the same pattern all questionnaires were filled by the researchers during face-to-face meetings with the sericulturists in each property. The data gathered was used to map the processes and related environmental aspects (inputs and outputs) of the cultivation of mulberry trees and the rearing of silkworms. The processes were modeled using the software tool Microsoft Visio to identify the physical flows.

4 Socioeconomic and Environmental Aspects of the Production of Silk Cocoons in Brazil

4.1 Socioeconomic Aspects

The socioeconomic aspects present in the production of silk cocoons contribute to several potential impacts, which are directly linked to the sericulturists. This section presents the main socioeconomic aspects concerning the production of silk cocoons in Brazil.

The production of silk cocoons incorporates two main activities, the cultivation of mulberry trees and the rearing of silkworms, which besides being arduous tasks require great physical effort and dedication from the silkworm rearers. In addition, sericulture has some characteristics that are specific to the activity, which can be seen in Fig. 1.

A few characteristics of sericulture overclass those of other similar activities. In sericulture, sericulturists have the advantage of counting on a monthly income, with greater profitability in a relatively small area of land, besides having autonomy over their work routine. Moreover, they also have the possibility of working with other parallel activities on the same property.

A few opportunities brought by sericulture also need to be highlighted, such as the expansion of labor through the entry of new workers, and the family succession in the business, in which sericulturists' sons and daughters carry on the activity previously performed by their parents. The sericultural activity also allows the establishment of sericultural associations and presents a vast field for research.

Nonetheless, sericulture also presents some disadvantages such as the off-season, in which producers spend an average of three months without income, due to zero production in winter. In such cases, they need to supplement their income seeking to cover the family's livelihood needs. Sericulturists also face difficulties regarding production costs, since a great part of their revenue goes to covering such costs. In addition, they feel the need for greater governmental assistance. Further challenges in their activities are long working hours, bad weather during leaf harvesting, and the risk of accidents involving fire and snakes, on top of little freedom, as they have almost no spare time available.

The production of silk cocoons is also faced with threats, such as diseases and pests that affect silkworms, climate instability that can affect the mulberry fields, and application of pesticides by large producers of other cultures. On top of those, sericulturists may also suffer from the lack of labor force and economic instability.

The main aspects addressed, as well as the categories and subcategories of socioeconomic impact, based on Unep guidelines [46], can be seen in Fig. 2.

In order to obtain a silk production in which socioeconomic aspects lead to minimum impacts, it is necessary to provide favorable conditions for those involved in the activity.

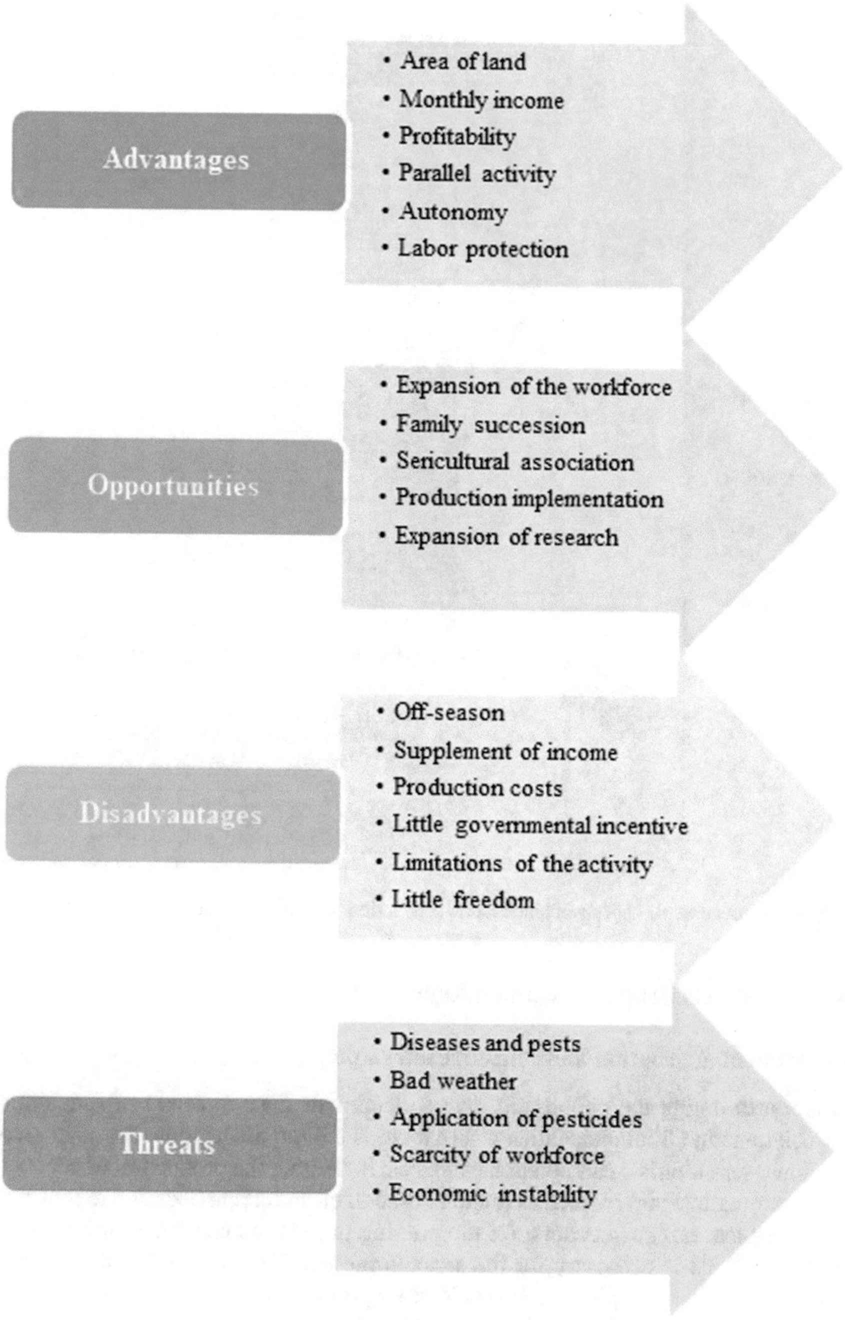

Advantages

- Area of land
- Monthly income
- Profitability
- Parallel activity
- Autonomy
- Labor protection

Opportunities

- Expansion of the workforce
- Family succession
- Sericultural association
- Production implementation
- Expansion of research

Disadvantages

- Off-season
- Supplement of income
- Production costs
- Little governmental incentive
- Limitations of the activity
- Little freedom

Threats

- Diseases and pests
- Bad weather
- Application of pesticides
- Scarcity of workforce
- Economic instability

Fig. 1 Characteristics of the sericulture in Brazil

Fig. 2 Socioeconomic aspects of silk cocoon production in rural properties in Brazil

4.1.1 Aspects Related to Human Rights

Freedom of association and collective bargaining

It is worth noting the opportunity that sericulturists have to establish sericultural associations in all municipalities where there is still no association. Through associations, sericulturists can obtain advantages, benefiting the production of cocoons, by obtaining machinery, such as tractors, used to clean rearing beds and to transport mulberry leaves; brush cutters, for the pruning of the mulberry trees, among other equipment that is necessary for the sericultural activities. That way, sericulturists with less economic capital can invest in other production needs.

There is also a need to develop cooperativism among sericulturists, in which it is essential for associations to be established and prosper, as well as the exchange of experiences and the establishment of partnerships. Through associations, several actions with socioeconomic benefits can be planned and implemented by the sericulturists themselves.

Child labor

In the studied region, the presence of child labor in sericultural activities was not observed. All children at school age were properly enrolled, according to parents' information. It is noted that only when in their adolescent age they start contributing to the family business, in a moderate way, starting with the feeding of silkworms.

Fair wages

Brazilian silk-related businesses accompany the international market, therefore sericulturists might face positive or negative oscillations in prices [36]. The monthly income of sericulturists coming exclusively from the sericultural activity is satisfactory to meet the basic needs of the family (mainly food); however, those who work as sharecroppers, go through financial hardships. Nonetheless, the profitability of the sericulture, considering the size of the property, is better than that of other types of crops.

Parallel activities are another reality for sericultural families. However, parallel activities should exist as a supplement for families to have an extra income, and not as a complement of income for the fixed monthly expenses.

There is a possibility of producing and selling other products through the sericultural association (creation of cooperatives), along with the production of cocoons, considering that they can develop the capacity to transform second-quality cocoons into textile products for the home, such as curtains, sofa shawls, pillows, among other products. This initiative can gain the attention of children, guaranteeing family succession. Sericulturists can also seek partnerships with other companies and/or projects for the second-quality cocoons, as an option for extra income.

Another possibility is the development of an annual productivity reward system for sericulturists. Considering the partnership system, the company needs sericulturists to obtain the raw material for the production of silk yarns, that is the silk cocoons, and the sericulturists need the yarn manufacturing company to buy their production of cocoons. Without the yarn manufacturer, the sericulturists do not have whom to sell their production to nationally. This award system must be well structured so that it can benefit sericulturists in a fair and equal way.

Moreover, the yarn manufacturer could anticipate the sericulturists' monthly production and dilute the annual payment made to each of them into 12 payments (considering the average of the previous harvest), paying the sericulturist in advance during the three months when there is no production, and discounting the amount paid in advance from the payments in the next nine months. This way, sericulturists can have better living conditions in the off-season.

Equal opportunity/discrimination

In the sericultural activity, no difference was identified between women and men in the price paid for the kilogram of cocoon.

Social benefit/social security

All cocoon production is sold to the yarn manufacturer and a sales invoice is issued to the sericulturist. In this case, the sericulturists are also affiliated with rural unions, being duly supported by social insurance in case they need it.

Education

Most sericulturists have basic literacy. However, shockingly, some are found in a state of illiteracy, that is, they never went to school. Considering children and young people, everyone has been or is being schooled.

4.1.2 Aspects Related to Working Conditions

Working hours

Work is necessary to move toward a sustainable economy, but a balance between working time and free time is also needed. The workload of sericulturists is considerably high because it is usually just the couple who owns the farm who works full-time, thus demanding them a lot of time.

One way to provide more free time would be the mechanization of time-consuming activities, such as automation of the silkworm feeding, taking into account that it is one of the activities that most demands time and physical effort.

Forced labor

There is no presence of forced labor in the sericultural activities in the studied region. Sericulturists work in partnership with the silk yarn manufacturer, by contract, and they can leave the activity at any time they wish. However, there is no such intention on the part of sericulturists, because the current arrangement is reported to be the best option for family farming at the moment.

Health and safety (working environment)

There is no data available on the exposure of sericulturists to chemicals such as formaldehyde, known to be carcinogenic by the International Agency for Research on Cancer (IARC), or to biological material, indicating an important gap in protecting the health of these workers [36].

Any serious accidents during the sericultural activities have not been observed or reported. However, there is a need for certain precautions, such as the use of personal protection equipment (PPE), considering that sericulturists handle chemicals such as herbicides, formaldehyde, chlorine, tools such as machetes, hoes, sickles, and may even encounter snakes in the middle of the mulberry fields.

The possibility of implementing a training program in health and safety at work is noted, which can be developed by the yarn manufacturer and offered to sericulturists through partnerships such as with the Paraná Institute of Technical Assistance and Rural Extension (EMATER—*Instituto Paranaense de Assistência Técnica e Extensão Rural*), the Agronomic Institute of Paraná (IAPAR—*Instituto Agronômico do Paraná*), and municipal governments, which are all already partners, among others.

It is also noticed that the equipment owned by the sericulturists is appropriate for their activities; however, the possibility of improving technicity is noted, expanding the types of equipment, machinery. and automation. One example would be installing air conditioning equipment for the sheds, which can be facilitated by the silk spinning company since there is a financing system for partner sericulturists.

Psychological working conditions

In general, it can be noted that sericulturists are professionally satisfied and are willing to receive training related to sericultural activities, in order to improve their capabilities and also the productivity of the cocoon production.

In this sense, there is a need to develop an intensive production training program, in which sericulturists can sane their doubts, as well as get updates on new production techniques, something that goes beyond the traditional monitoring of the rearers by the spinning company agents.

There is also a need for a program to value the sericulturist, where issues of social nature, collective leisure, lectures, among others, can be addressed in order to provide sericulturists with greater well-being and to emphasize the importance of the sericulturist as the social capital of sericulture.

4.2 Environmental Aspects

The silk cocoon production comprises several environmental aspects, which contribute to varied environmental impacts. This section presents the major environmental aspects embedded in the silk cocoon production in Brazil, taking into perspective model practices.

The major input flows, which contribute to extracting/consuming resources from the environment, and output flows, which contribute to releasing emissions to the environment (to air, soil, or water), are depicted in Fig. 3.

The two main sets of activities that can be observed in the rural properties, for the production of silk cocoons, are the cultivation of the mulberry trees, and the activities related to the rearing of silkworms.

4.2.1 Mulberry Production

Within the mulberry production, the main activities that can be pointed out are soil correction, soil preparation, mulberry planting, and soil maintenance.

Sericulturists usually practice a sort of rotation in the cultivation of mulberry trees in order to assure a year-long supply for feeding the silkworms. They usually have a certain area which is divided into three smaller areas and each one of them comprises one separate cycle of mulberry cultivation. That is given due to the need

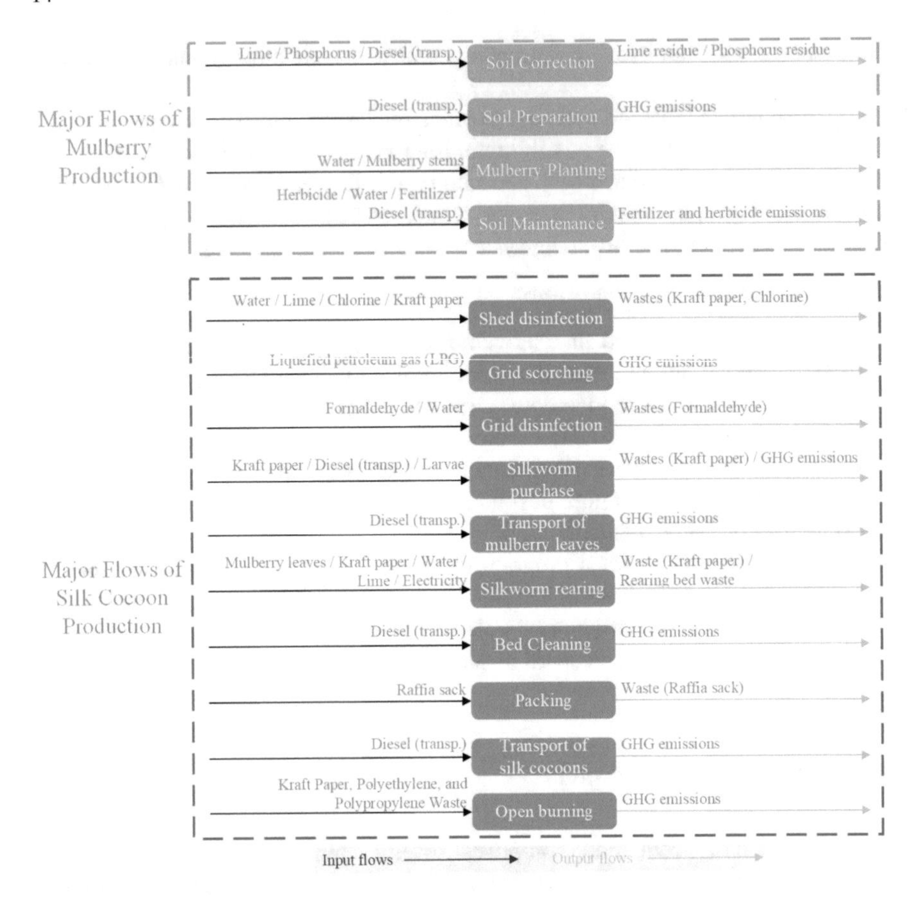

Fig. 3 Major flows for the production of silk cocoons in rural properties in Brazil

for the leaves to have adequate characteristics to be fed to silkworms, otherwise either the silkworms will reject them or they will not provide adequate quality feed. Moreover, it is recommended that every ten years the mulberry fields get renewed, in order to maintain both soil and crop quality.

Soil Correction

Part of soil correction activities takes place once a year, and part of it, every ten years, when the mulberry fields are renewed.

Lime and phosphorus are used in the soil to provide adequate conditions for the production of quality mulberry leaves for silkworms. Their use serves to supplement the soil and is given according to the technical analysis of the soil, or when renewing the mulberry fields.

For the application of both phosphorus and lime, a tractor is used, consuming diesel, accounting for different amounts of diesel consumption in their applications.

With those inputs, even though lime and phosphorus have positive effects on mulberry cultivation, their negative effects might permeate longer. Phosphorus is a nutrient largely used in agriculture, but it might cause eutrophication. As sericulturists use ground wells, this might lead to water contamination, which can be either consumed at the property or carried to other water bodies. Lime, in turn, can be a way to control soil acidity, but in excess can lead to carbonation and sulfate attack.

As outputs for this process, one can observe greenhouse gas (GHG) emissions, due to the consumption of diesel, contributing to air pollution and to aggravating global warming effects.

Soil preparation

This activity takes place only when the mulberry field gets renewed, thus every ten years, according to the technical recommendation. It consists of preparing the soil to receive the new mulberry crop.

Different equipment is used for this activity, as it is observed that the sericulturists have to till the soil. The main input and output flows in this activity account for diesel and GHG emissions, respectively.

Mulberry planting

For mulberry planting, the sericulturists basically use water and mulberry stems.

Water is obtained from ground wells and the mulberry stems are provided by the silk manufacturer. This activity also takes place only once every ten years, when the mulberry fields need to be renewed, or according to the technical recommendation. As this activity is conducted manually, no direct emissions are accounted for.

Soil Maintenance

Maintaining the soil means keeping it free of pests and also fertile. Therefore, the main inputs accounted for in this activity are herbicides and fertilizers, as well as water to dilute them, and diesel for their application on the fields.

Needless to say, herbicides are the causes of negative effects on both the direct environment and on human health. They can have long-lasting effects, permeating soil and water, contributing to concerning levels of toxicity. Besides, they might also reach non-target organisms and cause environmental imbalances in soil biology, even more because the products of their degradation can continue to be toxic in the long term.

Fertilizers, as for phosphorus, can be the cause of eutrophication. Eutrophication comprises a high load of nutrients (e.g. phosphates and nitrates). This mainly concerns the contamination of water bodies, affecting the fauna and flora of these habitats, and causing an imbalance in natural systems. Moreover, fertilizers also carry chemicals that contribute to GHG emissions.

4.2.2 Infrastructure and Other Environmental Aspects of Mulberry Production

Besides the input and output flows for the cultivation of mulberry trees, many of the activities mentioned previously need heavy machinery and equipment to be performed. These infrastructure aspects have not been taken into account and they lack addressing in environmental assessments of silk production systems. Agricultural machinery, such as tractors (as well as passenger cars and trucks used for transportation), also account for environmental impacts for their manufacturing, by extracting natural resources and accounting for varied emissions to air, soil, and water for their manufacturing and distribution.

All packaging systems of the products used for the mulberry cultivation also contribute to environmental impacts, for the same reasons mentioned above, since they require extracting resources from the natural environment, processing/manufacturing, and distributing.

Furthermore, manpower is also an aspect that can contribute to environmental impacts. It requires efforts that result in the consumption of resources (e.g. water and food) that would otherwise not be consumed or used to produce results for other systems. Nonetheless, manpower is (dare we say) never included in the accounting of environmental impacts due to the difficulty in tracking related inputs and outputs.

4.2.3 Silk Cocoon Production

Within the silk cocoon production, the main activities can be pointed to be shed disinfection, grid scorching, grid disinfection, silkworm purchase, transport of mulberry leaves, silkworm rearing, bed cleaning, packing, transport of silk cocoons, and open burning (paper and plastic packaging waste).

Shed disinfection

The sheds need to be disinfected before receiving a new lot of silkworms. It takes a short time between one batch of silk cocoons being delivered to the silk manufacturer and a new batch of silkworms coming into the sheds. In the meantime, the sheds need to be disinfected in order to prepare them for the new silkworms.

Inputs for the disinfection include chlorine, lime, water, and kraft paper. Chlorine and lime are diluted in water and sprayed over the walls, rearing beds, and floors. Besides, lime is dusted over the rearing beds and floors to prevent humidity, which can be hazardous to silkworms, thus preventing the spread of diseases. Moreover, the bottom of the rearing beds is covered with kraft paper, to help protect and maintain temperature, and to facilitate taking out the rearing bed waste.

As outputs, chlorine, kraft paper waste (from the lime package), and kraft paper waste from the bottom of the rearing bed (which gets mixed with silkworm litter and mulberry stems left behind by silkworms) can be observed. The chlorine is likely to evaporate with time, at varied rates depending on (e.g.) climate conditions, thus becoming a hazard to human health. Lime wastes are accumulated and eliminated (together with the kraft paper) along with silkworm litter and mulberry stems when the rearing beds are cleaned.

Grid scorching

When silkworms are ready to form cocoons, they are accommodated in grids. When the cocoons leave the grids, these are scorched, to eliminate the remains of previous cocoons, in order to prevent the proliferation of fungi that might cause diseases.

With adequate equipment, the grids are scorched using liquefied petroleum gas (LPG). Although little, compared to other potential sources, LPG is responsible for GHG emissions.

Grid disinfection

After being scorched, the grids are disinfected. Inputs for this disinfection are formaldehyde and water. Formaldehyde is diluted in water and sprayed over the grids.

As outputs, this results in air emissions of formaldehyde, which can also be hazardous to human health.

Silkworm purchase

The purchase of silkworms includes transportation of what the sericulturists call boxes, which are small packages with approximately 40,000 larvae each, a piece of kraft paper (where the larvae come wrapped in), and a negligible amount of mulberry leaves which are eaten by silkworms.

Therefore, inputs in this activity include the kraft paper, the larvae, and the diesel consumed for transportation. Outputs mainly include the kraft paper waste (which is burnt), generating GHG emissions, along with the emissions from the use of diesel.

Transport of mulberry leaves

This activity is labor-intensive, comprising the collection of mulberry leaves from the mulberry fields and their transport to the sheds where the silkworms will be fed. This transportation is done using a tractor. Thus, besides manpower, the main input for this activity is the diesel used in the tractor. Outputs, hence, comprise mainly GHG emissions. It is worth mentioning that in this activity most of the sericulturists harvest the mulberry leaves manually, with the help of a sickle.

Silkworm rearing

This activity comprises the main efforts employed during the rearing of silkworms. The most volume-intensive input in this activity is the mulberry leaves, which are the only food for the silkworms. Mulberry branches are taken from the trees and placed on the rearing beds. Silkworms feed on the mulberry leaves and leave the stems behind as silkworms climb them staying always on the surface.

Other inputs comprise lime, water, kraft paper, and electricity. Lime is dusted over the beds to absorb moisture from the decomposition of the mulberry leaves and silkworm litter. Kraft paper is laid on top of the rearing beds to both help control temperature and avoid the action of predators (e.g. birds). Water is used to water the floor, in order to avoid dust from floating and laying over the rearing beds and also to maintain the mulberry leaves always fresh in the storage room. Electricity is consumed for lighting and to power a few pieces of equipment (e.g. for disinfecting the sheds and for cleaning the silk cocoons before packing).

Outputs from this activity include a mixture of lime, silkworm litter, and the mulberry stems left behind by the silkworms. This mixture, named here as rearing bed waste, is used as a biofertilizer and laid on the mulberry fields. As the lime might be beneficial to control soil acidity, as commented previously, this excess of lime can cause eutrophication in the long term. One positive aspect is that due to the organic load present in the silkworm litter and the remains of mulberry branches, it decreases the need for mineral fertilizer.

A further output is the kraft paper waste (the paper used to cover the rearing beds), which is burned and thus responsible for GHG emissions.

Bed cleaning

This activity comprises the transportation of the rearing bed waste from the sheds to the mulberry fields. This is done with the help of a tractor. Thus, diesel is direct input. The output, hence, comprises mainly GHG emissions.

Packing

Once the cocoons are ready, they go through a classification process and are packed. The packaging comprises putting them into raffia sacks for transporting to the silk manufacturer. Thus, the main input in this activity is raffia sacks. The output comprises raffia sack wastes.

Transport of silk cocoons

After packing, the silk cocoons are taken to the yarn manufacturer using a pickup truck. Thus, the main input in this activity is diesel. The output, hence, comprises mainly GHG emissions.

Open burning

Most of the packages of the products used in the sericultural activities, along with the kraft paper, are subject to open burning. Inputs for such open burning include the wastes of raffia sacks (the ones that can no longer be reused), lime and chlorine packages, and any kraft paper waste. Outputs include GHG emissions derived from the open burning.

4.2.4 Infrastructure and Other Environmental Aspects of Silk Cocoon Production

Just as for the mulberry production, it requires manpower, and thus all related aspects are also tied to the activities.

Aspects of infrastructure are also to be accounted for, including the heavy machinery necessary for the transportation of mulberry leaves, as well as for transporting the larvae to the property and the silk cocoons to the yarn manufacturer. Equipment needed for disinfection of the sheds and grids, as well as for scorching grids also, contributes to consuming resources and generating emissions (considering extraction of raw materials, manufacturing, and distribution).

Packages also need consideration. Packages for chlorine and formaldehyde (used in the shed and grid disinfection) and the raffia sacks (used for packaging) are products that went through previous processing, and their environmental loads should be accounted for.

Moreover, further aspects that need to be considered regard the use of electricity, which depend on the mix that is delivered by the carrier.

5 Concluding Remarks

In Brazil, sericulture has developed in a way that has put Brazilian silk among the largest silk producers in the world, in addition to its increased demand for the quality it presents. However, it is necessary to consider the socioeconomic and environmental impacts of Brazilian silk production. It is important that efforts be made to improve the silk cocoon production processes, in order to achieve sustainable sericulture.

The socioeconomic impacts that directly and indirectly affect sericulturists are present, both in the activities of mulberry cultivation and in the rearing of silkworms. Sericulture requires a lot of dedication and physical effort from those involved, in addition to bringing potential health risks by the use of chemicals. For sericulturists, government incentive is important and is part of rural sustainability, since via this incentive they can develop further and with greater flexibility, being able to channel investment into production.

The sustainability of sericulture may be entering a state of instability due to the aging of sericulturists, the lack of new labor, and unsatisfactory family succession. There is an urgent need to plan actions toward changing this future scenario in Brazilian sericulture. Sericulturists fear that sericulture will not continue as a promising activity, and will not be sustained over the coming decades if there are no strategies to improve their socioeconomic status. Another relevant point is that sericulturists do not realize the risks they face, nor how much their activities impact their lives positively and/or negatively. This makes clear the need for further research that contributes to improving socioeconomic performance in sericulture.

Furthermore, all inputs and outputs for every process in the sericultural activities have their respective impacts either on the environment or on human health. Both the production of mulberry and silk cocoons account for environmental aspects (exchanges with the natural environment) and elementary processes which will somehow lead to environmental aspects in other stages of production, be it in the extraction of resources to produce raw materials (e.g. chlorine, formaldehyde, lime, and packages) or the manufacturing phase of these products. The set of all these aspects lead to the overall environmental impacts of the production of silk cocoons.

There are particularities for producing cocoons in different locations, under different (technological and climate) conditions. In this chapter, we set out to reveal the average conditions of the production of silk cocoons in the most representative region in Brazil, thus accounting for the general production of silk cocoons in the country.

Determining the complete environmental impacts of the production of silk cocoons calls for the use of quantitative tools, such as the life cycle assessment, which has been said to be the most complete and complex, but also time-consuming, tool for environmental profiling.

Regarding the expected impacts, a concerning issue is the lack of awareness of the sericulturists about the hazardous effects many of the chemicals and other products they use have both on their health and on the environment.

This chapter presented the impacts of the production of silk cocoons on the socioeconomic and environmental conditions, from which measures can be drawn toward improving those conditions to both provide a better quality of life for silk cocoon producers and improve the environmental conditions of the activity, in order to ensure continuity of the sericulture in the Brazilian territory.

Overall, the results show that sericulture has proven to be ahead of many other rural activities with regard to socioeconomic aspects, for being an alternative to diversifying small rural properties, generating monthly income to families in rural areas. Moreover, a fair background was given to enable the deployment of efforts to make sericultural activities less environmentally impacting. Nonetheless, further investigation and practical guidance are needed in order for the sericulture to reach more sustainable conduct.

Acknowledgements This study was financed in part by the Coordenação de Aperfeiçoamento de Pessoal de Nível Superior—Brasil (CAPES)—Finance Code 001, and the National Council for Scientific and Technological Development (CNPq) (Sponsored by CNPq 310686/2017-2), Fundação Araucária (Support for Scientific and Technological Development of Paraná) by means of CP20/2013, Doctoral Program in Textile Engineering in partnership with a Foreign Institution (Uminho/UEM/UTFPR), and the ERDF through the Operational Program for Competitiveness Factors—COMPETE and national funds by FCT—Foundation for Science and Technology.

The authors would like to show their gratitude for the support provided by the Universidade Tecnológica Federal do Paraná (UTFPR) (Brazil), University of Minho (Portugal), and the State University of Maringá (Brazil).

References

1. Abit - Brazilian Association of the Textile and Clothing Industry [*Associação Brasileira da Indústria Têxtil e de Confecção*] (2019) Sectoral profile [*Perfil do setor*]. https://www.abit.org.br/cont/perfil-do-setor. Accessed 05 Feb 2020
2. Abit - Brazilian Association of the Textile and Clothing Industry [*Associação Brasileira da Indústria Têxtil e de Confecção*] (2020) Foreign trade [*Comércio exterior*]. https://www.abit.org.br/cont/dados-comercio-exterior. Accessed 05 Feb 2020
3. Abraseda – Brazilian Silk Association [*Associação Brasileira da Seda*] (2018) Núcleos. https://abraseda.wixsite.com/abraseda/nucleos. Accessed 19 Mar 2020
4. AEN – Agência de Notícias do Paraná (2015) Paraná pode aumentar produção de bicho-da-seda em até 50%. Accessed 23 Feb 2020
5. Astudillo MF, Thalwitz G, Vollrath F (2014) Life cycle assessment of Indian silk. J Clean Prod 81:158–167. https://doi.org/10.1016/j.jclepro.2014.06.007
6. Astudillo MF, Thalwitz G, Vollrath F (2015) Life cycle assessment of silk production–a case study from India. Handbook of Life Cycle Assessment (LCA) of textiles and clothing. Woodhead Publishing, Cambridge, UK, pp 255–274
7. Babu KM (2015) Natural textile fibres: Animal and silk fibres. In: Textiles and fashion. Woodhead Publishing, Cambridge, UK. https://doi.org/10.1016/b978-1-84569-931-4.00003-9
8. Babu KM (2019) Silk reeling and silk fabric manufacture. In Silk, 2nd edn. Woodhead Publishing, Cambridge, UK. https://doi.org/10.1016/b978-0-08-102540-6.00002-4
9. Babu KM (2019) Introduction to silk and sericulture. In: Silk, 2nd edn. Woodhead publishing, Cambridge, UK. https://doi.org/10.1016/b978-0-08-102540-6.00001-2
10. Barcelos SMBD, Luz LM, Vasques RS, Piekarski CM, Francisco AC (2013) Introductory background for life cycle assessment (lca) of pure silk fabric. http://dx.doi.org/10.14807/ijmp.v4i1.67. Accessed 06 Apr 2020
11. Barcelos SMBD, Guedes MG, Salvador R and Francisco AC (2019) Introductory analysis for conducting life cycle assessment of silk cocoon in Brazil. In: 6th international scientific conference SWS on social sciences & art – ISCSS, 26 August – 01 September 2019, pp 179–186. https://www.sgemsocial.org/index.php/conference-topics/jresearch?view=publication&task=show&id=4103. Accessed 06 Feb 2020
12. Barcelos SMBD, Salvador R, Guedes MG, Francisco AC (2020) Opportunities for improving the environmental profile of silk cocoon production under Brazilian conditions. Sustainability 12:3214. https://doi.org/10.3390/su12083214
13. Baltar CS, Baltar R (2016) Caminhos da seda no Paraná: a convergência de diferentes processos migratórios na expansão da sericicultura de São Paulo até o Vale da Seda (PR). In: VII Congreso de la Asociación Latinoamericana de Población e XX Encontro Nacional de Estudos Populacionais, ABEP, Foz de Iguaçu-PR, 17 a 22 de outubro. http://www.abep.org.br/publicacoes/index.php/anais/article/view/2809/2695. Accessed 16 Feb 2020
14. Busch APBA (2010) Análise da conjuntura agropecuária. Safra 2010/2011. SEAB – Secretaria de Estado da Agricultura e do Abastecimento, DERAL – Departamento de Economia Rural. http://www.agricultura.pr.gov.br/Pagina/Boletins-Conjunturais. Accessed 13 Feb 2020
15. Bratac (2020) Nossa história. http://www.bratac.com.br/bratac/pt/index.php. Accessed 16 Feb 2020
16. Cirio GM (2014) Sericicultura no Estado do Paraná. Safra 2013/2014 - Relatório Takii. SEAB – Secretaria de Estado da Agricultura e do Abastecimento, DERAL – Departamento de Economia Rural. http://www.agricultura.pr.gov.br/Pagina/Boletins-Conjunturais. Accessed 15 Feb 2020
17. Cirio GM (2016) Sericicultura no Estado do Paraná. Safra 2015/2016 - Relatório Takii. SEAB – Secretaria de Estado da Agricultura e do Abastecimento, DERAL – Departamento de Economia Rural. http://www.agricultura.pr.gov.br/Pagina/Boletins-Conjunturais. Accessed 15 Feb 2020
18. Cirio GM (2017) Sericicultura no Estado do Paraná. Safra 2016/2017 - Relatório Takii. SEAB – Secretaria de Estado da Agricultura e do Abastecimento, DERAL – Departamento de Economia Rural. http://www.agricultura.pr.gov.br/Pagina/Boletins-Conjunturais. Accessed 15 Feb 2020

19. Cirio GM (2018) Evolution and Current Condition of Sericulture in Paraná [*Evolução e Condição Atual da Sericicultura no Paraná*]. In: Inovações na sericicultura do Paraná: tecnologias, manejo rentabilidade. Dimas Soares Júnior, Edson Luiz Diogo de Almeida, Oswaldo da Silva Pádua: orgs. – Londrina: ABRASEDA: IAPAR, 103 p. il. http://www.emater.pr.gov.br/arquivos/File/Biblioteca_Virtual/RedesReferencia/Livro_Sericicultura2018_2aED.pdf. Accessed 12 Jan 2020
20. Cirio GM (2018) Sericicultura no Estado do Paraná. Safra 2017/2018 - Relatório Takii. SEAB – Secretaria de Estado da Agricultura e do Abastecimento, DERAL – Departamento de Economia Rural. http://www.agricultura.pr.gov.br/sites/default/arquivos_restritos/files/documento/2019-09/sericicultura_2019_v1.pdf. Accessed 15 Feb 2020
21. European Commission (2013) EVALSED: the resource for the evaluation of Socio-Economic Development. https://ec.europa.eu/regional_policy/sources/docgener/evaluation/guide/guide_evalsed.pdf. Accessed 18 Mar 2017
22. Giacomin AM, Garcia JB Jr, Zonatti WF, Silva-Santos MC, Laktim MC, Baruque-Ramos J (2017) Brazilian silk production: economic and sustainability aspects. Proc Eng 200:89–95. https://doi.org/10.1016/j.proeng.2017.07.014
23. Iapar – Paraná Agronomist Institute [*Instituto Agrônomo do Paraná*] (2019) Paraná enters the international silk network [*Paraná ingressa na rede internacional da seda*]. http://www.iapar.br/2019/11/2607/Parana-ingressa-na-rede-internacional-da-seda.html. Accessed 05 Feb 2020
24. ISC - International Sericultural Commission (2019) Global Silk Industry. https://www.inserco.org/en/statistics. Accessed 06 Feb 2020
25. ISC - International Sericultural Commission (2019) Silk - an introduction. http://inserco.org/en/silk_an_introduction. Accessed 06 Feb 2020
26. ISO (International Organization for Standardization) (2006) Environmental management - life cycle assessment - principles and framework, 2nd edn. ISO 14040:2006. ISO, Geneva, Switzerland
27. ISO (International Organization for Standardization) (2006) Environmental management - life cycle assessment - requirements and guidelines, 1st edn. ISO 14044:2006. ISO, Geneva, Switzerland
28. Jain J, Ratan A (2017) Developing a Conceptual Model to Sustain Handloom Silk Industry at Sualkuchi, Assam, India. Eur J Sustain Dev 6(3):413–422
29. Kasi E (2013) Role of women in sericulture and community development: a study from a South Indian Village. SAGE Open, pp 1–11. https://doi.org/10.1177/2158244013502984
30. Lima AS (2018) Analysis of the contractual relations between producers and the silk industry in the municipality of Diamante do Sul, Paraná State [*Análise das relações contratuais entre sericicultores e indústria de seda no município de Diamante do Sul, Estado do Paraná*]. https://lume.ufrgs.br/handle/10183/197790. Accessed 04 Feb 2020
31. Mahan B (2012) Silk industry among the Tai-Ahom of Assam, India as an attraction of tourist. Int J Sci Res Publ 2(12):1–4
32. Makarov I, Sokolova A (2016) Coordination of the Eurasian economic Union and the silk road economic belt: opportunities for Russia. Int Organ Res Journal 11(2):29–42
33. Pennacchio HL (2016) Casulo de seda. Indicadores da agropecuária – Conab, Brasília, Ano XXV, n. 10, outubro, pp 01–114. Disponível em. https://www.conab.gov.br/info-agro/precos/revista-indicadores-da-agropecuaria?start=30. Accessed 14 Feb 2020
34. Pádua OS (2005) A origem da sericicultura. https://www.yumpu.com/pt/document/read/14574154/a-origem-da-sericicultura. Accessed 14 Feb 2020
35. Paul D, Dey S (2014) Essential amino acids, lipid profile and fat-soluble vitamins of the edible silkworm Bombyx mori (Lepidoptera: Bombycidae). Int J Trop Insect Sci 34(4):239–247
36. Pinto NF (2015) O processo saúde-doença dos/as trabalhadores/as da sericicultura no Paraná. Cascavel, PR: Unioeste, 128p. http://tede.unioeste.br:8080/tede/handle/tede/654. Accessed 06 Feb 2020
37. Pinto NF, Murofuse NT (2017) Processos protetores e destrutivos da saúde dos(as) trabalhadores(as) da sericicultura. Saúde debate 41(spe2). Rio de Janeiro. http://dx.doi.org/10.1590/0103-11042017s210. Accessed 04 Feb 2020

38. Porto AJ (2014) Sericicultura no estado de São Paulo. B. Ind Anim Nova Odessa 71(3):291–302. http://www.iz.sp.gov.br/pdfsbia/1412257067.pdf. Accessed 15 Feb 2020
39. Pratama AG, Supratman S, Makkarennu M (2019) Examining forest economies: a case study of silk value chain analysis in Wajo District. Forest Soc 3(1):22–33
40. Sakthivel N, Kumaresan PT, Balakrishna R, Mohan BV (2012) Economic viability of sericulture in southern Tamil Nadu - a case study. Agric Sci Digest 32(2):98–104
41. Siddappaji D, Latha CM, Ashoka SR, Raja MGB (2014) Socio-economic development through sericulture in Karnataka. J Humanities Soc Sci 19(10):24–26. https://doi.org/10.9790/0837-191052426. Ver. V
42. Silin Y, Kapustina L, Trevisan I, Drevalev A (2018) The silk road economic belt: balance of interests. Econ Polit Stud 6(3):293–318
43. Sunitha Rani D, Jayaraju M, Kannan M, Ashok Kumar K (2015) Socioeconomic influence on farmers in seri-business from Tamil Nadu. Int J Eng Bus Enterp Appl 11(2):136–140
44. Tikader A, Vijayan K, and Saratchandra B (2013) Muga silkworm, Antheraea assamensis (Lepidoptera: Saturniidae)-an overview of distribution, biology and breeding. Eur J Entomol 110(2)
45. UNEP – United Nations Environment Programme (2009) Guidelines for social life cycle assessment of products. https://www.lifecycleinitiative.org/wp-content/uploads/2012/12/2009%20-%20Guidelines%20for%20sLCA%20-%20EN.pdf. Accessed 15 Mar 2017
46. UNEP – United Nations Environment Programme (2013) The methodological sheets for subcategories in social life cycle assessment (S-lca). https://www.lifecycleinitiative.org/wp-content/uploads/2013/11/S-LCA_methodological_sheets_11.11.13.pdf. Accessed 15 Mar 2017
47. Vanclay F, Esteves AM, Aucamp I, Franks DM (2015) Evaluación de Impacto Social: Lineamientos para la evaluación y gestión de impactos sociales de proyectos. https://www.iaia.org/pdf/Evaluacion-Impacto-Social-Lineamientos.pdf. Accessed 12 Feb 2017
48. Wani KA, Jaiswal YK (2011) Health hazards of rearing silkworms and environmental impact assessment of rearing households of Kashmir, India. Indian J Public Health Res & Dev 2(2):84–88
49. Wichitsathian S, Nakruang D (2019) Knowledge integration capability and entrepreneurial orientation: case of Pakthongchai Silk Groups Residing. Entrep Sustain Issues 7(2):977–989

Adapting Sustainable Design Assessment Tools for Local Development: Some Insights into Argentina's Textile and Clothing Industry

Carola Ruppel, Rocío Canetti, and Celina Monacchi

Abstract Clothing value chain is fundamental to Mar del Plata (Argentina) as it holds a high rate of industrial added-value, is rooted in local history, and has shaped relationships and institutions. This means sustainable clothing practices are interwoven with Mar del Plata's social and economic development. However, fashion entrepreneurs must overcome many difficulties when talking about sustainable design. To begin with, some aspects of international labels and certifications are unsuitable for textile industry in a developing country. Due to the absence of measure tools, companies lack information about their production and cannot afford improvements. Furthermore, there is a gap between academic knowledge production and industrial production. Therefore, the following questions arise: How to engage firms in assessment processes to work on a sustainable culture? How to enrich relationships between companies and research groups? The Sustainable Design Research Group (GIDSu) belonging to Universidad Nacional de Mar del Plata, Argentina, has been working for local industries for more than ten years. GIDSu has been able to design a tool to evaluate sustainable design and management in local fashion and textile firms, which finally took the name of CeDiS (Spanish acronym of *sustainable design certification*). CeDiS was developed in order to offer a tailored solution for small and medium companies. It works as an ecolabel, based on self-declared claims from the firms and audited by the research group (GIDSu). The aim of this tool was to acquire information about each enterprise, detect problems to display it to managers and suggest sustainable changes for continuous improvement. In addition, it worked as a way to close the gap between private companies and public science. To design the CeDiS, researchers based on present methods and standards (D4S, LIDs Wheel, ISO 14006, ISO 14021) as well as field tests and interviews. As a result, the tool includes an interview protocol, a process matrix; a result display (wheel matrix); and a diagnosis and suggestions sheet. Finally, during September 2019, GIDSu verified the CeDiS, employing it over 20 textile and fashion firms from Mar del Plata, supported by a national program for local clusters (called *Programa de Apoyo a la Competitividad* or *PAC* in Spanish). The aim of this paper is dual. First, it describes the process of adapting and designing sustainable assessment methods

C. Ruppel · R. Canetti (✉) · C. Monacchi
UNMdP FAUD CIPADI, Mar del Plata, Argentina

M. Á. Gardetti and R. P. Larios-Francia (eds.), *Sustainable Fashion and Textiles in Latin America*, Textile Science and Clothing Technology,
https://doi.org/10.1007/978-981-16-1850-5_2

for industries in a developing country. Second, it shows the positive impact of this tools in local development. Through the CeDiS study, we share a possible blueprint for fashion and textile value chain (including businesspeople, scientists, politicians, consumers, and workers) to face sustainable development in complex—and often uncertain—settings.

Keywords Sustainable assessment · Design assessment · Assessment tool · Local development · Textile industry · Design

1 Introduction

In Argentina, the textile and clothing value chain covers fiber processing, the production of yarn and fabric, and the design and manufacturing of apparel and home textiles. Sub-sectors vary along with the country and show differences depending on geography. This paper focuses on the city of Mar del Plata.

Mar del Plata is a medium-sized city, located over the Atlantic sea 400 km south of Argentina's capital city—Buenos Aires (Fig. 1). It is a well-known destination for tourism; as a result, its population (regularly, around 1 million people) grows up to 300% in summer. Even though tourism is socially recognized as the main industry, fishing, textile, and building industries have been fundamental in Mar del Plata's economy and are rooted in its history.

In this way, two sub-sectors of the textile and clothing value chain have developed in Mar del Plata: the textile activity, oriented almost exclusively to knitwear produced with flat knitting machines, and clothing.

The knitting industry in the city has a long history. The first local experiences date from the mid-nineteenth century and engage migrating families, mostly Italian. Favero [10] analyzes the effect that those family and ethnic ties had on the activity. The consolidation of the local knitting industry took place between 1960 and 1975, coinciding with the establishment of Mar del Plata as a main holiday destination in the country. Knitwear thus became a "tourist souvenir" even though the industry failed to overcome the economic problems of the 80s and 90s [9]. Since then, the weaving activity for national or international companies has intensified due to the recognition of the local product and its quality [30].

Furthermore, clothing and garment production came up as a support activity for the knitting industry. However, by the end of the 70s, it became consolidated through the establishment of local companies. It continued to grow until the 90s upon the structure of small businesses. Between 1994 and 2005, the textile facilities decreased by 65% while the clothing ones raised 29% [27]. This growth could be related to the entrepreneurship fever that was prompted after the national currency devaluation process that also drove the establishment of small-sized enterprises focused on apparel design in Buenos Aires, which is commonly known as the "signature design" phenomenon [28]. Fashion entrepreneurship continues expanding nowadays, along with the design education offer.

Nowadays, the fashion and textile sector in Mar del Plata and its region consists of 55% knitting companies and 45% clothing [36]. These are mostly micro and small-sized enterprises.[1] There are also a few medium-sized enterprises, and all are mainly family businesses. Concerning entrepreneurship, records are showing that the highest percentage of local initiatives are related to the textile and clothing sector [5]. According to records, the companies in this sector add up to 13% of the local industry and 10% of the industrial activity of the regional SMEs [33]. The textile and clothing industry in Mar del Plata shows characteristics distinctive of industrial

Fig. 1 Mar del Plata is located in center-east Argentina, on the Atlantic seacoast. *Source* Google Maps. Photo from: Fermin Rodriguez Penelas, available at Unplash.com

[1] In Argentina, the SMEs segment is recognized as "PyMES". Through the resolution Nr. 220/2019, the Department of "PyMES" and Entrepreneurs, from the Production and Labor Bureau, classifies companies in the "PyMES" category. In the industrial sector, a micro-sized company has an estimated turnover of $21.990.000, a small-sized of up to $157.740.000; a medium-sized "tramo 1" up to $986.080.000 and a medium-sized "tramo 2" has a turnover of up to $1.441.090.000. Source: https://www.produccion.gob.ar/area/secretaria-de-emprendedores-y-pymes.

districts: a large proportion of SMEs with their own brands, high interaction with local sub-contracting workshops, production flexibility, high rooting on the region's history, a relevant flow of unencoded knowledge between the participants, and a growing presence of design professionals [14].

Another significant fact about the knitting sub-sector is its working method based on outsourcing for Argentinian clothing brands, called "fason" (toll processing). It is a method related to the putting-out system: the workshops and factories get commissions from other companies that outsource their production (usually called "comitentes"). In Mar del Plata's economy, toll processing signals both the outsourcing activity and the manufacturer suppliers. Furthermore, it is used to describe two types of service suppliers. Firstly, it is referred to small productive units, mostly domestic and individualized [41], where companies decentralize their activity and outsource their fixed costs. This paradigm makes invisible the crucial nodes, creating asymmetric relations where the workflow becomes irregular and there is no encouragement for job stability [40]. Secondly, toll processing is used to name knitting companies that allocate their resources to satisfy the demand of regional or international brands. "Fason" is clearly a complex concept. For the purposes of this paper, we call the working method "outsourcing", the small productive units as "domestic manufacturers", and knitting companies as "manufacturers". Similarly, those companies that outsource their production are being called "clients".

The local Textile Chamber estimates that 60% of the national knitting production is carried out in Mar del Plata. This indicates the city's acknowledgment associated with its economic activity, through the quality of its products and specific know-how. Oppositely, the clothing sub-sector consists of companies that produce under their own brands and occasionally outsource some processes to workshops in Buenos Aires. From the economic perspective, companies that work as manufacturers completely count on their clients and have reduced bargaining power, engaging so much as 70% of their production for the clients that sometimes even fix services' prices [7]. In this way, the manufacturers relegate their own brand production. This results in a lower investment in matters of branding, image, product communication, and trend analyses [31].

The local textile–clothing value chain relies also on supportive roles engaging in the traditional productive stages (Nutz and Sievers [32]. For instance, there are some educational organizations acknowledged, such as the National University of Mar del Plata and the National Technology University, along with political organizations such as the Textile Chamber of Mar del Plata and the Association of Clothing and Related Manufacturers (ACIA in Spanish). Additionally, since 2017 these organizations, along with the township of General Pueyrredón, through a support program for micro-, small- and medium-sized enterprises, constitute the Textile and Clothing Cluster of Mar del Plata. This organization outruns existing business institutions and involves new stakeholders in the sector. Its goal is to readjust this sector and promote cooperative relations between companies, the public, and academics. Its core matters are related to R&D, communication, and new niche markets. Accordingly, this unprecedented associativity fosters the local economic activity, although there are recognized differing interests about this program among the different social groups [29].

Along with this context, the Sustainable Design Research Group (GIDSu in Spanish), located in the Center for Research and Actions in Industrial Design (CIPADI in Spanish) within the National University of Mar del Plata, has a long history of working with local companies and their relation to sustainability. Its interest has been to introduce in the local manufacturing context the sustainability concept, considering its intrinsic peculiarities. In 2017, the Textile and Clothing Cluster of Mar del Plata, along with the National University of Mar del Plata, created a favorable space for spreading these studies. Furthermore, companies became interested in new markets and in adding value to their products, as an opportunity to improve their competitiveness in the unstable Argentinian economic context.

2 Sustainability in the Local Textile–Clothing Industry

2.1 *Opportunities and Limitations*

Since 2016, the GIDSu has made its central goal to raise SMEs' interest in sustainability, as well as to redefine the theoretical framework of the locally adapted sustainable design [17, 18]. To this end, the group has been focusing on developing a certification system politically agreed with every social stakeholder and technically defined by the group, to encourage sustainability (social, environmental, and economic) and improve the information given to the consumer.

GIDSu has chosen to focus on the textile and clothing sectors on account of its tradition and its effect on regional employability [15]. The project is addressed in two stages: (1) textile–clothing diagnosis regarding sustainable methods and strategies; (2) instrumental approach and testing.

The diagnosis resulted in a SWOT analysis of the industry from a sustainable perspective. As previously indicated, some strengths of the local companies are their tradition in production, skilled workforce, high-quality standards, and existing capacity, among others. Regarding weaknesses and threats, it can be inferred that:

- company owners showed low interest in fully incorporating the sustainability concept or had a deficient understanding of the matter. This means that even on the greener companies, sustainability strategies were associated only with the use of natural or local materials. The approaches that implemented design strategies such as multi-functional or zero-waste clothing, modular garments, etc. [39] got a score next to zero;
- the deficiency of specific data, particularly quantitative, encumbers the statistical analysis of the SMEs' environmental impacts [37, 46];
- there are detectable difficulties to adapt international sustainability assessment tools (such as SIMA-PRO software) to the local context, because of the related

costs (economy, time, and need of professional assistance) (Retamozo, loc. cit.); and
- the industry is vulnerable to the political and economic context, in which the agro-export activity gets fostered, the imports freed, the domestic market falls, and the subsidies withdrawn.

Moreover, the diagnosis allowed to find some opportunities for the sector:

- The implementation of sustainable design management tools could act as an encouragement for organizing the production and improving the products concerning social, environmental, and economic responsibility. Accordingly, it becomes an added value for companies. Design for Sustainability or D4S is a competitive management system [46]. D4S "*...is a globally recognized way in which companies can work to improve efficiencies, product quality, and market opportunities (locally and internationally) and at the same time, increasing environmental performance.*" [34], p. 23.
- The growing social sensibility toward sustainable consumption draws attention to consumers [20]. A survey carried out in Mar del Plata revealed that 67% of the respondents are willing to embrace new developments in durable consumer goods [15]. However, consumers explained that despite their interest, they do not consume eco-friendly products given the difficulty in recognizing and acquiring them [38].

In this context, it seems logical for companies to implement a continuous improvement process, such as the ISO 14.001. In fact, in the local textile sector, management improvements affect 41% of the product innovation and/or production processes. On the contrary, only 14% of companies adopt some type of ISO certification [36]. **Therefore, there is a need for designing tools that are adaptable to companies of all sizes, as well as it is important to reduce the obstacles related to fees and bureaucracy and to ease their implementation.**

From the perspective of local development [1, 44] and industrial design [25, 26], there is a need to adapt global methodological tools to the regional context considering not only the environmental and economic attributes but also the production, social, and cultural ones. Therefore, it is relevant to enhance and organize those variables to turn them into innovative strategies with local added value.

Also, there is a possibility to enhance the regional development through joint efforts of the academics and the industry who can integrate their knowledge to organize, simplify, and advance the implementation and management of eco-design [35]. Accordingly, the Quality Assurance Standards (ISO 9001), the Environmental Management Standards (ISO 14001), and the Guidelines for incorporating eco-design from the Environmental Management Systems (UNE 150301/ISO 14006 Ecodesign) involve valuable sustainable design concepts and measurements (Fig. 2). Particularly, ISO 14006 defines eco-design as follows:

> "(...) a process integrated within the design and development that aims to reduce environmental impacts and continually improve the environmental performance of the products, throughout their life cycle from raw material extraction to end of life." (p. 8)

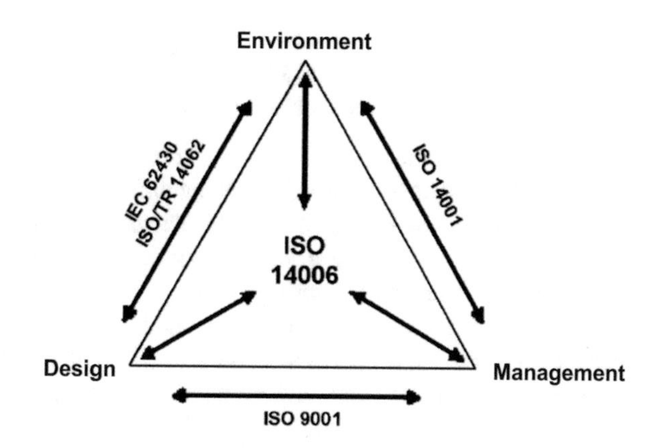

Fig. 2 Framework including ISO 14001, ISO 9001, ISO/TR 14062, IEC 62430 e ISO 14006. *Source* ISO 14006:2001 (es.) Sistemas de gestión ambiental—Directrices para la incorporación del ecodiseño

Finally, the relation university–industry–government was considered. To this extent, the instrument has to assess and analyze the sustainability within the company and over time, in terms of continuous improvement. Also, it must evaluate and compare the sustainability from a group of companies over time.

2.2 *Existing Norms and Instruments*

Nowadays, designers and companies rely on a variety of tools and methods to approach sustainability as a support to design processes and product development [8, 45]. Depending on the project's complexity and progress degree, the following tools can be recognized: checklists, Ashby charts, MET Matrix, D4S Strategy Wheel, and Life Cycle Assessment (LCA). They can be used along with other tools such as Eco-Indicators, Eco-Labeling (type 1, type 2, type 3), GRI Reports (Global Report Initiative), ISO 26000 (social responsibility), sustainable materials inventories, or "materialotecas" [21].

This project takes two of them as its main focus: D4S Strategy Wheel and Eco-Labeling.

- In the local context, Bernatene and Canale [3] focused on the Strategy Wheel [4] that stands out due to its ease of use and implementation. There are two ways to use it. The first one, more analytical, goes around in a clockwise direction leading to understanding the product life cycle—extraction of raw materials, production, use, and disposal. The second way goes counter-clockwise and allows us to get a project overview inviting us to think environmental strategies from three levels: product-system, product-structure, and product-component. However, this approach tends to have quite qualitative impact evaluations. To solve this, Canale [6] developed a method that blends the Strategy Wheel and the rating guidelines to measure the impact degrees.

- Also, regarding environmental labeling, many mechanisms are aiming at giving accessible information and identifying the goals needed to achieve sustainable production. Next, three labeling types can be recognized (Table 1).

GIDSu focused on the product self-declarations or Type II, which is based on the ISO 14021. They are developed by the manufacturer under their criteria, referring to specific stages of the product life cycle. The main responsibility of guaranteeing the data accuracy is of the person carrying out the self-declaration. They have to give all the information needed for verification.

This type of declaration is the most suitable for the stated requirements for local certification. Firstly, they would allow working with currently available resources at a lower cost than other labeling systems. Also, they involve public organisms and academics. They allow as well higher flexibility, as the company itself defines the criteria. Finally, self-declarations give simple and precise information to consumers. As a result, the GIDSu' project has been framed into environmental self-declarations.

Table 1 Environmental labeling systems comparison, Types I, II, and III

Environmental labeling		Type I	Type II	Type III
		Certified eco-labels	Product self-declaration	Environmental Product Declaration (EPD)
Charact eristics	Certification b y a third party	required	not required, but enhances credibility	not required, but enhances credibility
	Communicates	Better environmental performance	Improvement of one environmental aspect	LCA data
	ISO	ISO 14024	ISO 14021	ISO 14025, ISO 21930
Evaluation	Useful for	improved communication with the final consumer	improved communication with the final consumer	improved B2B communication and green procurement
	Standards	the company has to adapt to standardized criteria from certifiers	development of own criteria	requires LCA, so the company has to adapt to standardized criteria from certifiers
	Associated costs	high costs assigned to certifying body	could be lower when using own resources	high costs assigned to certifying body or experts
	Data availability	measurable and accurate data about the LC and production	verifiable data could be of only one production aspect	measurable and accurate data about the product

Source self-construct based on ISO 14024, ISO 14021, ISO 14025 and ISO 21930

2.3 CEDiS: Theoretical Aspects and Units of Analysis

The **instrument GIDSu developed was initially identified as Sustainable Design Certification, in Spanish: Certificación en Diseño Sustentable—CEDiS. However, finally, it was named after Self-declaration in Sustainable Design Management**. It is registered under the Argentinian Intellectual Property Law, with the reference number 17017885. Taking into consideration all the previously mentioned reasons, this instrument has been defined within the Environmental self-declarations, Type II which implements the ISO 14006 structure and the Strategy Wheel.

The CEDiS development took close to two years, in which GIDSu had to answer *what* and *how* to evaluate. This process was full of discussions and analysis. However, the present paper focuses on the *final form* that CEDiS adopted. Only two of the instrument's main focuses will be in-depth described: 4-Corporate Social Responsibility and 7-Innovation and expansion of useful life.

Presently, CEDiS uses eight units of analysis (henceforth UA) integrated into D4S strategies. The UAs significance was weighted upon a 100-point scale.

1. **Selection of low environmental impact materials, 10/100**. The materials used in products are analyzed by their environmental impact. Their use and selection criteria are verified (for instance, the origin of the materials). The raw materials selection is an essential process; however, the local industry shows difficulties regarding supply and demand. Therefore, this item was given a low weight compared to items where companies have higher bargaining power.
2. **Minimization of materials usage, 10/100**. Strategies for avoiding oversizing of clothing and accessories, while maintaining its functionality, are evaluated. This item also has a low weight because companies focus their actions on reducing the materials use in their whole production process (see UA 3).
3. **Optimization of the production system, 18/100**. This UA includes: (1) documentation of processes to find early problematics and make adequate decisions; (2) quality assurance, implementation criteria, and improvement strategies; (3) maintenance of own and third-party equipment, to avoid production delays and ensure quality; (4) optimization strategies to conceive alternative methods that allow lower environmental impact, efficient use of resources, and reduction of waste materials (digital pattern making) or use of efficient processes (integral garment knitting); (5) resources management (electricity, gas, and water); (6) waste and waste disposal management, taking into account the proper separation and final disposition (reinserting, recycling, among others); (7) health and safety standards and human resources management; (8) acknowledgment of the role that the design sector plays in creating added value and in strategic planning. This results in the highest weighted item, by cause of the variety of strategies included, and because most of the environmental improvement actions are focused on this approach.
4. **Social responsibility, 15/100**. In this item, the relation between the companies and the different local stakeholders is verified. This includes the companies'

employees, the outsourced workshops and suppliers, domestic workers, and other community sectors—in vulnerable situations or not. In the long term, the objective is that companies commit to environmental, social, and economic improvement, beyond the current legislation. Then, this item involves (1) labor conditions of outsourced workshops and domestic workers, taking into consideration legal aspects and communication between parties; (2) labor relations and human resources management (minority incorporation, skills training programs, motivation); (3) interaction with non-traditional production sectors by incorporating vulnerable social groups (through coaching, seminars, integration initiatives) and community engagement (responsible use of public spaces, working with NGOs).

5. **Optimization of the distribution system, 10/100**. Strategies are focused on lowering the environmental impacts of the distribution whether it is transportation inside the city (factory–stores–outsourced workshops) and transportation to wholesalers, franchises, and customers. These strategies could be focused on the packaging (reducing, reusing, or using biodegradable materials) and/or on the logistics (closed circuits, reverse logistics, among others). It represents an item with a high environmental impact; however, the local context and market conditions restrict the companies' taking of action. As a consequence, this item has a low weight compared to the rest.
6. **Reduction of the impact during use, 10/100**. Although the use phase has the highest impact on the clothing life cycle, it was complex to measure it during this project stage. Most of the difficulties were associated with the lack of quantitative and qualitative data about users' behavior during use. On CEDiS, the strategies are verified for reaching the minimization of the environmental impacts of products in their use and maintenance.
7. **Innovation and expansion of useful life, 17/100**. This takes into consideration strategies for improving product efficiency and expanding the product's lifetime. It focuses on reducing the planned obsolescence of textiles and clothing. It also considers developing disruptive innovations (such as making new yarns) or incremental innovations (blending yarns, combining fabrics, or changing garments to improve functionalities, changing machinery, among others). It results in a relevant item, because of its relation to the company's design management and the possibility of rearranging the available resources (materials and humans).
8. **Communication, 10/100**. It evaluates if the communication between the company and its stakeholders is clear, coherent, and appropriate to each profile. This item also evaluates how the company obtains information about its customer experience and preferences and how it conducts the research and implements the results. This UA is considered by taking into account that information, and communication can improve the product and the company's position, enhancing the value.

2.4 The CEDiS Aspects

GIDSu developed a **questionnaire** to collect information about the UAs, as well as an interview and data collection protocol. The collected data is then processed through an **evaluation matrix**. This matrix consists of a spreadsheet where a partial score, as well as an overall score, is calculated. As explained before, each UA has a different weight according to its relevance in the textile–clothing industry, the possibility to implement related strategies, and the different aspects involved (technological, economic, social, managerial, and environmental).

The UAs are divided into analysis variables that are individually classified according to the company's actions: zero/low, medium, or optimal/high. For each ranking and variable, there are criteria, examples, and cases established that allow the researchers to measure the company's action. For example, Table 2 illustrates a

Table 2 UA3 weighting criteria, Sect. 3.1. Use of follow-up tools

zero	medium	optimal
0	0,5	1
Not verified. / There is no standardized use of follow-up tools or process documentation.	There is an implementation of follow-up tools or process documentation for some products. / No control or improvement measures and actions are taken for documented processes.	There is an implementation of follow-up tools or process documentation for every product. This allows communication within different company' areas and the implementation of control and improvement measures.

Source CEDiS Evaluation Matrix (extract)

variable from UA3, optimization of the production system. In this particular case, the use of production documentation tools is evaluated (technical sheets, process follow-up sheets, among others). The standardized use of these tools allows the company to obtain reliable information about their situation in order to improve their system. The company not using this type of tool gets a 0-zero score. If the researchers cannot verify the use of production documents during the interview, the score is also 0-zero. The company that implements these tools but has an infrequent use, or does not take any improvement measure, gets a 0, 5-medium score. Lastly, if the company implements standardized follow-up tools and improvement measures, it gets a 1-optimal score.

The score calculation delivers two types of values: one for each UA and one overall result. The first type is displayed in a radial graph, which indicates the level of compliance for each UA, to the D4S strategies. Series 1 (in blue) shows the optimal results according to CEDiS. Series 2 (orange) shows the company's performance in each UA (Fig. 3).

The second type illustrates the overall result of the company's actions related to the D4S strategies, in a score from 0 to 100. CEDiS is thought of as a program with a progressive implementation for D4S compliance. There are three possible

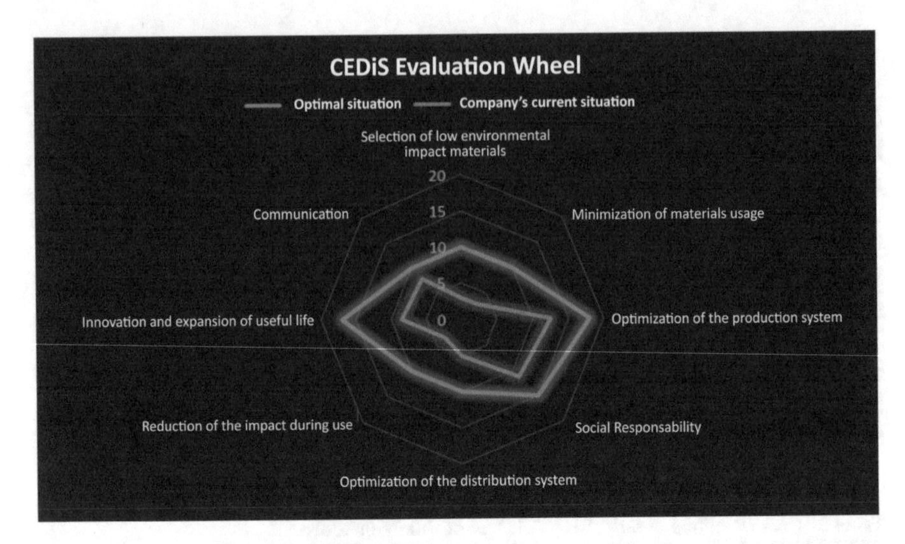

Fig. 3 Radial graph that compares the level of compliance: optimal situation (in blue) and the company's current situation (orange). *Source* CEDiS Evaluation Matrix (extract)

Table 3 Score rating for the different identification levels

Minimum score	rating	Maximum score
30,00	silver	45,00
45,00	gold	75,00
75,00	platinum	100,00

Source CEDiS Evaluation Matrix (extract)

levels of identification: silver, gold, and platinum that depend on the overall score (Table 3). The idea is to repeat the study in a period no longer than two years, to verify the improvement measures taken by the companies, many of which could be implemented in the medium term.

Finally, the analysis results and possible improvement measures are shown in the **Diagnosis Sheet**. This document is handed out to the companies with the results of the study carried out. It is presented as a table and summarizes the most relevant facts as follows:

- The first column describes the problem or the detected opportunities for improvement.
- The second column explains the actions to be taken to solve the problem or to capitalize on the detected strengths.
- The following columns evaluate the recommended actions for the company (Table 4). To easily communicate to the company the priority actions, it was chosen to use simple spreadsheets. For each firm and strategy, the following has

Table 4 Final rating and priority of the recommended action, considering its feasibility and benefits

Technical or economic		environmental, social, and managerial benefits		Priority of each recommended action
feasible	difficult /unknown	significant	limited	
X		X		1
X			X	2
	X	X		3
	X		X	4

Source CEDiS Diagnosis Sheet (extract)

to be considered: (1) the technical/economic feasibility; and (2) the environmental, social, and managerial benefits.

- The last column weights each recommended action.

2.5 *Units of Analysis: Item 4—Social Responsibility*

2.5.1 Background

The instruments studied in Sect. 2.2 are focused on the environmental impact. However, from a regional development perspective, it is also relevant to consider the social aspect to encourage corporate social responsibility (CSR) and Fair Trade.

The ISO 26000 defines CSR as the "responsibility of an organization for the impacts of its decisions and activities on society and the environment, through transparent and ethical behavior that contributes to sustainable development, including health and the welfare of society. The guideline and its implementation are voluntary and not a certifiable norm."[2] (ISO 26000, p. 2). The fair trade concept is incorporated due to its emphasis on the cooperation between stakeholders (state, consumers, and producers) and to focus on the ethical aspect of sustainability. The vision of the Fair Trade movement is "a world in which justice and sustainable development are at the heart of trade structures and practices so that everyone, through their work, can maintain a decent and dignified livelihood and develop their full human potential" (Manifiesto Comercio Justo [24], p. 1). Although it is not a concrete example describing specific actions, it is an encouraging movement that enhances values and ethical actions in search of labor and social equality.

Regarding specific certifications, the actions from *Fashion Revolution* an international NGO that works on creating conscience about the unsustainability of the

[2] ISO 26000 Guideline Social Responsibility Global STD, pp. 1–2. From http://www.globalstd.com/networks/blog/guia-iso-26000-responsabilidad-social.

current textile and fashion production systems can be noticed. They also created a campaign called *Who made my clothes?* to make visible the textile value chain creating conscience and connecting the consumer with the manufacturer in a way of respecting the environmental and ethical aspects of the production.[3]

Eco-labels are another example that contemplates the relation between consumer and manufacturer through product traceability. In the national context, the cases of "Sello Verde" in Rosario (green seal) and the "Ecosello de gestión salteña" (Salta's management ecolabel) can be recognized. They involve social aspects but do not develop them in depth. Another relevant case is B-System. According to B-Lab, the certifying organization, the B companies: "*measure their social and environmental impact and commit (…) to make decisions considering the consequences of their actions in the long term.*"[4] The B-System strictly evaluates the social and environmental impact of companies and proves that a business is meeting the highest standards of verified performance and transparency. However, none of these certifications have a comprehensive design foundation.

2.5.2 Adaptation to the Local Context

From the design perspective, the regional development and social sustainability aspects, the following threats and weaknesses of the local industry can be identified:

- There is a high level of informality in the outsourcing labor relation.
- Production units turn out as disjointed and invisible. There are not only workshops but also factories that do not belong to any chamber, which minimizes lobby opportunities.
- There is an unbalanced production system (higher production volumes for the winter season). This seasonal production system leads to inactivity periods and labor instability.
- There is a need for valuing craftsmanship and employee training and development. This relates to the need of keeping a stable and skilled workforce.
- There is a growth in business competition and diminished demand.
- There is a need to advertise the city as a high-quality production center, in the local and regional context, as a distinctive value.

Thus, GIDSu decided to focus on two areas of the CSR concept. On one side, previous local researches regarding CSR [46] were taken into account, related to the community (community engagement, minority incorporation), health, safety, and employee training (training programs, family support programs, volunteer programs, etc.). On the other side, outsourcing and sub-contracting systems, and the toll processing modality were considered. This last aspect proved to be the most relevant since it is a conflictive topic in the local context.

[3] From: https://www.fashionrevolution.org, last revised on 04.21.2020.

[4] From: https://www.sistemab.org.

To evaluate outsourcing in the local context, the concepts of outsourcing [2] and labor informality [42] were strictly defined. Law 12713 was also examined, which defines the conditions for textile home-based work taking into account the aspects of corporate responsibility. The configuration and employment conditions of the home-based workshops [41] were also analyzed. The resulting evaluation of these aspects is of interest since these types of workshops play a central role in the local textile value chain. They are also contingent on the relations built related to the "ways of doing" due to the local production tradition and craftsmanship.

With this analytical basis, the development of variables and measures was addressed. In the first CEDiS draft, outsourcing was not included due to its implementation complexity. Firstly, there was a debate around the assessment of the concept. From the social sustainability perspective, outsourcing arises as a positive aspect since the manufacturing activities allow the creation of jobs outside of the company's facilities enhancing local development. However, also in the local context, outsourcing has a negative aspect related to labor instability. This can be observed in the working relations that appear asymmetric regarding development and salaries when comparing home-based workshops and factories [41]. Consequently, it is a sensitive topic for companies, and even under the self-declaration nature of CEDiS, they were not able to disclose in-depth and verifiable information. GIDSu took into consideration the goodwill of the interviewees and the observations of the interviewers considering the certification protocols.

However, once the pilot cases were gathered, it was clear the difference between companies that centralized their production and those that outsourced most of their processes. The first ones incurred high personnel investment, while the second ones did not know the situation in the workshops. In this first approach, both types of companies got a similar score on the CSR unit of analysis. This showed the researchers the need for including the topic of outsourcing and sub-contracting into the certification. Firstly, as a sensitizing method and to open the "pandora box" of the local production; secondly, to put under the spotlight the sub-contracting modalities and to implement, in the future, improvement strategies.

In a second CEDiS round, companies were questioned about their engagement in outsourcing activities. The aim was to examine the company's responsibilities and their knowledge about the operational and production aspects of the workshops they worked with. Furthermore, GIDSu resolved that those companies not outsourcing any of their processes were going to be identified as *optimal*. Outsourcing is not considered here as negative per se. However, the idea is to reward those companies that make an extra effort to keep all processes inside their facilities. Considering the local context, it takes a significant investment to keep a stable workforce in the long term (salaries, facilities maintenance, employee training, among others). This can be interpreted, as an achievement from the private sector, in an unstable economic context and a seasonal production industry.

For the final CEDiS version, some other aspects were incorporated. Firstly, it was evaluated which reasons lead companies to outsource (if they do), such as the business model selection, bottlenecks, unusual processes, among others. Secondly, the employment situation in the workshops was evaluated. Therefore, three levels were defined: *zero* for those companies who do not know and do not adopt the legal framework for home-based workers; *medium* for the companies opting for the "Monotributo Social" (individual taxpayer category that allows entering the formal economy); and *optimal* for those who adopt the right legal framework (Law 12713).

Lastly, communication and exchange conditions are evaluated. For example, the companies that understand the external processes, and define those together with the workshops, get a positive score. To evaluate this situation, the adoption of standardized communication tools (technical sheets and production documents), delivery systems (of raw materials and end-products), quality standards agreed between parties, and product systematization is considered.

Some aspects of the relationship between the company and the internal employees are measured in the UA "Optimization of the production system".

2.5.3 CEDiS Diagnosis

The implementation of the instrument during 2019 allowed the researchers to test it and to obtain general information about the local industry. The most significant findings related to the UA 4—Social Responsibility, in particular those related to outsourcing, are the following:

- **Difficulties to define the production traceability**. Outsourcing shows different levels of sub-contracting. This means that there are different levels of intermediaries, which makes it hard to identify the responsible party in charge of manufacturing.
- **Persistency of labor irregularities**. The most common outsourcing modalities found are: (1) workshops enter the formal economy as "monotributo" (individual taxpayer category), and (2) there are operational workshops with no legal framework (small places or home-based productions, consisting of mostly women). The first case implies that the relation between the home-based worker and the company-owner is about business, when in fact they have a work and production relationship [40]. A dependence on the home-based workers can be observed, in terms of prices, product conditions, and lead time.
- **Lack of information about the work processes in the outsourced workshops**. This directly affects the quality of the end-product and its traceability.

The opportunities and improvement measures detected are:

- **Finding the correct legal framework for each case**. There are some gray zones between the company-owners and the home-based and outsourced workshops. However, the revision of the Law 12713 is suggested, which explains the acknowledgment of the dependence relation between an employer (company-owner) and the worker [42]. This law considers items such as social security and pension fund of the textile home-based workers. Although this legislation fell into disuse, the idea is to make an interdisciplinary revision meeting (state organisms, businesses, and labor unions) to evaluate the best way to enforce it. This would also benefit other areas such as traceability, making visible the production processes, eco-labeling, and an increase in product added value.

- **Appraisal of skilled workers in terms of trade**, by taking into account the craftsmanship behind textile and clothing manufacture. The fares proposed by the union of home-based seamstresses as a basic price criterion are recommended for use. It is also recommended to use communication strategies to sensitize and make visible the existence of these types of production units (through campaigns such as Who made my clothes?). These strategies could foster in the medium and long term, the regional development, and restructuring of the textile sector. Moreover, these could create a context of enhanced quality, added value, and industry access for vulnerable social groups.

2.6 Units of Analysis: Item 7—Innovation and Expansion of the Useful Life

2.6.1 Background and Adaptation to the Local Context

The structure of this item is based on two dimensions of the D4S Strategy Wheel [34]: optimization of the initial lifetime and development of a new product. In terms of D4S, the optimization of a useful life involves those expected product features that allow the extension of the product life cycle. The conditions to analyze are related to reliability and durability, ease of maintenance and repair, modular product structure, classic design, a strong relation between user and product, and involvement of local systems. Additionally, new product development involves those eco-design variables related to conceiving a new object. Some of its proposals are dematerialization possibilities, shared use of a product, function integration, and functional optimization of components. The D4S wheel also includes a third dimension: optimization of the product's end-of-life, which refers to reuse, remanufacturing/refurbishing, recycling of materials, in connection to local systems.

From the beginning of CEDiS design, GIDSu decided to address the dimensions of optimization of the initial lifetime and new product development together in one UA: Innovation and expansion of the useful life. At first, the item "optimization of the product's end-of-life" was evaluated separately from that UA. However, in consecutive iterations, the researchers decided to approach the item tangentially due to the difficulty to analyze it with the information provided by companies. For example, recyclability and compostability are extremely specific and rely on given physical and chemical conditions [43]. Moreover, the local waste disposal system, particularly the textile one, is diffuse. Then, many companies define their products as recyclable or compostable, only by their material composition without knowing, and so not being able to inform, about the right guidelines for the end-of-life processes. In the future, this variable is expected to be gradually incorporated into CEDiS.

As a result, UA-7 incorporates a systemic perspective, without downplaying its two composing dimensions. This proposal aims for a comprehensive vision of the design process, related to sustainability and the possibility of innovation to extend the product lifetime from its projection. It is in line with the idea of Fletcher and Grose [11] that in the clothing and textile sector, sustainable transformations should be made in the system more than in the products. Also, Gwilt [19] proposes to explore new strategies when designing and producing: to think about innovating systems that could, for example, include renting and returning, shared use, product combinations. This holistic view aims to change the perspective of product-only sustainability. When observing the local value chains, characterized by small- and medium-sized productive organizations, innovation appears widely and is fostered by the interaction between different local stakeholders.

In CEDiS, UA-7 is one of the most relevant variables assigned with a high weight. This decision refers to the importance of innovation in small- and medium-sized companies that allow them increased competitiveness in new markets. The focus is given to innovative strategy design to guarantee the maximal expansion of the product lifetime. This approach takes into account that company owners still consider sustainability as an obstacle and an expense. When talking about expanding the product lifetime, it sounds contradictory to the main goal of selling that companies have. Accordingly, the same global debate between capitalism and sustainability is reflected in the local value chains.

Following the D4S variables, this dimension is focused on evaluating:

- product durability strategies;
- design for maintenance and easy repair from local systems;
- modular product structure, transformable or disassemble;
- classic design;
- a strong relationship between the user and the product; and
- disruptive product development cooperating with scientific and technical institutions.

This dimension enables a wide range of possible answers, aligned with the company's possibilities and its context. To avoid mistaken associations between innovation and technology—mainly related to hardware and software—there are open-ended questions with some explanatory examples. Those strategies that help to reduce textile and clothing obsolescence are considered here, such as durability, ease of maintenance and repair, customer after-sale services or quality assurance systems, multi-functional or modular structures, classic design, or the relation with the user (signature design, capsule collections, inclusive design, among others). There is also a question about disruptive design (such as developing new yarns), or incremental design (blending yarns, combining fabrics or changing garments to improve functionalities, changing machinery, among others). This dimension measures the interactions between organizations, under the framework of the National Innovation System (Sistema Nacional de Innovaciones in Spanish), and the involvement in other institutions to develop new products, enhance the quality, or optimize processes. For this dimension, it is highly valued the demonstration of intention or future projection, even when it is not implemented at the time of evaluation.

2.6.2 Found Problem Areas and Recommendations for Continuous Improvement

The findings related to the UA 7—Innovation and expansion of useful life define certain problem areas and potentialities for the clothing and textile sector. Among them:

- **Difficulties to engage with scientific and technical institutions, R&D, universities**. Although some companies demonstrated their link with this type of organization, when looking to achieve product or process improvements, it is mostly about a one-time relation. The possible access barriers to these institutions are the lack of dialogue and knowledge/research dissemination and the lack of articulation between science and technic with the industry.
- **Deficiency of product maintenance and repair strategies**. Most of the companies mentioned that they have an after-sale service for their customers, in cases of damaged goods. Although, beyond this option, there are no other strategies implemented to innovate in the user's care and repairment of the product.
- **Limited relation between the user and the product**. Some strategies to strengthen this relation were detected, although quite incipient. While some durability strategies are implemented, the idea is to focus on a comprehensive user experience related to a symbolic aspect.

Design acts as a creator of innovative ideas, comprehensive projects that enable the implementation of sustainability across the company. The difficulties to implement such innovations are attributable to the economic adversities or the lack of investment, according to companies. Respectively, there are some suggested improvement measures, which require low investment and would work as relieving actions for the periods of declined production:

- **Articulation with differential propositions**. Innovation is fostered in varied scenes, such as in participation in the "Good Design Seal" (Sello de Buen Diseño in Spanish), association with business chambers, the involvement in technological actions as the hackathon, among others.
- **Involvement with scientific and technical institutions, and R&D organizations**. Multiple university research centers work on a wide range of topics that could assist the development and improvement of troubled areas. Also, companies could present their inquiries and research topics to research institutions. The strengthening of the link between the private and public sectors and academics acts as a central driving force for innovation.
- **Design creative strategies for process and product innovation**. Each company has specific features and implemented systems that could be efficiently improved. This is where the designer's creativity plays the role of a unique strategies' developer. Examples of these could be art or brand collaborations, renovation or adaptation of machinery for new products, yarn blending, among others.
- **Design linking strategies, between the user and the product, in terms of symbolism and functionality**. However, the project aims to the local value

chain, consumers' choice answers to global social and cultural trends. The strategies comprehend the functional and symbolic dimensions. The first ones refer to propositions that relate the consumer with product use and care. For instance, the DIY concept by giving assembling or customization instructions, garment care instructions, or even online tutorials about these aspects could be implemented. Regarding the symbolic strategies, it could be implemented as a genderless capsule collection or inclusive collections also with a wider size curve, among others. All of the mentioned examples are intended for achieving consumer empathy with the product and thus, creating brand loyalty.

3 Final Considerations

CEDiS is the result of a series of researches over many years and the engagement of a diversified team involved with sustainable design in Argentina, particularly in the city of Mar del Plata. This project has been nurtured by different sustainability areas and the academic journey of the involved researchers and scholars. Industrial design has been entangled in new ways of innovation. Social innovation, particularly, depends on specific design actions: experimentation, replication, and connection, both in expert design and diffuse design, are fundamental to global development, initiatives, and goals (loc cit). According to Manzini, "*is it possible to outline a design scenario built on a culture that joins the local with the global (cosmopolitan localism), and a resilient infrastructure capable of requalifying work and bringing production closer to consumption (distributed systems)*" (loc. cit, p. 2). That is why, designers' hardcore differentials should include a refexión-action capacity which allows people to develop in a symbolic environment, diverse and complex (Galán 2018, p 71).

In this way, the article first focuses on the strength to develop a specific method for the local context. Previous studies lead to identifying the local difficulties for the implementation of eco-labeling and other sustainable certifications, into the clothing and textile value chain. These difficulties are related to economic aspects, disregard for the topic, and bureaucratic management challenges. After all, local sustainability is a relegated topic for clothing and textile SMEs. Therefore, the CEDiS project focuses on a self-declaration model introduced by a simple audit that is conducted by the researchers and experts, who contrast the information through prior knowledge and observation during the interview. Its structure was designed to analyze those sustainable design aspects quickly and systemically, evaluating the results, identifying problem areas, and proposing improvement measures. This structure enables local companies to reach sustainability processes and gives them the chance to gradually incorporate green actions into their processes and products.

Furthermore, CEDiS helps to prevent the blind spots that a firm could incur when performing green actions. Such blind spots could be innovative strategies not correctly informed to the customers, sensitizing about their product traceability, or outsourcing in poor conditions. It is also acknowledged the interest of micro-, small- and medium-sized enterprises, in the sustainability issue as a way of capitalizing the specific know-how in the region. The structure of micro and SMEs develop an enhanced human connection and particular knowledge about the local value chain, where territorial processes are easier to control, and where quality is valued over quantity ([13]; Thackara, *op.cit*).

It is also worth emphasizing that GIDSu's aim has always been to replicate CEDiS for other industries and has been making progress around it. The group has already analyzed sustainability and design among metal-mechanic and plastic local firms; however, this is not enough to scale the assessment method. On the one hand, each industry has its own impacts: differences translate into specific impacts and strategies. This would mean a careful redesign of the CEDiS, to adapt the tool and maintain its simplicity and usability. On the other hand, CEDiS's development did not only rely on theoretical work. On the contrary, CEDiS success is entangled with the development of the Textile and Clothing Cluster of Mar del Plata. CEDiS work was based on a common comprehension about what sustainable clothing meant, common goals and values, shared along with firms and institutions.

At this point, relationships between state, university, and private sector regarding sustainability become more relevant. Clearly, a joint action to transition toward sustainable projects in the city is fundamental for projects like CEDiS to thrive.

This experience shows the need for the implementation of academic projects involving community transference. For social innovation, design action is fundamentally based on social work along with the territory. These research-action scenarios [12] are highly valuable for the discipline and allow designers to identify new intervention roles, for both the professionals and their context. This strategic perspective is larger than product design and leads it to new management areas, coordination, and action in the territory and its actors. In the case of CEDiS, its implementation led to active linking the sector considering the empiric experience, iterating about inconsistencies that could complicate the tool implementation. Also, it turned the researchers into managers with the need to make CEDiS attractive for companies and make them participants, linking public and private institutions, managing resources [29].

Finally, this certification method was developed to be a flexible solution that encourages constant feedback between stakeholders and the creation of new improvement and cooperation scenarios. At GIDSu, this pilot experience lead to think about consumers' awareness, how to further implement CEDiS in Mar del Plata, and how to make this certification attractive for companies by its autonomy and advantages.

References

1. Alburquerque M (1999) Manual del Agente del Desarrollo Local. Ediciones SUR, Santiago de Chile
2. Alonso Herrero J (2002) Maquila domiciliaria y subcontratación en México en la era de la globalización neoliberal. Plaza y Valdés, S. A. de C. V.
3. Bernatene Ma R, Canale G (2018) Innovación sustentable en Diseño a partir de la integración del análisis de Ciclo de Vida (ACV) con Cadenas Globales de Valor (CGV). Cuadernos del Centro de Estudios de Diseño 69:151–174
4. Brezet H, Van Hemel C (1997) EcoDesign: a promising approach to sustainable production and consumption. UNEP, France
5. Cabut M, Mocerla O A, Bertone B, Petrillo J D (2019) Estudio sobre las incubadoras de la Universidad Nacional de Mar del Plata y de la Unión del Comercio, la Industria y la Producción. Observatorio Tecnológico – OTEC. FI UNMdP, Mar del Plata
6. Canale G (2010) La Caja de Herramientas del Diseño Sustentable. El Diseñador como generador de consecuencias antes que Productos. 5to Encuentro Latinoamericano de Docentes de Diseño UNC, Córdoba
7. Canetti R, Retamozo E, González Trigo M, Vuoso V, Zimmermann M (2014) Análisis de la Cadena de Valor de un Producto de la Industria Textil/Indumentaria Marplatense. 1er Congreso Latinoamericano de Diseño: Diseño, Producción y Región. San Juan, Argentina
8. Charter M, Tischner U (eds) (2001) Sustainable solutions: developing products and services for the future. Routledge, UK
9. Costa E, Rodríguez G (1998) La industria textil marplatense ante la globalidad y el cambio. FACES 5:45–75
10. Favero B. (2011) Las tramas de una identidad: el trabajo textil doméstico entre las mujeres inmigrantes italianas de posguerra en Mar del Plata a partir de entrevistas orales. Jornadas Internacionales Sociedad, Estado y Universidad, Mar del Plata
11. Fletcher K, Grose L (2012) Gestionar la sostenibilidad en la moda. Diseñar para cambiar. Editorial Blume, España
12. Galán B (2007) Diseño y territorio: Transferencia de diseño en comunidades productivas emergentes. In: Díaz-Granados F (ed) Diseño & Territorio. UNC, Bogotá
13. Galán B (comp) (2011) Diseño, proyecto y desarrollo. Miradas del período 2007–2010 en Argentina y Latinoamérica. Wolkowicz Editores, Buenos Aires
14. Gennero de Rearte AM, Liseras N, Baltar F, Graña F (2005) Creación de empresas por graduados universitarios. Comunicación presentada en X Reunión Anual de la Red PyMEs-MERCOSUR, Neuquén, Argentina
15. Graña F, Bachmann F, Liseras N (2016) Análisis de los determinantes de la innovación bajo un enfoque sistémico. XXI Reunión Anual de la Red PyMEs-MERCOSUR, Tandil
16. Graña F et al (2019) Estudio global de las empresas del Parque Industrial y Tecnológico Mar del Plata-Batán. FCEyS UNMdP, Mar del Plata
17. Grupo GIDSu (2016) Informe Final Proyecto de Investigación GIDSu 2015–2016. Mar del Plata (unpublished)
18. Grupo GIDSu (2018) Informe Final Proyecto de Investigación GIDSu 2017–2018 Mar del Plata, (unpublished)
19. Gwilt A (2014) Moda sostenible. Una guía práctica. Gustavo Gili, España
20. IHOBE (2014) Manual Práctico de Compra y Contratación Pública Verde. Modelos y ejemplos para su implantación por la administración pública vasca. Basque country, Spain
21. INTI-Diseño Industrial (2018) Introducción a la Innovación Sustentable. Herramientas y enfoques para diseñadores. Cuadernillo de trabajo
22. ISO 14.006 (2011) Sistemas de gestión ambiental. Directrices para la incorporación de ecodiseño. https://www.iso.org/obp/ui#iso:std:iso:14006:ed-1:v1:es:sec:4.2
23. ISO 26000, Responsabilidad Social. http://www.globalstd.com/networks/blog/guia-iso-26000-responsabilidad-social

24. Manifiesto por un Comercio Justo Elecciones al Parlamento Europeo (2014) http://comerciojusto.org/wp-content/uploads/2014/03/manifiesto-castellano1.pdf. Accessed 21 Apr 2020
25. Manzini E (2015) Design, when everybody designs. An introduction to design for Social Innovation 2. MIT press, Massachusetts
26. Margolín V (2002) Las políticas de lo artificial: ensayos y estudios sobre diseño. Designio, México
27. Mauro LM, Graña FM, Liseras N, Bosch FB, de Rearte AMG (2012) El sector textil-confecciones en la región de Mar del Plata. In: XIII Encuentro Nacional de la Red de Economías Regionales del Plan Fénix. Facultad de Ciencias Sociales, Universidad de Buenos Aires, Argentina
28. Miguel P (2013) Emprendedores del Diseño: Aportes para una sociología de la moda. Eudeba, Buenos Aires
29. Monacchi MC, Canetti R (2018) Teorizando el mapa del diseño. Caso CEDiS, Mar del Plata, Argentina. REVISTA DAYA Diseño, Arte y Arquitectura 5:87–101
30. Monacchi MC (2017) Destejiendo historias y recuperando la identidad del tejido de punto marplatense. Revista I + A Investigación + Acción I 20(19):61–78
31. Monacchi C (2020) Diagnóstico de la cadena de valor textil marplatense a partir de la herramienta analítica del Design Thinking. Informe Final beca de investigación UNMdP tipo A 2017–2020
32. Nutz N, Sievers M (2016) Guía general para el desarrollo de cadenas de valor. Cómo crear empleo y mejores condiciones de trabajo en sectores objetivos. OIT, Ginebra
33. Observatorio PyME. Gennero de Rearte A, Graña F, Liseras N (2009) Informe regional. Industria manufacturera. Evolución reciente, situación actual y expectativas de las PyMES industriales. Observatorio PyME Regional General Pueyrredón y zona de influencia de la Provincia de Buenos Aires. Universidad Nacional de Mar del Plata, Argentina
34. PNUMA, Programa de las Naciones Unidas para el Medio Ambiente (2007) Manual en Español D4S. http://www.D4S-de.org. Accessed 24 Mar 2020
35. Portales L, García de la Torre C (2009) Modelo Penta-dimensional de Sustentabilidad Empresarial (MOPSE). Primer Congreso Internacional de Micro, Pequeña y Mediana Empresa: Cumex, Pachuca, Hidalgo, México., Pachuca, Hidalgo, México
36. Rech L (2019) Exposición Estudio global de las empresas del sector textil y confecciones Mar del Plata-Batán. Grupo de Análisis Industrial FCEyS UNMdP, Mar del Plata
37. Retamozo E (2017) Análisis del ciclo de vida de un indumento y su packaging mediante el uso de un software específico. Revista I + A Investigación más Acción 19:79–98
38. Retamozo E, Bengoa G (2015) Empujando el carrito de compras: el consumidor como motor del ecodiseño. DISUR VIII, Buenos Aires
39. Retamozo E, Bengoa G (2016) Producción y sustentabilidad en Argentina: Estrategias de diseño de indumentaria. Revista I + A Investigación más Acción 18:77–106
40. Ruppel, C (2017). Tercerización Textil, la Otredad de la Producción. Caso Mar del Plata. DISUR IV, Mendoza
41. Ruppel C (2019) Innovación social aplicada a las pequeñas esferas productivas: Caso de las costureras a domicilio en la región de Mar del Plata. Tesis de maestría en Diseño orientada a la estrategia y la gestión de la innovación, dependiente de la UNNOBA
42. Salgado P (2015) Deslocalización de la producción y la fuerza de trabajo: Bolivia - Argentina y las tendencias mundiales en la confección de indumentaria. Universidad Nacional de Tres de Febrero, Buenos Aires
43. Thackara J (2015) Cómo prosperar en la economía sostenible. Diseñar el mundo del mañana. Experimenta, España
44. Venturini EJ (2011) Diseño para un mundo sustentable: reflexiones teóricas y experiencias en diseño industrial. https://rdu.unc.edu.ar/handle/11086/15074
45. Wrisberg N, Udo de Haes HA, Triebswetter U, Eder P, Clift R (eds) (2002) Analytical tools for environmental design and management in a systems perspective. The combined use of analytical tools. Springer, UK

46. Zimmerman M (2015) Ecoetiquetas en el mercado: Contribuciones a una conciencia ecológica en la ciudad de Mar del Plata". In: Luis Rodriguez G (ed) Actas XXIX. Jornadas de Investigación. XI Encuentro Regional. Si + TER. Investigaciones territoriales: experiencias y miradas. FADU UBA, Buenos Aires

Non-governmental Organization Support for Sustainable Artisan Business

Elizabeth Davelaar and Marsha A. Dickson

Abstract The purpose of the study was to determine how training and support provided by non-governmental organizations (NGOs) impact the capabilities of artisans and their craft groups' internal constraints and business sustainability in the context of external constraints that may influence the long-term viability of artisan businesses. Awareness of social and environmental responsibility was explored as one potential internal constraint, among others. Utilizing a multiple case study design with embedded units of analysis, five case studies were conducted with artisan groups and their NGO partners during a 10-day field study in Guatemala. Additional interviews were conducted for context within the Guatemalan artisan section with 3 NGOs, a trainer, and an apparel executive. Two main conclusions can be made from this study. First, the product development trainings have an impact on artisan capabilities, but have a positive impact on business success only when two things are present: demand for the product and a strong NGO partner. A second conclusion from the study is the strategic choice of imported raw materials for product differentiation contributes to artisan group sustainability. This differs from previous research. Lastly, it was determined that artisan had little awareness of environmental and social responsibilities.

1 Introduction

The purpose of the study was to determine how training and support provided by non-governmental organizations (NGOs) impact the capabilities of artisans and their craft groups' internal constraints and business sustainability in the context of external constraints that may influence the long-term viability of artisan businesses. Awareness of social and environmental responsibility was explored as one potential internal constraint, among others.

Research questions were

E. Davelaar (✉) · M. A. Dickson
University of Delaware, Delaware, USA

M. Á. Gardetti and R. P. Larios-Francia (eds.), *Sustainable Fashion and Textiles in Latin America*, Textile Science and Clothing Technology,
https://doi.org/10.1007/978-981-16-1850-5_3

(1) What type of support have NGOs provided to expand artisan capabilities for addressing the internal constraints they face?
(2) What is the relationship between changes in artisan capabilities and artisan group business sustainability?
(3) How aware are the artisan groups about social responsibility and environmental responsibility?
(4) What external constraints are influential to artisans and their business success?

2 Literature Review

2.1 *Overview*

Artisan crafts have been created and sold within the local communities around the world involving traditions of daily life that are passed down through generations. They have become useful tools for development because they allow families and communities to make an income from things that they already know how to do, using materials that are easily accessible or grown locally [3, 14, 19]. Yet, traditional markets for artisan products have shifted because of globalization [19]. Innovations used in western goods, such as synthetic fabrics and industrialized manufacturing, create substitutes for traditional crafts that can be made cheaply and quickly. This has created less of a market for artisan goods, forcing many artisans into financial crisis because their crafts are no longer a viable option for making a living [3, 19]. Although cautious of creating neocolonial relationships and criticized for their efficacy, many NGOs work with artisan groups to strengthen their skills, support them in pursuing better methods of sales, and innovate new products and processes for producing and marketing goods for new markets [2, 6, 19].

2.2 *Chamber's Web of Responsible Wellbeing*

When looking at any form of development solutions, it is vital to consider the people involved (artisans) and put their needs first; a perspective that is incorporated into a conceptual framework developed for the study. Chamber's Web of Responsible Wellbeing contains five elements that work together to describe the development and was the starting point of the conceptual framework. The elements in Chamber's Web that were incorporated in this study were capabilities, the abilities or skills of the individual, and sustainability, the overall long-term viability of the solution. Chambers explains that capabilities can be elevated through training, practice, and education [5], such as the training and workshops offered by NGOs for artisan groups they work with [15, 21].

NGO interventions range in the degree of their involvement in the development and sale of products made by artisan groups. Some organizations do not intervene with product development, only buying what they find, while others are deeply involved with the product development process of the artisan groups. MarketPlace India, for example, has designers from the United States involved in much of the product development process [22].

When looking at NGO intervention, cultural appropriation and related issues need to be addressed. Many skills are region-specific, giving that region a comparative advantage for those skills [31] and many artisans keep to the traditional methods of production, much of which is by hand. The resulting culturally significant products have been knocked off by Western designers and other craft makers [19]. Intellectual property for cultural crafts is difficult to define because these crafts are based on the generational traditions of the region [8]. In 2003, UNESCO instituted a convention regarding the safeguarding of intangible cultural heritages, as well as respect for communities that had strong intangible heritage [30]. Safeguarding of the heritage does not mean freezing time for the community but instead giving them support so they can continue and grow in their traditions (UNESCO n.d.).

2.3 *Internal and External Constraints*

There are various internal and external constraints threatening the sustainability of businesses in the artisan sector, and these were incorporated into the conceptual framework. Internal constraints are the ones that are within the artisan's control and can be impacted by expanding their capabilities. These include technology, production levels, access to local markets, and education. External constraints are the ones that are not within the artisan's control and include economic policies, legal framework/industry formation, and market structure, and raw material availability [3, 7, 15, 16, 23, 26].

Previous literature has shown that hyper-competition is prevalent in the artisan sector because of low barriers of entry, which creates over-saturated markets making it hard to remain competitive. The informal structure of the market also makes it difficult for the artisan to obtain financing to buy raw materials [3, 16, 18, 19]. NGOs also have been known to engage in "compassion exports" or export items that are of subpar quality [18]. Social and environmental sustainability and cultural relevance also represent external constraints faced by artisan businesses.

2.4 *Understanding Artisan Business Success*

It is important to remember that artisans are business people and want to create products that will sell [1]. Likewise, because artisan enterprises are businesses, understanding what success means in both a conventional sense as well as for the artisans themselves was important for measuring the impact of NGO support and the long-term sustainability of the business.

Criteria for a successful small business were addressed in a study done by Gorgiexski et al. [10], of 150 Dutch business owners. The criteria that are useful for determining "entrepreneurial success" for this study are discussed below.

Profitability and growth are two traditional measures that are popular criteria for measuring business success [10]. Because artisan businesses are typically income-generating activities, with all money going back to the families [14], income for artisans for this study will be more important than the overall profitability of the business.

The other criteria important for this study include innovation, which is defined as "introduction of new products or production methods"; firm survival/continuality, which can either mean that the firm is financially successful enough that it can be passed on from one generation to another or sold; and contributing back to society, which means that the business is "socially conscious [and uses] sustainable production methods" [10]. Many artisan businesses who work with NGOs are concerned with fair and socially conscious labor practices, creating better lives for the community through honoring culture, building relationships, and having environmentally conscious products [11, 20].

Personal satisfaction is another criterion for measuring success and is defined as "attaining [the] important things in life, such as autonomy, challenge, security, power, creativity, etc." [10] this is similar to Chambers concept of wellbeing. The wellbeing of artisans in India was analyzed by Littrell and Dickson [22] and self-respect, confidence, better healthcare, respect, changes in housing, and increased social interaction were some of the biggest benefits to wellbeing that resulted from working with the organization that was studied.

Satisfied stakeholders are "satisfied and engaged employees [and] satisfied customers" [10]. Satisfied stakeholders of artisan businesses include the organization buying and selling artisan products, the customers around the world that purchase the products [20], and the artisans themselves [22].

Public recognition means that the organization/business has a good reputation and is well known [10]. For artisan groups, this may come from being number one in the community in sales [1] or from international sales and the recognition their countries get from the sales [12]. The last measure for business success is the utility or usefulness of the organization in the context of the society where it is present and that it fills a need for a product and a service [10]. Artisan groups fill the need for work for the artisans that work within them, ultimately creating income for the artisan through sales.

2.5 Guatemala and Artisan Crafts

One area of the world where craft-based development can be found is Guatemala, where field research was conducted for this study. Poverty is widespread in Guatemala, with over half of the population living in poverty [28]. Cloth and clothing have played a large role in Maya culture, which is the ethnic population found in the Guatemalan highlands. Traje is the traditional clothing that is worn by the Maya population. It is handwoven primarily by women on backstrap looms as a part of daily life, with each region having patterns specific to them [17, 25, 27].

While traje is a source of pride and Mayan unity [25, 27], the indigenous population of Guatemala experiences discrimination, especially the women, when wearing their traditional clothing [24]. During the civil war, which lasted from 1960 to 1996, people abandoned traje for fear of getting targeted and potentially killed. Over 200,000 people were killed in the war and over 80% of the victims were of Mayan descent [13]. However, the creation of traje is still a skill that many people possess and use for income generation.

3 Method

The research method used for this study was a multiple case study design with embedded units of analysis. Each case focused on one of five artisan groups and the NGO that was working with the group, with the artisan group being a unit of analysis and an NGO they work with being the other. A case study research was useful because it answered the "how" and "why" of the impacts of product development training in the field [32].

For each case study, individuals from the NGOs and the artisan groups they worked with were interviewed to understand how training was approached and the outcomes of the training. Because fieldwork was conducted in Guatemala, a retired NGO director, who is originally from the United States but has lived and worked in Guatemala for 16 years and is an expert in weaving and artisan craft, was recruited to assist with the research. She arranged local logistics and recruited participants from different organizations and groups based on sampling requirements, set up interviews during the ten days of field research in Guatemala, and acted as the Spanish to English translator and travel guide during the field research.

The research for each case study involved tours of the NGO offices and interviews with the NGO employees, which typically took between two and three hours. Then interviews were conducted with artisans who worked with them, which lasted between one and three hours (see Table 1). Additional interviews with a trainer, three other NGOs, and an apparel executive and board member of AgExport were conducted for further insight into the context of the artisan work in Guatemala. One interview was conducted totally in Spanish, three were conducted mostly in Spanish but with some Kachiquel translation where needed, and one was conducted using entirely double translation from Kachiquel to Spanish to English.

Upon returning to the United States, recordings of the interviews were transcribed and then analyzed using an inductive approach, reading and rereading the data, and making connections through this process to form themes associated with the conceptual framework and research questions.

Table 1 Case Study Overview

Case Study #1: Building Baskets, Building Opportunity
NGO Location: Guatemala City and Panajachel Artisan Group: 12 women total, 3 interviewed Location of Artisans: Small village west of Sololá Products Created: Baskets Organization Responsible for Training: IXOQI'
Case Study #2: "Compassion" Training
NGO Location: Panajachel Artisan Group: 10 women, 6 interviewed Location of Artisans: Patanatic Products created: Baskets, randas, neckties, clay necklace beads, tassels Organization responsible for the training: Randas training from NGO Also sold to two other NGOs in the area
Case Study #3: Weaving Opportunity
NGO Location: Santiago Atitlan Artisan Group: 30 women, 10 men, 6 women interviewed Location of Artisans: based in and around Santiago Atitlan Product created: Woven goods Organization responsible for training: Fundap
Case Study #4: Weaving in the Mountains
NGO Location: Chimaltenango Artisan Group: 5 women, all of which were interviewed Location of Artisans: Agua Caliente Products created: Woven products (backstrap/foot loom) mirrors, and crowns Organization responsible for training: NGO
Case Study #5: Friends for Life
NGO Location: Chimaltenango Artisan Group: 5 women, all of which were interviewed Location of Artisans: Comolapa Products created: Woven products (backstrap/foot loom) Organization responsible for training: NGO and AgExport partnership

Inductive analysis was also conducted on notes from observations made at local markets and the transcripts of additional interviews. The case studies and other information were then compared and contrasted to each other and with previous literature under the structure of the conceptual framework developed for the study. This cross-case analysis answers the research questions and develops a theoretical model that can be used to guide future research.

4 Results

The results for this study are presented as five separate case studies. In the final section, the results of the additional interviews and observations from the markets are presented.

Case Study #1: Building Baskets, Building Opportunity

Case Study #1 included an NGO founded in 1989 that had locations in Guatemala City and Panajachel and sold home good products ranging from baskets to felt animals. They sold wholesale to stores in Guatemala and exported to different vendors in the United States. The NGO had extremely high standards for quality control and imported different materials to elevate the quality of their products (Fig. 1).

The artisan group that the NGO worked with comprised 12 women ranging in ages from 17 to 41. The group was formalized in 2015 after receiving basket training at a local institution, and three women in the group found the NGO to partner with. They recruited nine other members to learn informally from them. They created baskets of grass that were gathered from around Lake Atitlan, pine needles that are grown in a specific region of Guatemala, and raffia that was sourced from Madagascar via the United States. This group was led by a 21-year-old, who held the group to the same high-quality standards the NGO expected.

Fig. 1 Baskets created by artisan group

Because they had the opportunity to work with the NGO, the artisans were able to develop skills in business and production. They had a hierarchy of responsibility, with a president, vice-president, secretary, and treasurer, and these four worked together on handling money. The group also developed a production schedule, splitting up the work evenly each month and assisting each other in meeting deadlines if needed.

Internal constraints were impacted because of product development training that was provided by a local training facility and the partnership with the NGO. The group had actively worked to find an NGO partner to create a demand for the products they learned to create through the training, thus increasing access to local markets. Their ability to create products efficiently was positively impacted through the training and practice because the women were able to execute to the high-quality standards their NGO partner desired. Interestingly, new financial constraints were added because of the training as the women had to purchase the pine needles to make the baskets.

Business success was expanded due to the increase in individual income from consistent work-making baskets. The artisans also enjoyed their work and were proud to create high-quality products. The group worked well together and respected each other.

The women associated sourcing the grass and the pine needles to be environmental impacts from the products they created. The grass needed to be pulled up from the root to ensure the fibers were long enough, leading the women to wonder whether eventually there would be no grass left to harvest.

Access to raw materials and transportation for delivery was the biggest two external challenges faced by the artisans. The pine needles can only be found in a certain region of the country and can only be harvested from January to April and can only be picked after it rains. The raffia for the baskets also had to be imported because it is not available in Guatemala. Transportation for delivery was an external constraint because of costs and time required. Delivering to Panajachel from the artisans' village took either 150 quetzals to hire a driver or 27 quetzals for a bus ride that had multiple buses and stops and took more time.

Market access was also an external constraint for this group. Baskets designed by the NGO differed from traditional ones because of the incorporation of imported raffia instead of the traditional omega and modern designs (flat coasters, placemats, and chargers). To maintain a competitive advantage, the NGO asks that the artisans they source from not sell any baskets to anyone other than them, including baskets that do not make it through quality control and are not accepted. This restricts production opportunities for the group, and tightly ties their business success to the NGO. The products created by the group had interesting cultural implications because they drew on traditional skills prevalent in the area, but the artisans had not traditionally made baskets. The design and the addition of raffia changed the product into something that was not traditionally found in the area.

Case Study #2: "Compassion" Training

Case Study #2 included an NGO founded in 1996 in Panajachel. The NGO was directed by an American woman, who was interviewed, and relied heavily on international volunteers to run the organization. They had their own private line of textile

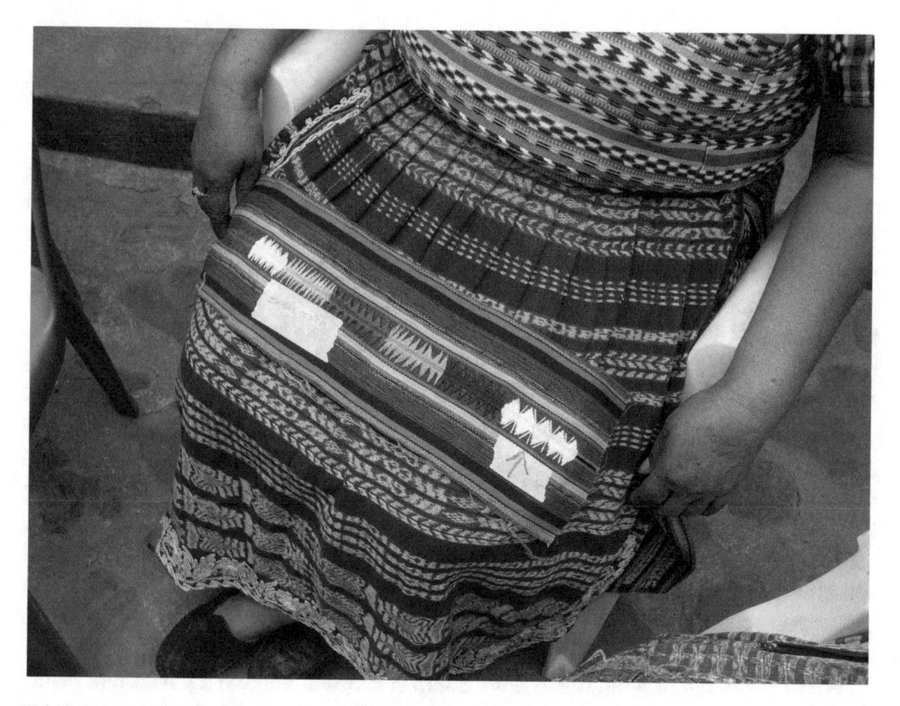

Fig. 2 Example of Randas created by artisan group

goods (bags, table runners, etc.) they exported wholesale, but also worked with designers around the world to create specific products for their clients. The artisan group included six women ranging in ages from 28 to 70. The group was founded in 2007 and originally included 12 members. They began with a basket training in 2007 provided by the NGO but had requested additional training form because of the need for more work (Fig. 2).

The group had the skills to create baskets, randas, tassels, hook rugs, neckties, clay beads, and embroidery. The group sold not only to the NGO in the case study but also to two other NGOs in the area. They had received randas training from the NGO but had only two orders for the products. Each artisan was able to do one stitch well but had failed to learn all of the stitches taught because orders normally only included one stitch. This further segmented a group that lacked cohesion.

Even with the additional training in randas, the group faced the internal constraints of limited finances, had issues with quality control, access to local markets, and limited production levels. There was not a strong demand for any single product; instead, each of the women made several different products (baskets, hook rugs, and other embroidery products) that used completely different skills for sales to at least three different NGOs. For all of the skills the women had, there wasn't a single product that had strong consistent demand from any NGOs.

Efficient production levels were an internal challenge, especially in regard to the randas. Because each woman had only learned one stitch, orders needed to be fulfilled by one person who could do the stitch. For very large orders for just one stitch, it would have taken substantially more time than if they all knew how to do all the stitches. Through practice they were improving their skills and had hopes that more orders would come.

Financial constraints stemmed from the women's responsibility for harvesting and paying for the pine needles used for making the baskets. Additionally, while the group is a legally formed business, they entrusted payment of taxes to the NGOs they work with. This internal constraint could have been lifted by additional education.

Lastly, quality control was an internal constraint because the group did not have high standards for their work, and the NGO studied would accept and sell most products. This resulted in the group not being able to meet the quality control standards of other NGOs, causing them to miss out on the opportunity to have consistent orders.

While the artisans appreciated that the work provided them some income, the group overall was not financially successful. The whole group had earned a total of 2000 quetzals working for the NGO studied, did not have monthly orders from the NGO, and the training itself had not provided a regular income. Although the artisans said that they liked the work, they indicated they would continue it only until something better came along, suggesting the artisans were unsatisfied stakeholders in the partnership. The group lacked cohesion and every woman was working to make money for herself.

When asked about what made their group successful and what gave them pride, there was a very long pause in the conversation as the women thought about what this meant to them. The women were proud to be part of a group that their products were sold internationally and that they had opportunities to meet international visitors, which had opened them up to new experiences. These responses point to the personal satisfaction of the group, which was a positive finding for business success.

The group did not see any harm done to the environment in their production methods because if not used for the baskets, the pine needles that they use would decompose, leaving no detrimental impacts on the earth. In regard to social sustainability, the women in the group liked what they did and wanted to continue to make baskets and randas until something better came along. They wanted their children to learn the skills they had to supplement their education from school.

Raw materials were an external constraint because the women must travel and pay to collect raw materials for the baskets. Hyper-competition was a prevalent external constraint for this group; the women had vast skills, but no distinguishable products or markets to sell them in.

Case Study #3: Weaving Opportunity

Case Study #3 involved a unique organization whereby the artisans were not a separate entity; instead, they were employees of the NGO. The NGO was founded in 1983 and was located in Santiago Atitlan. The NGO focused on creating upscale products that would be exported to the United States, Canada, and Hong Kong, sold in Antigua, or sold in their own store at their headquarters. Thirty women and 10 men worked for the NGO, all having different tasks on different products, from winding warps, weaving, to sewing. The women interviewed had received jaspe and brocade

Fig. 3 Examples of products created

training, both given by a local training facility. Both trainings were given because the NGO saw a need for more efficient or expanded production or an opportunity to reach a new market (Fig. 3).

The NGO had a much different internal division of labor than the rest of the groups interviewed. Each artisan had their own specialty, with only two or three being skilled in each area, and orders were given based on what the women could do. Every product created for orders had two to three people working on it. The organization was moving away from this for jaspe fabrics because of the difficulty in weaving the final product. This was a reason the NGO had offered training.

Production planning for the group is done by one of the founders of the organization. The organization also takes care of money for the artisans and all business expenses. Quality control is handled by the weaving teacher, production manager, and designer if needed.

Access to local markets was an internal constraint that was lessened by the training and by how the group worked. The women do not weave for other organizations but do weave for themselves and their neighbors. One woman said she would not copy or duplicate anything that the NGO made but originally opened her shop because of uncertainties about the NGO's business sustainability. In addition to being able to sell products to friends and neighbors, the training gave the women another skill that they could use to sell in the local market.

The ability of the artisans to efficiently create each product differs for every product and technique but the training that had been used had been actively trying to improve the group's output. Training has addressed how to reduce the time it takes to weave certain designs. Another way that production levels have increased is that more women are involved in jaspe, making the work more evenly distributed throughout the group. The training, overall, had positively impacted the internal constraints the NGO faced.

Sales for the whole group (40 people) were approximately 300,000 quetzals a year, with each person paid approximately 450 quetzals a month depending on skill. It was also reported that the NGO itself was self-sustaining and not reliant on external funds, making the NGO at least a breakeven company and the only one studied that did not rely in part on grants to stay in business. Sales were the biggest challenge for the organization overall, but quality control was a key challenge for production and for satisfying stakeholders and was implemented at every step of production.

The women like working for the organization because it gives them different opportunities and the opportunity to meet different people. They also appreciated that the work was steady, they knew they were going to get paid, and they were able to support their children. Success for the women was on a personal level, feeling loved, and respected and that they were admired for the products they created. The typical buying/selling practices on the street devalue their products and what they are capable of.

In the early years of the NGO, at least two of the women's mothers had been weavers for the group and the women would weave with their mother who would sell their products to the organization. At least one of the women had started working for the organization at 13 years of age. The NGO has since changed this practice, whereby products cannot be made by anyone who is not employed by the NGO in an effort to stop the practice of selling things made by children. While one of the women still had her son help her on occasion, keeping ribbons flat for weaving and winding bobbins, the majority do not have their children help them. The women work on weaving four to five hours a day, on average. The women want to continue weaving as long as they can.

The NGO has incorporated an upcycled leather that is imported from the United States in efforts to improve environmental sustainability. The women believed that most of the products they create did not have an environmental impact. However, the NGO said that those involved in the jaspe training had wondered where the dye goes after it has been used.

Communication was a constraint for this group and the NGO staff because most of the artisans only spoke the local dialect, not Spanish. The administrator who translated for the interview spoke the local dialect, so she could communicate with the weavers easily. When discussing if other people made the same products as the organization, the women said that weaving was the tradition of the pueblo. It was normal for the women that everyone weaves, so while there may be extreme competition, they did not believe they were impacted by it.

Transportation was a clear constraint for this group. They sourced many of their closures from Guatemala City, which requires an entire day and hiring someone to provide transportation. The alternative would be going across the lake to Panajachel and taking public transportation into Guatemala City, a quicker but not always a safer method. During the fieldwork, stories were told about artisans being raped and having money stolen from them on public transportation. Women never went alone, normally in pairs of two, but this was consistently a worry for the NGO and the artisans.

Case Study #4: Weaving in the Mountains

A fourth NGO facilitated interviews with two separate artisan groups they partner with for cases 4 and 5. The NGO had been founded in 1990, in Chimaltenango. Three of the main employees were interviewed for context. This organization had been affected by the recession in 2008, which had caused them to re-evaluate their work. They had not been able to consistently order from the groups interviewed, and there was a lack of trust on the NGO. Still, the organization was working toward creating new opportunities for the artisan group. They exported textiles goods (yamikas, prayer shawls, bags, etc.) to the United States, Vietnam, and Germany (Fig. 4).

The artisan group in this case study was located in a mountain village called Agua Caliente, about 10 km from Comolapa on very twisty dirt roads. The five women in the group could produce products on backstrap looms, foot looms, and they also could sew. The group lacked sales and had not had an order from the NGO in over a year and a half. They also created crowns and mirrors, which were made from cardboard, cloth, and figurines that the women bought from the market in Comolapa.

The group lacked sales for all of the products they created causing them to not work at capacity despite desiring more work. The product development training that most of the women had taken was to learn how to make mirrors with figures tied around the edges. The NGO paid for the training and had sent a woman who worked

Fig. 4 View of road on the trip to Agua Caliente

Fig. 5 Example of wall hanging made by group

for them to do the training ten years prior to the interview. Initially, women were not interested in making the product the NGO wanted to create, so the training did not happen at that time. Six months later, the women saw the mirrors for sale in Antigua and then asked the organization for the training. However, the process was difficult and not all of the women could create the product after the training. There had been very little demand for the new product, and this meant little to no orders from the NGO.

The women were able to weave fabric for products on foot looms, which was typically a man's job. When there are orders for sewn goods, the women who sew cut their own products from the woven fabric, even though only three of the women sew. The two women who could make the mirrors could make two a day.

One main internal constraint that was faced was the lack of access to local markets. The group of women had a broad skill set that would seem to create more opportunities, but the women had very few outside sales other than small quantities to other customers or in the local town.

The women tied much of their success to whether or not they had their orders. The group was legally registered and could issue receipts, which gave the group recognition from the local government, but the group had very limited, if any income, making their business barely functional. The fact that there had not been orders from

Fig. 6 Example of the mirrors created through training

the NGO in over a year was a cause of strife between the women and the three NGO employees that joined the interview, illustrating that the artisans were unsatisfied stakeholders but the NGO workers were also unsatisfied because they could not support the artisans through work. But the women did take pride in the fact that they knew how to do many different things.

The women wanted to continue the work they were doing because it was their life. One woman had her son helping cut the molds for the mirrors and crowns and others let their children help. The women did not think about the implications of their products on the earth.

Transportation and location were large external constraints for this group given their location in a remote village. Public transportation only runs three days a week and picks up from a stop that was a ten-minute car ride from the village. This means that any time orders had to be delivered, it had to be the correct day and it would be very time-consuming for the women. This situation constrained the women from traveling to and from regional markets as well.

Case Study #5: Friends for Life

Case Study #5 consisted of the relationship between the fourth NGO and an artisan group from Comolapa. The original group consisted of 12 women; only five of whom remained active had been together since they were children. They had learned how

Fig. 7 Products created by group in Case Study #5

to foot loom weave and sew from a woman in Comolapa. They were also able to backstrap weave. They had learned how to sew baby hats and booties from the NGO, although there had not been orders from the NGO for some time. They tried to sell their products in the local market but were not always successful.

The sewers were the only ones that had received training, and that was how they created the children's items for the NGO. They had been given two workshops for the baby products, each meeting three times over a week; one was taught by an American woman and the other by a Guatemalan woman. Six artisans attended the training and they learned to make two hats and baby booties. The most valuable thing that they took away from the course was their ability to create more items. In addition to sewing, the women also weave the fabric. For production, all of the women cut their own products and have a meeting to split up orders based on what they can do at certain points in time.

Three of the women were in charge of the finances, keeping records, and quality control. The weavers could get their yarn in Comolapa, but the sewers had to go into Guatemala City to purchase their supplies, specifically zippers. They would go in pairs when they needed to get supplies and they went whenever they had orders.

Access to markets was an internal constraint faced by this group. In addition to selling to the NGO when there were orders, the women would also sell huipiles in town at the local market. While they had a wide variety of skills and had the ability to produce products from weaving to sewing, there was not a large demand for their products from the NGO and their products were not very different from others in the community, only using higher quality thread, meaning they faced competition.

The women linked their success with the fact that they did not have work, thus when looking at overall business success this group was not successful. Yet, the women were very proud that they were able to support their families through the work they were doing. The women appreciated that they were able to take a customer's design and translate it into a product and that their work does not get rejected by the NGO.

This group was interesting because they had all started working at approximately the age of ten. The women wanted to continue the work they were doing because they felt that they were too old to switch jobs and this was something that they knew and could support their families with. The women do not think about how their products affect the earth. They use basic commercial yarn for their products, so may not have the opportunity to think about its origins and any environmental impacts.

One of the biggest external constraints for the group was the reliability of raw materials. The group sourced yarn from the higher quality yarn producer in Guatemala, but the company was very unreliable. Randomly, colors would become unavailable and stay unavailable for months. This was an external constraint because it was a sector-wide problem, affecting many artisan groups.

4.1 *Insights from Additional Interviews*

Interviews with a trainer, three other NGOs, and an apparel executive and board member of AgExport provided further insight into the context of artisan work in Guatemala. Key themes from these interviews are related to the quality of artisan goods, materials used in products, and education/literacy of the artisans.

The quality of the products was addressed in several interviews. The trainer said that the materials locally available are of low quality and yarn bleeds, making everything created for local consumption disposable. She believed that improving quality would help things sell better, not create as much cut-throat competition, and give people the opportunity to sell items at a fairer price.

To elevate the quality and distinctiveness of the goods, several of the interviews addressed innovative solutions to materials and importing of materials. One NGO in Guatemala City had differentiated its products by using naturally colored cotton, which was locally grown in Guatemala. Another NGO in Panajachel created hooked rugs out of old t-shirts. Everything used in producing the rugs, the base fabric, the hooks, and the embroidery hoops used to hold the base fabric steady, and all had to be imported until recently. Now, the NGO is able to locally source the base fabric but still relies on imports for the rest, except for the t-shirts, which are available in local second-hand markets. Importing raw materials limits the amount of competition they have.

A company in Antiqua also addressed the need for differentiating materials for Western markets. They had an artisan switch from making rugs from a traditional material that was coarse to making the same product out of softer cotton. The company was able to sell the product as a blanket and increased the marketability and sales of the product.

Having education-level constraints for the artisans was a key theme as well. When talking about training and quality control with the business in Antiqua, one insight the business gained was that whoever does the training requires the artisans in the training sessions to repeat back what they did in order to make sure that they understand it. The education level of the artisans was discussed in the context of hyper-competition and culture by the Guatemalan apparel executive, who observed that artisans lacked understanding of how new products add value. The artisans see that their neighbors are making placemats and being successful, so they started making placemats. He also said that a possible impediment to training or business success is related to literacy. Women know how to write their names and how to answer basic questions but often do not know how to read and write. This leads people to say they understand and demonstrate the basics, but then not actually understand the basics, which can exasperate communication barriers further.

5 Cross-Case Analysis and Discussion

5.1 Overview

This section includes a cross-case analysis that addresses the research questions and discussion of the findings in the context of previous literature. Based on the findings, a model of Product Development Training Impact on Artisan Capabilities and Sustainable Business Success is presented depicting how training affects capabilities and business success, and the role of internal and external constraints.

Research Question #1: What type of support have NGOs provided to expand artisan capabilities for addressing the internal constraints they face?

The artisan groups from all of the case studies have learned a new skill, for example how to weave jaspe. In this aspect, the training could be seen as positively expanding the capabilities of the artisans [5]. However, the training for some artisan groups did not impact the capabilities as much as they should have, with the artisans successfully learning only part of the skills taught or only a few artisans completing the training. Business capabilities developed through the trainings included accounting for splitting up money and production planning based on skills and available time of the individual artisans needed to produce the orders.

Despite the expanded capabilities provided to the groups through training, the results of those on internal constraints differed, reducing internal constraints for some, leaving them unchanged for others, and in some cases adding new constraints. Internal constraints were similar to those described in the literature [3, 15, 16, 26] and included access to local markets, efficient production levels, and quality control.

In general, financial constraints were reduced because the NGO partners assisted in supplying materials, which reduces the overall financial risk required by individuals [6].

The most successful trainings were based on need—either to impart skills needed to pursue a new market or a need for increased production capacity; thus, they reduced those internal constraints faced by the groups.

In contrast to the needs-based training, the NGO from Case Study 2 trained the artisan group on a skill in hopes of increasing their income, but without clear market demand or standards for quality. Similar to the ILO's [18] reference to compassion exports, this situation could be called "compassion training", one that the trainer interviewed indicated must stop in order for the industry to grow.

Training did not influence the internal constraints quality control or access to markets for some groups because of lack of demand from the NGO. *Not having a partnership with a strong NGO seemed to have increased the internal constraints of the groups.* For example, the NGO in Case Study #2 had only limited orders and lenient quality standards, so the women did not have opportunities to continue to improve their skills.

While the trainings offered did expand the capabilities for artisans, the nature of the partnership between the NGO and the artisan groups had a bigger impact on expanding capabilities that lessened the internal constraint of product quality. Quality control, while not addressed directly in many of the trainings, was a main concern for some NGOs and the artisans were directly involved in the execution of the quality control. This was particularly seen in Case Study #1, where the quality control demanded by the NGO was very high and the artisans were expected to understand and execute the high standard. The finding is similar to that of Littrell and Dickson [21] who found that artisans working with an NGO in India had increased capabilities because of the partnership.

Research Question #2: What is the relationship between changes in artisan capabilities and artisan group business sustainability?

Business success was analyzed based on several criteria. One element that was consistent across all of the cases was that being part of the artisan group brought positive personal satisfaction for the women. They could all find one element that made them proud, no matter how successful the business was overall.

The artisan groups in Case Study #1 (basket makers) and Case Study #3 (jaspe) seemed to be the most successful overall. Both artisan groups saw an increase in income from the trainings because they were able to sell the items that they produced. Both cases had satisfied stakeholders in the NGO partners and their international consumers, mainly because of the high-quality control executed by the artisan groups. Ultimately both businesses did what they were formed to do, which was to generate income and had positive usefulness/utility.

Even though all the women in the cases could find positive personal satisfaction, the other three artisan groups (2, 4, and 5) could be considered less successful. The trainings had not provided substantial additional income, and the groups had experienced negative growth, having fewer members than they had started with.

Additionally, because of the lack of substantial income, the artisan businesses were not meeting their intended outcomes, making them less useful than they should have been.

One criterion that was not listed as the ten found by Gorievski et al. [10] but that was a key criterion for success observed in these case studies was the demand for the product created by the groups. An unanticipated finding was that an increase of capabilities through diversifying skills (e.g., basket weavers learning tassel making) did not necessarily mean that the business would be successful. In contrast, the two that seemed the most cohesive and happiest with the partnerships (Case Study #1 and #3) had learned one new skill (basket weaving) or had built off existing skills. The group from Case Study #2 that had developed the most diverse skills was the most fragmented, which was observed in their body language during the interview and in how many NGOs the various individuals in the group worked with. The relationship between training and business success revolved around the demand for the product and the overall success of the training. If there was a strong demand or need for the product, then it was more likely the business would be successful because they had a sales outlet.

Research Question #3: How aware are the artisan groups about social responsibility and environmental responsibility?

NGOs, specifically those belonging to the Fair Trade Federation, work toward the goals of having "environmentally sustainable production practices… [and] assurance of safe and healthy working conditions" [11]. There had not been previous research on how artisans are aware of these concepts. The artisans in this research had limited awareness of both social and environmental sustainability.

Research Question #4: What external constraints are influential to artisans and their business success?

The external constraints observed in this research were similar to those described in the literature [3, 15, 16, 26]. Transportation was a challenge previously observed by rural artisans [23] and was a constraint that impacted the artisans in Guatemala. Distance to and from the NGOs and also to the garment district in Guatemala City was not only time-consuming but also could be very dangerous for the women, even in pairs. While the war ended in 1996 [4], extreme poverty meant that violence was still a reality for the people in Guatemala. Being afraid of violence was a constant for many people and dictated how the artisans traveled in support of their businesses.

Access to local raw materials for production was a sector-wide problem for the artisans in the case studies, and sometimes made delivering correct orders on time difficult. The findings here contrasted to those of Durham and Littrell [7] regarding the importance of local raw material availability when starting a handicraft enterprise. Some NGOs in these case studies were very reliant on imported raw materials, using them to differentiate products from traditional local products. The imported materials added an additional element of quality and distinction that was not available in locally sourced materials and allowed for strong export sales.

Finally, sector formation and practices were constraints for the artisans and NGOs, specifically hyper-competition and quality control. Hyper-competition has been found to be prevalent within artisan sectors because of low barriers of entry and low skill levels needed for production [16]. Hyper-competition was observed in all of the markets visited, with the majority of the artisans selling very similar products, where the product differentiation observed by the artisan groups had more business success. In contrast, when training taught skills that were already in the community or other regions in Guatemala, hyper-competition was prevalent in markets.

The practice of training the artisans' skills used by others in nearby regions also brings up questions about cultural appropriation. Traditionally, each region in the highlands of Guatemala had a different huipil style and the skills that when along with that style were specific to that region [17, 25], giving that region a competitive advantage for those skills [31]. Teaching artisans from nearby areas or regions can give less significance to the skills and erodes the competitive advantage of the community that traditionally claimed that skill.

The sector practices around quality control also were an external constraint felt by the NGOs and artisan groups. There was a disconnect between Guatemalan quality and the expectations of Western buyers throughout the whole sector. Artisans would create products to the wrong specifications or with the wrong colors, causing a loss of production capacity and sales. While some of the NGOs would pay a lower price or delay payment for lower quality goods, they would still accept most of the products. These findings provide support for past research that has shown that quality control had created problems for NGOs [6] and that NGOs who engaged in compassion exports, specifically those who were more interested in helping people than buying high-quality crafts, had the potential to be hurting the sector more than helping the artisans [18].

6 Conclusions

Two main conclusions can be made from this study. First, the product development trainings have an impact on artisan capabilities, but only have a positive impact on business success when two things are present: demand for the product and a strong NGO partner. Without demand, NGOs engage in compassion training, which would contribute to the over-saturation of skills and ultimately does not help the artisan group improve their skills or businesses. The quality of the NGO partnership, along with the continued practice, and rigorous quality control were the three factors that positively influenced the internal constraints of the artisan businesses. High-quality goods differentiate the products from what is found traditionally in the markets, which helps sales to international clients. NGOs focus on business also continues to ideate and help the artisans use their skills learned in trainings for long-term production. The artisans who had spent a considerable amount of time learning the skills and had opportunities to perfect them over continued orders were more successful.

A second conclusion from the study is the strategic choice of imported raw materials for product differentiation contributes to artisan group sustainability. Traditionally, it has been recommended that NGOs use raw materials sourced locally, but one of the partnerships that was observed to be the most successful in business differentiated their products with imported materials. Other NGOs interviewed also found that this was the best method for differentiating products and for creating the highest quality items.

But through differentiation of products through raw materials and demanding high-quality products, the NGOs and the artisan groups could make positive changes in the industry.

From these conclusions, as well as the findings from the research questions, a model of Product Development Training Impact on Artisan Capabilities and Sustainable Business Success was developed. As seen in Fig. 8, strong NGO partnership, production capacity needs, and new market opportunities are important inputs for product development training. Product development training impacts capabilities and both the training and capabilities then impact internal constraints of market access, quality control, financials, and production levels. Being able to successfully address these internal constraints then influences the long-term viability of businesses. A strong NGO partner also directly impacts sustainable business success by helping create demand for products and providing support. A feedback loop from sustainable business success to internal constraints and capabilities shows the ongoing influence of successful business. All components of the model are surrounded by external constraints and cultural considerations, which are out of the control of the artisans themselves but impact the overall success of artisan businesses.

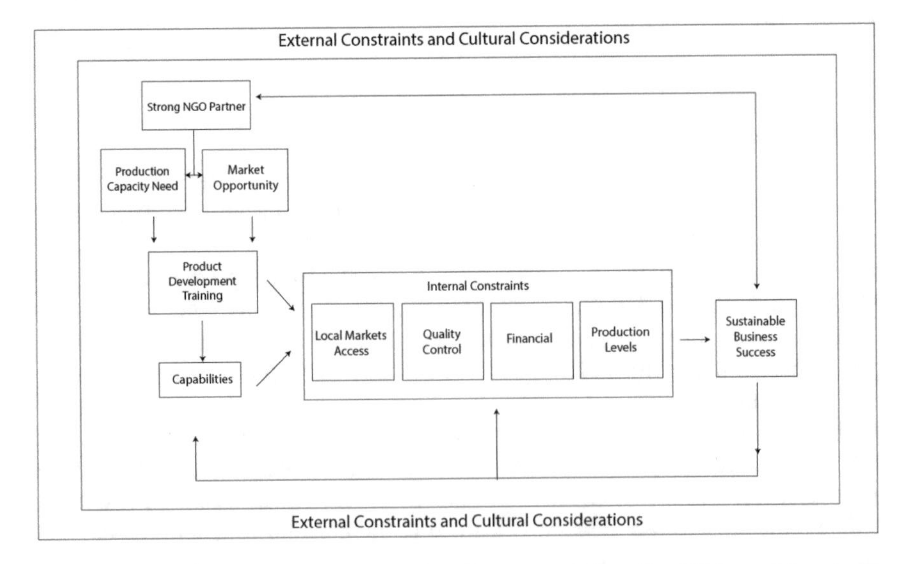

Fig. 8 Model of Product Development Training Impact on Artisan Capabilities and Sustainable Business Success

Like all research, this study had limitations that could be addressed by future research. One opportunity would be to analyze the economic impact that choosing imported raw materials rather than locally available materials has on the local community, the artisans, and NGOs, and how consumers react to the products. Because of the extent to which different NGOs used imported materials, additional research needs to be done on the impact these materials have. Another opportunity, based on a limitation of this study, is to more fully research how environmental and social sustainability is understood by artisans with the goal of understanding whether limited awareness is specific to Guatemala or is an international theme.

References

1. Antrosio J, Colloredo-Mansfeld R (2015) Fast, easy, and in cash: Artisan hardship and hope in the global economy. The University of Chicago Press, Chicago
2. Arvelo-Jimenez (2004) Kuyujani Originario: The Yekuana road to the overall protection of their rights as a people. In: Finger JM, Schuler P (eds) Poor people's knowledge: Promoting intellectual property in developing countries. The International Bank for Reconstruction and Development/The World Bank, Oxford University Press, Washington, DC., pp 37–51
3. Bhatt ER (2006) We are poor but so many. Oxford University Press, New York
4. Calderon MJ (2002–2011) Timeline: Guatemala's history of violence. WGBH Educational Foundation http://www.pbs.org/frontlineworld/stories/guatemala704/history/timeline.html
5. Chambers R (1997) Whose reality counts?. Intermediate Technology Publications, London, England, Putting the first last
6. Dickson MA, Littrell MA (1998) Organizational culture for small textile and apparel businesses in Guatemala. Clothing and Textiles Research Journal 16(2):68–78
7. Durham DE, Littrell MA (2000) Performance factors of Peace Corps handcraft enterprises as indicators of income generation and sustainability. Clothing and Textiles Research Journal 18(4):260–272
8. Finger JM (2004) Introduction and overview. In: Finger JM, Schuler P (eds) Poor people's knowledge: Promoting intellectual property in developing countries. The International Bank for Reconstruction and Development/The World Bank, Oxford University Press, Washington D.C., pp 1–36
9. Fluitman F (1989) Training for work in the informal sector. International Labour Organization, Geneva
10. Gorievski MJ, Ascalon ME, Stephan U (2011) Small businesses owners' success criteria, a values approach to personal differences. J Small Bus Manage 48(2):207–232
11. Grimes KM (2000) Democratizing international production and trade. In: Grimes KM, Milgram BL (eds) Artisans and cooperatives: Developing alternative trade for the global economy. University of Arizona Press, Tucson, pp 11–24
12. Gobagoba MR, Littrell MA (2003) Profiling micro apparel enterprises in Botswana: Motivations, practices, challenges, successes. Clothing and Textiles Research Journal 21(3):130–141
13. Gordon B (1993) Identity in cloth: Continuity and survival in Guatemalan textiles. Helen Louise Allen Textiles Collection, Madison, WI
14. Haan HC (2001) Training for work in the informal sector: New evidence from Eastern and Southern Africa. The International Training Center of the International Labour Organization (ILO), Geneva
15. Haan HC (2006) Training for work in the informal micro-enterprise sector: Fresh evidence from Sub-Sahara Africa. UNESCO-UNEVOC Book Series, Volume 3. Dordrecht, The Netherlands: Springer

16. Harris J (2014) Meeting the challenges of the handicraft industry in Africa: Evidence in Nairobi. Development in Practice 24(1):105–117
17. Hecht A (2001) Textiles from Guatemala. University of Washington Press, Seattle, WA
18. ILO (2006) Easing the barriers to formality: Registration procedures for microenterprises and handicraft exports in Cambodia. Artisans' Association of Cambodia http://www.ilo.org/wcmsp5/groups/public/—asia/—ro-bangkok/documents/publication/wcms_bk_pb_131_en.pdf
19. Liebl M, Roy T (2004) Handmade in India: Traditional craft skills in a changing world. In: Finger JM, Schuler P (eds) Poor people's knowledge: Promoting intellectual property in developing countries. Oxford University Press, Washington DC, The International Bank for Reconstruction and Development/The World Bank, pp 53–74
20. Littrell MA, Dickson MA (1999) Social responsibility in the global market: Fair trade of cultural products. SAGE Publishing, Thousand Oaks, CA
21. Littrell MA, Dickson MA (2006) Employment with a socially responsible business: Worker capabilities and quality of life. Clothing and Textiles Research Journal 24(3):192–206
22. Littrell MA, Dickson MA (2010) Artisans and fair trade: Crafting development. Kumarian Press, Sterling, VA
23. McGrath S, King K, Leach F, Carr-Hill R, Boeh-Ocansey O, D'Souza K, Messina G, Oketch H (1995) *Education and training for the informal sector*. vol 1: ser 11
24. Macleod M (2004) Mayan Dress as Text: Contested Meanings. Development in Practice 14(3):680–689
25. Miralbes de Polance R (2013) Maya attire and the process of its massification. In: Shaughnessy R (ed) Ancestry and artistry: Maya textiles from Guatemala. Textile Museum of Canada, Toronto, ON, pp 88–99
26. Rogerson CM (2000) Rural handicraft production in the developing world: Policy issues for South Africa. Agrekon 39(2):193–217
27. Shaughnessy R (2013) Ancestry and artistry: Maya textiles from Guatemala. In: Shaughnessy R (ed) Ancestry and artistry: Maya textiles from Guatemala. Textile Museum of Canada, Toronto, ON, pp 12–49
28. The World Factbook (2016–17) Central Intelligence Agency, Washington, DC https://www.cia.gov/library/publications/the-world-factbook/index.html
29. UNESCO (n.d.) Safeguarding without freezing http://www.unesco.org/culture/ich/en/safeguarding-00012
30. UNESCO (2003) Convention for the safeguarding of the intangible cultural heritage http://unesdoc.unesco.org/images//0013/001325/132540e.pdf
31. Wherry F (2008) Global markets and local crafts: Thailand and Costa Rica compared. The Johns Hopkins University Press, Baltimore, MD
32. Yin RK (2014) Case study research: Design and method, 5th edn. SAGE Publications, Thousand Oaks, CA

Flax Fibre Extraction to Textiles and Sustainability: A Holistic Approach

Sanjoy Debnath

Abstract One of the most ancient fibre is flax fibre, found in different archaeological studies and documented in history. It is supposed to be the oldest natural plant fibre found in different parts of the globe. It is a seasonal crop normally grown in different parts of Europe, America and Asian countries. This chapter covers the overall aspects of flax textile starting from cultivation to end product, disposability, and different sustainability steps involved in the value chain of flax. Apart from these, applications of different components of flax biomass and plant residue in some other areas like medicinal, industrial, etc. for diversification have been discussed. Further, the chapter focuses on minimum use of artificial resources (man-made products) for making the product and process environmental-friendly and sustainable.

Keywords Flax · Agriculture and textile · Sustainability · Textile and non-textile product diversification · Holistic approach

1 Introduction: History of Flax Fibre and Its Use in Different Parts of the Globe

The scientific name of flax is *Linum usitatissimum*, and also known as linseed. It is a member of the genus *Linum* belonging to the family of *Linaceae*. The flax fibre belongs to bast fibre category, and the fibre is extracted from the bark of the linseed/flax plant. It is one of the ancient fibres and being used since the human civilization. It is found from archaeological evidence that flax, nettle, cotton were used since ancient times. Egyptians used flax textiles to wrap the mummy during 2325 BC or even earlier to this. This fibre seems to be one of the oldest natural fibre in the earth used for textile applications. Back to many thousands of years, dyed flax textiles found in a prehistoric cave in Georgia (a country in the Caucasus region of Eurasia and located at the crossroads of Western Asia and Eastern Europe,

S. Debnath (✉)
ICAR-National Institute of Natural Fibre Engineering & Technology, 12 Regent Park, Kolkata, West Bengal 700040, India
e-mail: sanjoydebnath@yahoo.com; sanjoy.debnath@icar.gov.in

M. Á. Gardetti and R. P. Larios-Francia (eds.), *Sustainable Fashion and Textiles in Latin America*, Textile Science and Clothing Technology,
https://doi.org/10.1007/978-981-16-1850-5_4

bounded to the west by the Black Sea, to the north by Russia, to the south by Turkey and Armenia, and to the southeast by Azerbaijan) evidence the use of woven linen fabrics from wild flax grown date back over 30,000 years. Dyed flax fibres found in a prehistoric cave in Georgia suggest that the use of woven linen fabrics from wild flax may date back over 30,000 years [16, 29]. It was also found that linen textiles were used since ancient civilizations, including Mesopotamia [33] and ancient Egypt [30, 31]. There are many pieces of evidence of linen throughout the Bible, reflecting the ancient textile's entrenched presence in human cultures. Literature revealed that natural dye extracted mostly from different parts of plants (root, leaf, stem, flower, etc.) and mineral is used to dye/colour the textile materials. With the passage of time, in the eighteenth century and beyond, the linen industry got important in the economies of several countries in Europe [17, 38] and American colonies [27]. Flax was first introduced to the United States by colonists, primarily to produce fibre for clothing [2]. As the United States grew, more buildings were constructed and the need for flaxseed oil also increased. This flaxseed oil commonly called linseed oil used in paint making industries. The demand for flaxseed, or linseed, meal for livestock and poultry feed also increased. As a result, the flax processing industry was developed in the late eighteenth century. By the 1940s, however, cotton had replaced flax as commonly used fibre crops in the United States, and flax became nearly extinct as a commercially grown crop [2].

It is the flax plant wherein the fibre part is used for high-value textile product and the seed popularly known as linseed oil or flaxseed oil is mostly used in industry. This oil is a common carrier used in oil paint. It can also be used as a painting medium, making oil paints more fluid, transparent and glossy. The fibrous part of this plant (Fig. 1) is flax fibre, which is very strong and good absorbent; at the same time, it dries faster than cotton. Because of these characteristics, linen is comfortable to wear in hot and humid weather and is valued for use in apparel. It also has other distinctive characteristics, notably its tendency to wrinkle which gives an esthetic look. There is a long list of textile applications, including home furnishing items, and technical applications made from linen. During the nineteenth century and early twentieth century, flax/linen was contributing to the major economy of Russia [1].

Fig. 1 Flax plant and cultivation

At one point in time, flax/linen was the country's greatest export item, and Russia produced about 80% of the world's fibre flax crop.

Very recent documentation discloses that unlike India many countries are now promoting dual-purpose crop grown annually as a renewable source and are being popularly cultivated in different parts of the world as food and fibre crop [19]. Normally, when the flax fibre is converted into textile material, it is popularly known as linen. The flax fibrous material is traditionally grown to use for bed sheets, underclothes, and table linen. On the other side, the flaxseed used to extract oil, popularly known as linseed oil, is used in industry. This flax-seed/linseed oil, extracted from flax seed, is one of the most useful natural oils, used as a preservative for wood, concrete, and an ingredient in paints, varnishes, and stains. The flaxseeds contain between 35 and 45% oil [35]. This flax oil contains around 10% of saturated fatty acids (palmitic and stearic), about 20% monounsaturated fatty acids (mainly oleic acid), and more than 70% alpha-linolenic fatty acids acid. The protein content in seeds of flax varies from 20 to 30%. The chemical composition of flaxseed has identified areas in the study of preventive and functional properties. PUFA omega-3 family, dietary fibres and phytoestrogen lignans determine hypolipidemic and antiatherogenic actions of flaxseed. Flaxseeds under the conditions of storage and processing technologies are harmless food product. Consumption of 50 g/day of flaxseed showed no adverse effects in humans. Furthermore, the flaxseed cake (after extraction of oil from seed) are being mixed in certain proportion for preparation of poultry feed and to feed the other domestic animal to enrich omega-3 protein [28]. These flaxseed mixture feed produce omega-3 content in the poultry egg. However, this chapter mostly restricts about different aspects of flax fibre, its extraction to textile application and sustainability.

2 Flax Value Chain: Cultivation to Product Disposal After Subsequent Usage

Flax fibre is used to produce linen fabric that is comfortable to wear and use. It is a winter crop and cultivated in almost different parts of the world. A survey report [3, 4] revealed that following ten countries in descending order as the largest flax producer

1. Canada
   Canada has dedicated about 750,000 hectares of land for flax cultivation. This country produces the majority of the flax available on the global market. A part of the flax produced in this country is used to extract oil which is later used by the paint industry. Flax produced in this country is used to manufacture papers for the cigarette. Production in tonnes—872,000 [5].
2. Kazakhstan
   Kazakhstan is the second-largest producer of flax in the world. This country exports 46% of its total flax produced to Europe. Poland and China are the other

two main countries that import flax from Kazakhstan. This country cultivates flax mainly for extracting the oil from its seed and is expected to increase the area under cultivation to increase the production. Production in tonnes—419,957 [6].

3. China
   China is the third-largest producer of flax in the world. The world's largest populated country has been growing flax for more than 90 years; however, this country has been using oil from flaxseeds for more than 2000 years. This country uses flax for manufacturing linen cloth material and has the world's largest linen fabric manufacturing plant. Flax production in this country is the major supporting fibre source; however, the flax fibre straw yield is just half of the European flax fibre.
   Production in tonnes—387,088 [7]
4. Russia
   The USSR was once the world's largest producer of flax; however, the flax industry collapsed as USSR collapsed. 70% of flax was produced by Ivanovo and couldn't recover after the collapse. But, today the Russian government has taken measures to increase the production of flax in the nation and it is the fourth-largest producer of flax. This country is now rebuilding its flax industry by inviting investors to invest in flax production and linen manufacturing. Production in tonnes—365,088 [8].
5. United States of America
   Flax production in the United States of America began before colonization. By the seventeenth and eighteenth centuries, production increased. Today, the USA is the fifth-largest producer of flax. Majority of the flax produced by this nation is grown in North Dakota. A part of flax cultivated in this nation is used to produce oil, and the by-products are used to feed the livestock. This country also manufactures linen, furniture padding, and cigarette paper. Production in tonnes—161,750.
6. India
   India is the sixth-largest producer of flax in the world. The majority of the flax cultivated in India are used for oil, the source material for the manufacturing of paints, printing ink, and varnish etc. The linen industry in India does not use the flax produced in India, but export the raw materials from Europe because the quality of the Indian grown flax does not match European quality standards of flax. However, Chandra Shekhar Azad University of Agriculture & Technology (CSAUAT) has produced dual-purpose flax that can be used for both oil and fibre and matches the European quality flax. Production in tonnes—141,000 [9].
7. France
   France is the seventh-largest producer of flax in the world. However, this country has a reputation for producing one of the world's best quality flax in just 55,000 hectares of flax cultivated land. France along with the Soviet Union were the largest producers of flax once but collapsed with the end of World War. Production in tonnes—64,000 [10].

8. Ukraine
   Ukraine is the eighth-largest producer of flax in the world. Studies indicate that flax production in this country has been increasing since 2011 annually. This country has an average yield of 10–12 hundred kg of flax per hectare area cultivated in Ukraine. Ukraine has notably intensified the processing of their flaxseed by 15%, and this is much greater than in the past years. Vietnam and Poland are two of the main importers of flax from Ukraine. Production in tonnes—25,000. [11].
9. Argentina
   Argentina is the ninth-largest producer of flax in the world. But, once, Argentina had flax production more than nations like India and Russia. However, their production started collapsing since 1999. Today, the majority of the flax cultivated in Argentina is exported as linseed oil. The quantity of production is expected to increase in future as they have implemented more efficient cultivation methods. Production in tonnes—16,000 [12, 13].
10. Italy
   Italy is the tenth largest producer of flax in the world. Cultivation of flax in Italy is very old. This country compared with other countries has low quantity production. However, Italy is one of the three main producers of the world's best quality flax. The flax cultivated in Italy has high global demand. Production in tonnes—15,300 [14].

Apart from these major flax growers in the world, some quantity of flax is also grown in, Pakistan, China and Africa with little inferior (coarser) in nature. As far as the cultivation of flax is concerned, almost all soil types are suitable. However, alluvial kind and deep loams containing a large amount of organic matters are most preferred for this flax cultivation. This crop is often found growing just above the waterline in cranberry bogs. Heavy clays are unsuitable, as are soils of a gravelly or dry sandy nature. Farming flax requires few fertilizers or pesticides. Within eight weeks of sowing, the plant can reach 10–15 cm (3.9–5.9 in) in height and grows several centimetres per day under its optimal growth conditions, reaching 70–80 cm (28–31 in) within 50 days. The crop is harvested around 120 days of age. [2], studied the habitat and agronomic factors which differently affected the length of the flax plant, amount of straw and seed yield. They found that flax straw yield was the highest from the flax cultivation on the soil of cereal-mountain and when rapeseed was the fore-crop and nitrogen fertilization applied at a dose not exceeding 40 kg/ha. This study also concludes that the soil pH has no effect on straw yield; higher rainfall and lower temperatures throughout the growing season had a positive effect on both yields of straw and seed, which has good commercial value. Unlike other countries, Kazakhstan has recently developed few varieties (line) having higher production of both linseed and fibre (flax) which can reduce the cost of cultivation economic benefit to the relevant industries [36].

As far as the flax value chain (cultivation to product disposal) is concerned, the following figure describes the details of it:

Fig. 2 Value chain of flax

From Fig. 2 it is clear that flaxseed, flax fibre and woody particles were separated after suitable harvesting and extraction. The seed has its own application apart from cultivation for the next season. On the other hand, the fibrous part after suitable processing through textile/spinning machinery yarn/fabric can be manufactured. In one word, nothing is wastage from flax cultivation. Even the twigs/woody parts have the potential to convert into particle board. Even during harvesting and extraction, apart from fibrous component, woody part and seed, the remaining leaf and other part are used as compost to enrich the soil.

As far as the disposal issue is raised, the oil cake used for animal/poultry feed and the litter after processing can be treated as good manure as well as feed material for fish. The linseed oil used as natural oil has different medicinal values, used as nutritional food material, as paint making application, wood varnish, metal coating to avoid oxidation, etc. Since the oil is from a natural source and does not contain any harmful chemicals, it can be disposed of safely. The textile materials are prepared from the extracted fibres from the flax plant and can be disposed of safely after certain usages.

Overall we can see almost all parts of the plant is useful for the preparation of a value-added product. Starting from the retting (due/water), the residual component acts as manure in the field in case of due retting, and in the case of water retting, it can be consumed by the microbes in the water and subsequently useful for aquatic plants and animals present over. Flax fibre like pineapple [23], jute, ramie, etc. is annually produced, so no such issue of depletion of natural resources arises (Fig. 3). Figure 3 also depicts that except for the availability of flax fibre, other aspects like disposability, durability, colouration etc. are acceptable, compared to other natural fibres. Based on some of these important factors, flax/linen can well be considered as sustainable luxury textiles. In Fig. 3, the variables defined by the polar diagram are proportionally represented and approximate and no such industry standards represent.

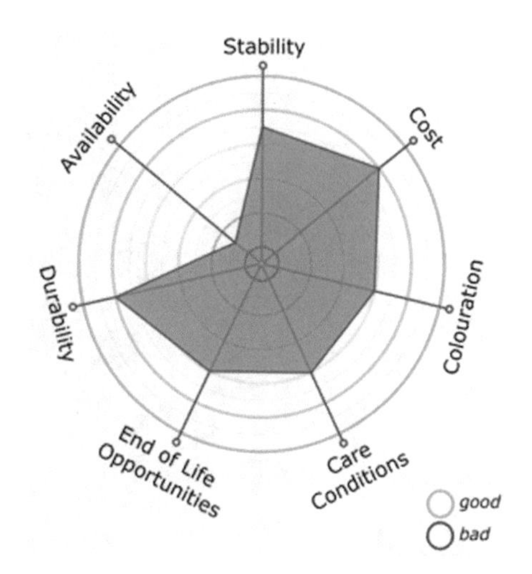

Fig. 3 Polar diagram of comparative aspects of nettle fibre [19, 24]

Apart from these factors, fabric construction, weight and blend composition with other fibres (in the case of blended products) may influence the perceivable ranking.

Renewable fibre like nettle, after subsequent usage of the flax-based textiles, can be disposed of using all end of life opportunities (Fig. 2). It can be handled in a similar way unlike other natural cellulosic fibres materials are disposed of. Many of the time either fall flax or cotton flax blended textiles are popularly manufactured by the leading manufacturers. Hence, there is a good prospect to be re-used or re-manufactured and can also be used as a source of cellulose feedstock for regenerated cellulose products. Being naturally biodegradable, after the end of usage the flax textiles can be disposed of safely or can be composted in some cases, if required as found in the case of nettle fibre [24] shown in polar diagram (Fig. 4).

Water is one of the major sources in any agriculture/technological intervention. Water-saving during agriculture can be achieved using advance/controlled irrigation system and bio-mulching application. Although of these interventions involve additional costs which is negligible if we consider the environmental and soil water conservation aspects. A significant amount of water is again required during chemical processing (scouring, bleaching, dyeing etc.) of flax textiles. Optimum use of environmentally safe chemicals, reuse of chemicals to a maximum extent and proper treatment of the treated liquor after chemical treatment prior to disposal may lead to saving of water resource. This may be a holistic approach, which leads to sustainable goals for future prospects in this flax industry.

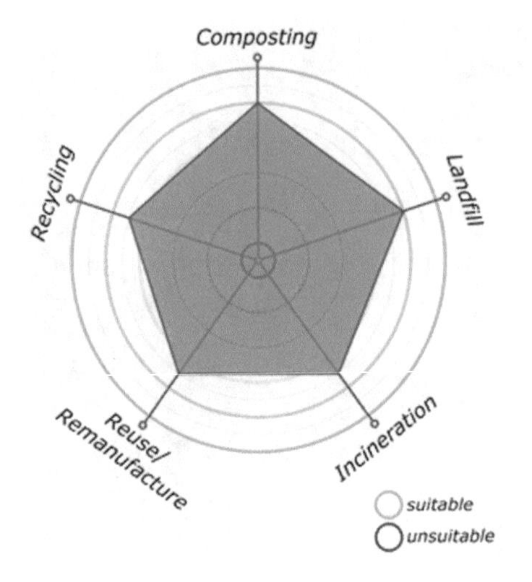

Fig. 4 Disposal of natural textiles [19, 24]

Fig. 5 Manual scutching and processing of flax waste fibre for spinning of high tenacity yarn [20, 21]

3 Product Diversification of Flax and Flax Biomass: Sustainable Possibilities

With the passage of time human civilianization takes place and there has been a lot of contribution of science and technological interventions in product diversifications in every sphere, including flax. Today, in many countries, this linen/flax is cultivated not only as a cash crop but also generates good revenue for country growth. In other words, the country's economy depends on the cultivation of this crop to some extent. As far as degradability is concerned, since it is an annually renewable natural fibre from plant origin, the majority of cellulose base can decompose easily without causing environmental hazards. In today's world, fine flax is used for fashion clothing in the summer climate [19, 37]. It has the wonderful property of absorbing moisture very fast compared to other natural fibres. This property is very important from a comfort point of view. When the body sweats, it absorbs the sweat quickly and keeps the body dry and cool; as a result, the wearer feels comfortable. Apart from its look the material has high moisture absorbing capacity and can be dyed in good shade. Based on the properties of the flax/linen fibre, fibre material gets swell and gains its strength in wet condition. That is why linen/flax spinning is normally preferred to spin in wet condition rather than in dry condition. As far as international market is concerned, there exists lot of pure/blended linen products. Some of them like ramie-linen *Kurti* for women for casual wear, linen blend trousers (50% viscose, 35% cotton, 15% linen. White: 65% viscose, 25% cotton, 10% linen) by Joanna Hope, linen mix trousers and short (55% linen, 45% cotton) by SOUTHBAY, single-breasted linen mix blazer (55% linen, 45% cotton. Lining: Polyester) by Williams and Brown, lightweight fabric will keep you cool but stylish in the warm weather (54% cotton, 46% linen) by Black Level Jacamo, linen mix 3/4 pants (55% linen, 45% cotton) by Southbay, ladies fashion linen bow decor straw braid summer sun hat (linen 100%), linen summer women's wide brim sun hat wedding church sea beach (linen 100%) by Kentucky Derby, ladies pleated crisscross fashion sexy linen dress, women's white linen dress (linen 100%), female slim blazer short jacket/linen blazer/ladies coat, ladies thong linen underwear/briefs/(85% linen, 12% nylon, 3% spandex), women linen shocks (85% linen, 15% nylon), etc. This shows there is a good fashion market that exists internationally. Apart from these, suitable chemical/enzymatic treatment can be given to flax textiles in different structures (fibre, yarn and fabric) for various product diversifications [18, 22, 39].

Flax is one of the oldest fibre plants that does not need specialized attention during the production of the fibre. Because of this reason, many of these fibres are being used for several centuries together prior to much developments in science and technologies. The main advantages of these natural fibres are that they are from plant sources and production to fibre extraction is sustainable [24]. With the advances in fibre technology, many synthetic/artificial fibres are being developed and used for the past 60 decades. However, in the last 15–20 years, special attention has been given to use lesser quantities of synthetic fibre and more emphasis has been imparted on the use of natural fibres at different fibre applications areas, and flax is one of them. With this, production of flax fibre got special attention and application of these fibres in apparel textile and technical textiles fields including fashion areas also increased. Due to scarcity of agricultural land and urbanization, in limited areas more amount of fibre is being produced for which more chemical fertilizers are being used, which may hinder the future crop due to loss of soil fertility in due course of time. However, with this improved productivity, more importance is being used in the overall sustainability of the production system.

Many of the cottage and small-scale industries flax fashion products are being produced directly from raw plant fibres, without much use of chemicals and machinery. Some of the fancy items are also being produced in decentralized sectors. However, large industry including composite plants is used to produce fashion products in huge quantities. There will be huge value addition if we switch to the production of diversified fashion and lifestyle items from conventional products. Many of the industries processing natural fibres worldwide are now concentrating on value addition in product design using fashion as a major component. Hence, product diversification from conventional products is the order of the day for sustainable development in the fashion industry.

Talking about further product diversification, industrial textiles from flax waste (tow fibre), also known as hackling waste, have good potential for making high tenacity yarns [21]. These high tenacity yarns are used for producing defence textiles. Many literature also covers the use of fibre reinforcement composite to make a natural fibre-based composite. These flax composites not only have higher strength but also lighter in weight and found suitable to replace many of the automobile components to make vehicle towards enviro-friendly without affecting the safety aspects. The particle boards prepared from the woody components of the flax plant after suitable extraction have immense potential in the use of medium density boards. These medium density particle boards prepared from flax wood/twig are suitable for furniture, indoor partition wall, false ceiling, etc. application. These boards also have good thermal insulation as well as sound-absorbing capacity, hence also suitable for wall panel in the auditorium to reduce reverberation effect, outside sound and maintain inside temperature. This specific area of application can reduce deforestation, where forest wood is used conventionally to prepare those application products.

The poultry/animal feed to introduce the omega-3 content in the human food chain is already discussed. A lot of successful business is being grown worldwide in this area wherein the raw material is the linseed oil cake (process waste material). The linseed oil has an established market and is a profit-making industry.

Holistically, if we consider each and every product, by-product and process waste, in totality, the industry will sustain on its own. This crop is annually renewable and with minimum support (water, pesticide and fertilizer) grows in various parts of the globe. Hence the production of raw material for textile and non-textile products out of the flax is not a constrain. The quality variation of fibre and oil content in flaxseed across the globe is not uniform. In the case of fibre, Canada, Russia and parts of Europe are potential producers as far as quality fibre is concerned. However, the flax fibre grown in India, Pakistan, China, Latin America and other countries is comparatively coarser in nature. The coarser fibre blended with other fibres or solo can be consumed in producing bed linen, table cover, furnishing, etc., fashionable products for a niche market [26]. The chemical requirement in converting the flax fibre into textile is also lesser compared to other natural fibre processing. Blending of other natural fibre with flax can open up a new direction, though few of the blended products exist in the market as discussed earlier.

4 Conclusions and Future Prospects

Flax being an ancient natural fibre resource have promising potential for future fashion and handicraft industries [25]. Due to the scarcity of natural petroleum-based reserve resources, synthetic fibre production will decline in near future. Optimum use of water and energy will produce a good amount of natural fibres which are still not explored potentially. Focus may be imparted towards wider application of eco-friendly processing technology and product diversification to meet the sustainable goals. It can also be concluded from this chapter that there are many textile and fashion/utility, and non-textile materials based on flax fibre, its extraction waste and by-products (not explored much) have immense potential with proper marketing, advertising and appropriate technologies to convert those fibres into textiles [20, 21] and non-textile products for different diversified applications. Minimum use of hazardous chemicals and maximum use of natural resources during cultivation to product development will lead to a greener process and can meet sustainable goals. Finally, flax/linen fibre has unique properties and thus blending with different natural fibre or blending with a minimum quantity of synthetic fibre will also provide diversified end uses as far as sustainable goals are concerned. Overall, the people involved in producing from fibre cultivation to end products (textile and non-textile) from flax fibre and its extraction waste will be remunerated and a good amount of people will be benefited out of it apart from its environmental impacts and carbon footprints. Hence, more emphasis needs to be made in the coming days to consider in a holistic manner for making the whole process sustainable so that the disposal to nature or cradle-to-grave will be of no problem for the future generation.

References

1. Akin D E (2013) Linen Most Useful: Perspectives on Structure, Chemistry, and Enzymes for Retting Flax". ISRN Biotechnology. 186534. https://doi.org/10.5402/2013/186534
2. Anonymous (2020a). https://www.agmrc.org/commodities-products/grains-oilseeds/flax-profile, Dated 30.06.2020
3. Anonymous (2020b). http://textilefashionstudy.com/top-flax-growing-countries-of-the-world-linen-fiber-production, Dated 30/06/20, 9:53 PM
4. Anonymous (2020c). https://www.worldblaze.in/largest-flax-producing-countries, Dated 10.10.2020
5. Anonymous (2020d). http://www.thecanadianencyclopedia.ca/en/article/flax/, Dated 10.10.2020
6. Anonymous (2020e). https://bnews.kz/en/news/ekonomika/apk/kazakhstan_flax_and_colza_in_demand_in_poland_and_china
7. Anonymous (2020f). http://unifiedcommunity.info/hemp/hemp-articles/pauls-articles/overview-of-hempflaxlinen-production-processing-in-china/, Dated 10.10.2020
8. Anonymous (2020 g). http://www.voiceofchina.com.au/history-linen-china/, Dated 10.10.2020
9. Anonymous (2020 h). https://www.csmonitor.com/2006/0208/p07s02-woeu.html, Dated 10.10.2020
10. Anonymous (2020i). http://agropedia.iitk.ac.in/content/flax-fibre, Dated 10.10.2020
11. Anonymous (2020j). http://www.swicofil.com/products/003flax.html, Dated 10.10.2020
12. Anonymous (2020 k). http://www.proagro.com.ua/eng/news/market/12845.html?t=5, Dated 10.10.2020
13. Anonymous (2020 l). https://www.ceicdata.com/en/argentina/agricultural-production/agricultural-production-oilseed-crop-flax, Dated 10.10.2020
14. Anonymous (2020 m). http://www.nationsencyclopedia.com/Americas/Argentina-AGRICULTURE.html, Dated 10.10.2020

15. Anonymous (2020n). https://www.statista.com/statistics/758135/production-volume-of-flax-fabrics-in-italy/, Dated 10.10.2020
16. Balter M (2009) Clothes Make the (Hu) Man. Science 325(5946):1329. https://doi.org/10.1126/science.325_1329a
17. Belfanti M (2006) "Reviewed Work: The European Linen Industry in Historical Perspective by Brenda Collins Philip Ollerenshaw. Technology and Culture. 47(1):193–195. https://doi.org/10.1353/tech.2006.0056
18. Chattopadhyay DP, Samanta AK, Nanda R, Thakur S (1999) Effect of caustic pretreatment at varying tension level on dyeing behavior of jute, flax and ramie. Indian J Fibre Text Res 24(1):74–77
19. Debnath S. (2020) Flax Fibre Extraction to Fashion Products Leading Towards Sustainable Goals. In: Gardetti M., Muthu S. (eds) The UN Sustainable Development Goals for the Textile and Fashion Industry. Textile Science and Clothing Technology. Springer, Singapore, pp. 47–57. Print ISBN: 978-981-13-8786-9; Online ISBN: 978-981-13-8787-6; DOI:https://doi.org/10.1007/978-981-13-8787-6
20. Debnath S, Basu G, Mustafa I, Mishra L, Das R, Karmakar S (2018) Flax Fibre Extraction to Spinning – A Holistic Approach. Indian Journal of Natural Fibres 5(1):69–75
21. Debnath S, Basu G, Mustafa I, Mishra L, Das R, Marmakar S (2019). Flax fibre extraction to spinning – A holistic approach, Book Title: Natural Fibre Resource Management for Sustainable Development', Dr. A. N. Roy, Editor in Chief, The Indian Natural Fibre Society, Published by Creative Incorporate Garia, Kolkata-700096, India, pp. 69–76
22. Debnath S, (2016a). Unexplored vegetable fibre in green fashion. In: Muthu S S, Gardetti M A (eds) Green fashion, environmental footprints and eco-design of products and processes. Springer Science + Business Media Singapore, pp 1–19.
23. Debnath Sanjoy (2016b). Pineapple leaf fibre–A sustainable luxury and industrial textiles. In: Gardetti M A and Muthu S S (eds) Handbook of Sustainable Luxury Textiles and Fashion, Environmental Footprints and Eco-design. Springer Science + Business Media Singapore, pp. 35–49
24. Debnath S (2015) Chapter 3: Great potential of stinging nettle for sustainable textile and fashion. In: Gardetti MA, Muthu SS (eds) Handbook of sustainable luxury textiles and fashion, environmental footprints and eco-design of products and processes. Springer Science + Business Media, Singapore, pp 43–57
25. Yunus Dogan, Nedelcheva Anely M, Obratov Petkovic Dragica, Padure Ioana M (2008) Plants used in traditional handicrafts in several Balkan countries. Indian Journal of Traditional Knowledge 7(1):157–161
26. Foulk JA, Akin DE, Dodd RB, McAlister DD (2002) Flax fiber: Potential for a new crop in the Southeast. In: Janick J, Whipkey A (eds) Trends in new crops and new uses. ASHS Press, Alexandria, VA, pp 361–370
27. Keegan T A. (1996). "Flaxen fantasy: the history of linen". Colonial Homes. **22** (4): 62. Retrieved 4 June 2020
28. Khan SA (2019) Inclusion of Pyridoxine to Flaxseed Cake in Poultry Feed Improves Productivity of omega-3 Enriched Eggs. Bioinformation 15(5):333–341. https://doi.org/10.6026/97320630015333
29. Kvavadze E, Bar-Yosef O, Belfer-Cohen A, Boaretto E, Jakeli N, Matskevich Z, Meshveliani T (2009) 30,000-Year-Old Wild Flax Fibers. Science 325(5946):1359. https://doi.org/10.1126/science.1175404
30. Lobell J. (2016). "Dressing for the Ages". Archeology. 69 (3): 9. ISSN 0003-8113
31. Warden AJ (1867) The linen trade, ancient and modern, 2nd edn. Longman, Green, Longman
32. Roberts & Green. p. 214. hdl:2027/hvd.32044019641166
33. McCorriston J (1997) Textile Extensification, Alienation, and Social Stratification in Ancient Mesopotamia. Curr Anthropol 38(4):517–535
34. Mańkowska Grażyna, Mańkowski Jerzy (2020). 'The Influence of Selected Habitat and Agronomic Factors on the Yield of Flax (*Linum usitatissimum* L.)', Journal of Natural Fibers, https://doi.org/10.1080/15440478.2020.1810843

35. Martinchik AN, Baturin AK, Zubtsov VV, Iu Molofeev V (2012) Nutritional Value and Functional Properties of Flaxseed. Vopr Pitan (Russian) 81(3):4–10
36. Sheng QC, Stybayev G, Fu WY, Begalina A, Hua LS, Baitelenova A, Yuan G, Arystangulov S, Hua KQ, Kipshakpayeva G, Lin ZX, Tussipkan D (2019) Flax Varieties Experimental Report in Kazakhstan in 2019. Journal of Natural Fibers. https://doi.org/10.1080/15440478.2020.1813674
37. SinclairR(2015)Textilesandfashion:materials,designandtechnology.WoodHeadPublishing Limited, Cambridge, UK
38. Takei A (1994) The First Irish Linen Mills, 1800–1824. Irish Economic and Social History 21:28–38. https://doi.org/10.1177/033248939402100102
39. Tyagi GK, Kaushik RCD, Dhamija S, Chattopadhyay DP (2000) Effect of alkali treatment on the mechanical properties of flax-viscose OE rotor spun yarns. Indian J Fibre Text Res 25(2):87–91

Strategic Design for Social Innovation In The Fashion System: The Sustainable Fashion Ecosystem Case

Karine Freire

Abstract This chapter explores a possible answer to the question: what kind of innovation will be required for a truly sustainable change in the fashion value chain? It presents cultural innovation as a path towards a sustainable society. This cultural innovation is based on Edgar Morin proposition of reform of thought, grounded on the worldview of Buen Vivir, conviviality and complex thinking. From this perspective, we explore a strategic design process for cultural innovation and sustainability that underpins the development of the Sustainable Fashion Ecosystem project

Keywords Strategic design · Cultural innovation · Buen vivir · Sustainable fashion ecosystem

1 Introduction

This text was written in the year 2020. It is important to contextualize it in time, as it was a time when the global COVID19 pandemic alerted us about our forms of living in society. It was a period when we were confined to our homes to protect ourselves and give health systems a chance to save as many lives as possible. And in my view, one of the most obvious issues that have emerged is how much social inequality is the biggest pandemic that we need to fight; and in the wake of it, racism, misogyny and all forms of violence and oppression. These are the reasons that lead me to reflect on sustainability in the fashion value chain. What kind of innovation will be required for a truly sustainable change in the fashion value chain? This text explores a possible answer: cultural innovation as a path towards a sustainable society. And

[1] Buen Vivir or Vivir Bien, "are the Spanish words used in Latin America to describe alternatives to development focused on the good life in a broad sense. The richness of the term is difficult to translate into English. It includes the classical ideas of quality of life, but with the specific idea that well-being is only possible within a community. It embraces the broad notion of well-being and cohabitation with others and Nature" [11].

K. Freire (✉)
PPG Design, Universidade do Vale do Rio dos Sinos, Porto Alegre, Brazil
e-mail: kmfreire@unisinos.br

M. Á. Gardetti and R. P. Larios-Francia (eds.), *Sustainable Fashion and Textiles in Latin America*, Textile Science and Clothing Technology,
https://doi.org/10.1007/978-981-16-1850-5_5

the sustainable society presented here is understood from an integral perspective, based on theories of integral ecology, *Buen Vivir*[1] and conviviality [2, 3, 10, 11, 15]. Here, we seek to present concepts and paths capable of leading towards cultural innovation that is really capable of modifying the unsustainable forms of production and consumption today.

One of the first aspects to be questioned is the rationality of modernity. There is the need to look for another possible rationality to produce knowledge and produce fashion: rationality that replaces modern rationality is linked to dominance, destruction and the sovereignty of man over nature. And in this search, we find a possible path in the epistemologies of the south, which draws upon the knowledge of the original Latin American peoples, especially as to their worldview linked to *Buen Vivir* [1, 15]. It is through a new humanism that reconnects us to the core values of human life in connection and balance with other forms of life that we can produce cultural innovation for a sustainable society.

And in this regard, it is necessary to discuss fashion. Fashion, as a cultural expression, is more than a clothing industry. This allows us to search within the culture itself for the elements necessary for its renewal and reframing. And for this, it is essential to discuss the importance of things. I look for inspiration in Manoel de Barros' poem, "On importance", to reflect on importance. The poet asks us if anyone knows for sure how to measure the importance of things, revealing that importance is defined by the point of view of the person evaluating it, depending on subjectivity and values. So, I reinforce my epistemological perspective: there is no way to separate the observer from the way of producing knowledge. And here I present myself as a researcher. I am a Brazilian and I received my education in Rio Grande do Sul and Rio de Janeiro. They are two places with completely different cultures, but both colonized by dominant Eurocentric thinking, with significant influence from Anglo-Saxon rationality. In the last 10 years, the strategic design for social innovation that I practise has undergone a transition from the influence of Anglo-Saxon thinkers to Italian thinkers. And specifically, in the last two years, I have been trying to decolonize my views as a researcher, studying epistemologies of the south as paths towards design.

Along this path, I follow the call of Edgar Morin [11] for reform of thought, seeking in complex thinking a way to change the paradigm of society towards a more sustainable society. In his proposal, for the reform of thought, Morin [11] seeks to reconnect knowledge as a paradigm alternative to disjunction. It is necessary to understand that a living being is only recognized in its relationship with its environment, from which it extracts energy and organization. Complex thinking is the answer to reconnect and at the same time separate human beings from nature and the cosmos, re-establishing the dialogue between humanistic and scientific culture. It is a way of producing knowledge that operates based on four principles: the systemic principle, circular causality, the hologram principle and the dialogical principle. It also allows us to understand man based on his three-fold nature: biology, psychology and social nature. Affected by and affecting nature and society, man can be an agent of cultural change and innovation. It is a way of thinking that enables us to discuss new ethics, in which responsibility and solidarity emerge as fundamental values in our relationship with each other and with nature.

Complex thinking and its principles have a closeness with the worldview of the original Andean peoples. These are the peoples who have resisted colonization by Western thought for more than 500 years in the Americas and have benefited from what science is capable of producing, without losing sight of the ancestral knowledge that enables them to engage in *Buen Vivir*. The founding aspects of *Buen Vivir* are the vision of the whole, co-existence in a multipolar world, the search for balance, the complementarity of diversity and decolonization [15]. For *Buen Vivir*, the goal of humans is to take care of nature. Society needs to understand itself as a community that has nature and the Whole (Pacha Mama) as its center. In the words of Solon [15]: "recognition and belonging to the group are the keys to *Buen Vivir*, justifying the principle of 'wholeness' as the core of the Andean worldview" (p. 25).

It is based on this worldview that we will revisit the work of the renowned researcher in the field of design, Ezio Manzini, who has been reflecting on the necessary reforms in design culture to make a change towards sustainability since the early 1990s. Manzini (1992) reflects on an ethical basis for design that considers environmental problems, caused by the industrial design itself and that finds a new sensitive horizon for design, capable of generating a wide range of cultural transformations and social practices; design that relates to a value system consistent with the awareness of relativity; and design that seeks in the models developed by ecology the bases for ethics: "in the context of general attitude built on the principles of responsibility and solidarity" [6]. Manzini [6] also highlights that it is the responsibility of design to contribute to the production of a habitable world, in which human beings can express and expand their cultural and spiritual possibilities, more than merely surviving. The author states that the design culture produced by the West needs to change; an idea that is connected to qualities. And quality is associated with complexity. The design needs to develop new models to understand reality without reducing complexity. And as a way forward, Manzini [6] draws on the understanding of doing held by the native peoples of America who believe that the answer is: "to produce and reproduce their cultural world - and therefore artificial world, by seeking to be in tune with the natural environment" [6].

Based on authors who discuss ecology and complexity, Manzini proposes a reconfiguration of doing, not as how to produce, but as how to reproduce, which presupposes: [1] the circularity of material transformation processes; and [2] a new aesthetic that gives value to materials and products that somehow incorporate a trace of their previous existences; this only being possible based on the participation of, care for and attention to all the parties interested in the project. Based on this perspective, Manzini proposes that designers should become a cultural figure in the process of creatively connecting that possible with that expected in visible ways. Design culture must know how to point out paths towards potential possibilities, thereby visualizing aspects of what the world could be like while showing the characteristics that need to be supported for there to be complex ecological balance and culturally attractive and socially acceptable alternatives for the change to occur. In this scenario-building process, there is the need for "social imagination to redesign new systems of values [...] Social imagination emerges from dynamic complexes of socio-cultural innovation in which a plurality of actors play a part" [6].

It is based on the worldview of *Buen Vivir*, conviviality and complex thinking that we will seek to imagine a new society, with a new value system, which does not have the economy at its center, but rather a dynamic balance for living. We need a new society that combines ancestral practices and knowledge with technological advances, whenever they contribute to a balance with nature and strengthen a sense of community. We need a new society that effectively promotes cultural patterns of sustainable consumption, strengthening local communities and ecosystems. We need a new society featuring respect, balance and complementarity between the different parts of the Whole and its bases. As Solon [15] argues, in order to achieve the necessary balance, it is important to depatriarchalize society. Patriarchy, with its forms of oppressing power relations, is one of the main obstacles to the balance between humans and nature [15], that is, to *Buen Vivir*.

In the following sections, we will address the concepts of design for cultural innovation and sustainability that underpin the development of the sustainable fashion ecosystem project. The results will be discussed in light of this knowledge to propose changes in the ways of producing fashion.

2 Buen Vivir: A Sustainable Alternative to Pursue

Buen Vivir is a concept under construction, based on relearning the practices and worldviews of the original peoples from Latin America. It is a philosophical conception and worldview that conceives the relationship between time and space and between man and nature in a different way from the dominant Western worldview. It considers that the universe is in constant evolution based on an indissoluble understanding between time and space. As part of *Buen Vivir*, everything is interconnected and forms a whole: humans, animals, plants and physical and spiritual universe. In this space, the past, present and future live side by side and relate to each other in a dynamic way. Time and space are not linear; they are cyclical. Time passes in a spiral form and the future is intertwined with the past. Every step forward involves going backward; everything is transformed.

The Whole has a spiritual dimension in which the conceptions of the self, of the community and of nature merge and are cyclically linked in space and time. Living considering the Whole implies living with affection, care, self-understanding and empathy towards others. The Whole needs to be favoured, understanding the multiple dimensions and interrelationships of all parts. Learning how to interrelate everything is a challenge. Existence depends on a set of relationships [15].

To engage in *Buen Vivir* it is necessary to value all experiences. Material life is only one aspect of the life cycle. Eating well, drinking well, dancing well, sleeping well, working well, meditating well, thinking well, loving well, listening well, dreaming well, walking well, speaking well, giving and receiving well are important dimensions of life [2]. *Buen Vivir* seeks a balance between human beings, between humans and nature, between the material and the spiritual, between knowledge and wisdom, between different cultures, and between different identities and realities [15]. It pursues a dynamic balance as opposed to growth. Humans are caregivers, cultivators and facilitators of what Pacha Mama gives them. *Buen Vivir* is the meeting of diversity. It practises multiculturalism. It is to recognize and learn from differences without arrogance or prejudice.

To overcome the systemic crisis in which we live, it is necessary to seek systemic alternatives to the dominant model that complement each other and that project other forms of cultural, economic and social organization. *Buen Vivir* instigates us to decolonize our territories and our being; to free our minds and souls captured by consumerism as a path to well-being. *Buen Vivir* must be a factor that contributes to the empowerment of communities and social organizations, in processes that promote their emancipation and self-determination.

3 Strategic Design as an Agent of Change

By understanding the world according to the epistemology of complexity, we start to think in terms of encounters, interconnections, flows and occurrences. Complex thinking replaces the standard of Cartesian thinking, of linear reasoning and even of systemic thinking, as part of a "post-modern" project, thereby accepting change and uncertainty as essential elements of complex and adaptive systems [12].

From this perspective, strategic design as an area of knowledge changes its process to question the status quo, to discover emerging factors, indicators of change to the environment, and to develop strategies in support of reorganizing the system, in such a way that it adapts and continues to exist. Designers contribute to the rearrangement of the significance of systems in such a way that they maintain their identity and persist, that is, "design as sense maker". As Manzini [8] asserts: "[design] collaborates actively and proactively in social construction of meaning; and therefore, also, of quality, values and beauty". We understand strategic design as a design process that is contaminated by the paradigm of complexity to develop, adapt and evolve organizational strategies, which will allow organizations to adapt to changes in the context, thereby sustaining themselves in the long term [9, 18]. The capacity of design to promote dialogue and collective construction is at the core of this process. In this way, strategic design can be understood as a process of social learning that creates apparatus capable of fostering changes in the culture of organizations and society. It is a process that generates knowledge and that internalizes the strategies of adaptation and even evolution [4].

From this perspective, strategic design is an approach that takes a step towards the study of strategies intended to guide design action and, above all, organizational action towards innovation and sustainability. In this regard, the ability to read and interpret signals emitted by the ecosystem, coupled with planning based on scenarios, is at the heart of design processes, since it makes it possible to consider what is regular, evident and possible, but also unpredictable, a chance, a deviation and a mistake. Understood in this way, strategic design makes it possible to give shape to the form, function, value and meaning of overall proposals for actions that give shape to societies and organizations run by people. In this way, it transcends the supply of unique products or services, and considers as a systemic whole the values of social groups, the structures of organizations, the differences in socio-cultural contexts, the potential of technologies and networks, the desired meaningful effects and the communication of processes and results [4].

Strategic design has the scope of thinking and doing committed to life: engaging in the development of projects for social and cultural innovation whose meaning is the harmonious interconnection between different ecosystems; identifying potentialities, weaknesses, opportunities or threats, mapping the context and trends; designing scenarios; and lastly by developing artefacts, services, experiences and culture, with the structuring purpose of allowing organized collectives to move forward with their plans to bring more quality to their life and its contexts. As a consequence, we propose that strategic design is a process for creating strategies that generate value for the different actors of a creative ecosystem. Design can contribute towards the creation of socio-technical apparatus, which, based on the constant creative reconstitution of existing technologies, results in the production of new meanings.

The design process is a creative process capable of encouraging the development of relations between the different actors of the ecosystem of innovation. The objective is to support the collective construction of necessary knowledge, through strategic dialogues, with actors and groups holding different roles. Here, strategic dialogues are understood as ways of thinking and action explained through the construction of possible future scenarios, in which it becomes possible to evaluate the different paths for the construction of the solutions [17]. By way of this collective knowledge construction process, it is possible to generate new ideas to foster the targeted systemic changes. It is a process based on transdisciplinarity and, for this reason, one needs to nurture interpersonal relations, prizing to the maximum the diversity of human qualities [5]. It is a nonlinear process, open to interactions with the environment of which it is part, considering the circular nature of the actions and retroactions. It is a process that accepts and aims to deal with uncertainty, randomness, unforeseeability and contradictions. For this reason, it is open to constant exchanges of information with the environment (in this case, the network of interpreters), through dialogical cooperation. This seeks learning through the comprehension of contradictions and differences.

As it turns out, research on design activity focuses on the creative processes allowing designers to critically anticipate future developments in society in order to be capable of creating a device able to affect it from the perspective of the qualification of life contexts. The ways in which designers perceive and experience the world, through their aesthetic and poetic sensitivity or scientific and operational expertise, lead to a specific form of designing. Designers undertake a qualitative and interpretative reading of realities and seek to identify the elements that will be the basis of solutions taking into account aesthetic, cultural and social values besides economic value.

This study considers strategic design as an agent for change for the entire socio-technical system when operating based on social innovation. Manzini [8] presents a definition for design for social innovation that places expert design as an agent that can be triggered to sustain and guide the processes of social change towards sustainability. In the author's opinion, social innovation is a creative recombination of existing resources for the development of new ways of thinking and doing capable of generating systemic discontinuities in the dominant models. They produce solutions based on new social forms and new economic models.

Manzini [8] points out that strategic design skills enable those involved in the problems to seek multiple perspectives and points of view and to reformulate the issues based on this critical thinking. In addition, strategic design activities can form design coalitions (a project network orientated towards a shared vision and results), through the identification of an appropriate group of partners and the co-creation of common values and converging interests.

Furthermore, the author adds that design specialists collaborate to create an environment favourable to coalitions, that is, to (social, economic and technological) ecosystems, in which the potentials of diffuse design can arise. Design experts must use their design culture (and their characteristic critical constructivism) as a driver to deal with turbulence in the environment in projects aimed at sustainability. In these projects, design experts and diffuse designers establish a social dialogue interacting in different ways (from collaboration to conflict) and at different times (asynchronously and synchronously), with creative and proactive activities. These processes are referred to as co-design.

Manzini [8] calls on design schools to be involved in design research activities, promoting socio-technical experiments, empowering individual communities, institutions and companies towards invention and encouraging the improvement of new ways of being and doing things.

The case discussed in the following section is an answer to this call; based on an exploration of design as activism, in which design activities offer solutions to problems in a provocative way, thereby bringing different views to dominant models. It is a search for results to promote cultural innovation, in which new values and behaviours emerge, thereby breaking with the dominant patterns: Really significant changes are supported by the value of collaboration and improving the quality of life. A theory that underlies this change is *Buen Vivir*.

4 The Case: Sustainable Fashion Ecosystem

Is it possible to trigger an ecosystem? If an ecosystem takes shape in the relationships and practices that are established, how can we trigger the relationships? These questions concerned us at the beginning of this project. We had a clear view that collaboration was a possible way to strengthen local micro-entrepreneurs, who chose to embark on sustainable fashion businesses as an alternative to the dominant system.

In 2016, we identified signs of change in Rio Grande do Sul linked to sustainable production in the fashion industry; more specifically, based on a small set of activist brands that sought in their proposals to promote a culture of sustainability, by producing devices from waste textiles and by valuing the workforce. Some of them had some peculiar characteristics: they had an organizational model identified as a social business and were members of collaborative houses. We began to propose a series of collaborative encounters, led by design processes to imagine ways of enhancing these fashion initiatives, which, through their discourses, advocated sustainability in production processes. After a series of workshops with these initiatives it was determined that it was necessary to give visibility to the initiatives and use the existing collaborative houses in Porto Alegre as dissemination hubs. 2016 was a difficult year for Brazilian democracy and the group eventually broke up, with some collaborative houses closing their spaces. At the same time, we realized that the message of sustainable fashion was growing in Brazil and brands from Rio Grande do Sul were recognized nationally because of the sustainability of their proposals. Some collectives had emerged to stimulate knowledge exchange and increase the visibility of these sustainable brands (coletivo 828 and coletivo viés).

In our study of some of these initiatives, we understood that for strategic design to contribute to the development of a culture of sustainability through the designing of business models in the fashion industry, designers should pay attention to the ecosystem relationships that are established and to the potential environmental impact that the business causes. In addition, designers should pay attention to the political side of their design work and should consider what they fight for and the effects of meaning for sustainable cultural transformation, both in terms of direct delivery to users (product and accompanying elements) as well as the different forms of the business organization.

Another relevant factor in the context is that the State of Rio Grande do Sul was looking for a way to reduce the deficit in the balance of trade for the fashion sector. Our state has had manufacturers along every step of the fashion value chain, but with the opening of markets, globalization and tax incentives from other Brazilian states, some manufacturers have closed and others have transferred their manufacturing plants to other locations. In 2016, we imported a lot of clothes from other Brazilian states and from other countries. The Federation of Industries and unions for the clothing, textiles and trade industries came together to seek solutions in order to boost local business. Together with the Secretariat for Economic Development for the State of Rio Grande do Sul, they called on Universities and other social actors (SEBRAE, SENAC) to think of a way to give visibility to products from Rio Grande do Sul. They tried to bring back a fashion event that had already existed (RS Moda) as a way of giving visibility to products manufactured in the state and stimulating a boost to local businesses. The hypothesis was that local products did not have visibility and that other states promoted fairs so that retailers could learn about products. They attempted a "business as usual" solution that served small and medium-sized manufacturers, within the same dominant system.

As a stimulus for the discussion on paths forward for the state, we proposed an alternative vision, in which sustainability is the vision to stimulate trade in local products. The high labour costs complained about by employers' unions could be re-signified as the value of being socially sustainable. And from an environmental point of view, the least that could be thought about is a reduction in the carbon footprint related to logistics costs. There was already a large retail store in the state that has included sustainability in its mission and that has already worked with its network of suppliers to reduce the environmental impact on its production processes. The proposal presented to the group was to co-create a space for social learning about sustainability in fashion, in order to change from a vision of fashion value chain to a value network; a value network with values including ethical and sustainable fashion, fair trade, transparency and the enhancement of local production and small businesses. The proposal was inspired by the innovation design model [14] in order to give rise to an ecosystem providing multidimensional (economic, cultural, environmental and social) value at different levels (users, organizations, ecosystems and society). The intention was to form a design coalition [8], in which members share a vision of what to do and how to do it and decide to do it together. This design coalition, set up by the network of interpreters of the sustainable fashion design discourse, could trigger a collaborative organization that could experiment with other ways of producing fashion: A space encouraging openness towards the possible, the imagination, the courage to risk making mistakes and learning from mistakes and a vision for entrepreneurship based on the empowerment of women. This was a proposal to discontinue traditional ways of doing business and ended up not being accepted by the group that was part of the government meetings.

Understanding that collaboration and social learning are foundational aspects of the design processes that seek the creation of social and technical devices to transform the world, we sought inspiration from organizations that have re-invented themselves to promote social well-being and the transformation of the ecosystem in which they operate to give rise to a creative ecosystem for sustainable fashion in Rio Grande do Sul.

For this, it was necessary to have clear drivers [16] and to experiment in a social innovation laboratory on possible organizational models to stimulate the transition towards a culture of sustainability in the state's fashion industry.

Over the course of six months, we identified different organizations and people related to the culture of fashion sustainability in the primary, secondary and tertiary sectors of the economy and invited them on a design trajectory to create scenarios and visions of a sustainable innovation laboratory capable of experimenting on a model to stimulate sustainable fashion in Rio Grande do Sul.

Considering that the idea was promising, we sought to connect different actors in society to develop a proposal based on the provocative vision of a social space for learning, collaboration and empowerment of women entrepreneurs. Thus, a strategic design process was initiated, with the various actors who shared this vision, to imagine possible scenarios and paths to implement this idea. Three workshops were held from September to November 2017, with different guests connected to the fashion value chain and sustainability to think about scenarios and visions in order to put the proposal into effect: it was necessary to think about ways of operating, governance and work rules, to disseminate the culture of fashion sustainability and to encourage growth in local production, fair trade and conscious consumption.

Fig. 1 Codesign sessions

In these workshops, we had different groups, with about 50 men and women from different social classes, working in different spaces of the fashion value chain: manufacturing, retail, micro-entrepreneurs, production cooperatives, teaching, creative collectives and researchers (Fig. 1). This diversity of views had the aim of developing visions for future scenarios based on the proposals by Manzini [7]: to bring the future to the present. This means understanding what needs to be done in the present in order to anticipate a possible and appreciable tomorrow; bring forward in the present a way to discontinue the current system of unsustainable production and consumption. Manzini [7] sees in the "activity of strategic design - whose objective is to create new business ideas: a mix of products and services situated in the social and spatial context"—a path towards this discontinuity. For the author, the effects of strategic design processes need to be seen as entrepreneurial initiatives with results in which everyone wins: the producer, the user and the environment.

Based on this methodological premise, we generated two scenarios: sustainable concatenation and *EntreLAÇOS*. The first scenario has a vision of a physical and collaborative environment integrating the different links in the value chain, with governance based on strategic design. Innovation was a necessary premise for the relations between the actors in public–private partnerships, with governance through strategic design; a space thinking about sustainability from an economic, social and environmental point of view, which would operate in a FABLAB format, combining the waste of manufacturing with the possibilities of professional training, entrepreneurship and cooperativism. The second scenario had the vision of being a space for "fashion metamorphosis to change the world": a space for social learning to develop human and social capital through the dynamics of exchanges of abundant knowledge between residents and partners; a space linking a human, transparent and sustainable network, empowering people who pass through it and making feasible initiatives that would not exist on their own. One of the premises was to strengthen small brands, seeking to popularize the products of creative sustainable brands through ecosystem co-creation, thereby establishing a place where success is measured by the number of lives impacted, people empowered and re-integrated into productive life.

Considering the possibilities of discontinuities in the current ways of interpreting sustainability, beyond the triple bottom line, the second scenario was chosen to be carried forward and to strive ahead as a project proposal based on the potential to develop a new understanding of wealth and well-being, through new forms of organization, founded on collaboration.

A group of strategic designers worked on developing a proposal based on this project vision. It established the following principles of the new organizational model to be created: collaboration, openness, experimentation, social learning, a relationship with universities, companies, civil society and the government, cultural and social innovation as the base, and expansion conceived through dissemination strategies.

We set a vision for creating an experimental, collaborative space capable of popularizing sustainable products of brands with creative designs, through ecosystem co-creation. For this, all the actors would need to understand what contribution they would make to the ecosystem through their resources and capabilities and the relationships established in collaborative design encounters. The ecosystem would create social and technical devices capable of transforming fashion based on three main focal points: production, creation and communication.

Based on these design principles, a third workshop was held to co-create a manifesto expressing the values of the space, capable of connecting all the agents necessary for systemic discontinuity. As a result, the following values were established: relationships of mutual trust, true interpersonal relationships, the strengthening of people, the promotion of diversity, the reduction of environmental impact, co-creation and co-production. The co-created manifesto can be seen in Fig. 2. The manifesto was interpreted by local artists and, based on the illustrations created, cards were produced to be distributed in lectures presented at sustainability events, combining online and offline strategies. The strategy of making a card with an illustration was conceived to prevent the material from going straight to the trash. The manifesto contained a challenge, which encouraged people to share their ways of changing fashion on social media, using a hashtag. This strategy was conceived as a way to evaluate the educational effects of the lectures. The #modamudamundo [#fashionchangestheworld] hashtag was created for two purposes: to decentralize the issuance of the message and to evaluate the dissemination of the ecosystem values. As such, any participant in the ecosystem could send messages representing the collective. The emphasis in these actions was activism in design, in order to spawn openness for questioning so as to spread the culture of sustainability.

After months of activist activities, in the monthly monitoring and co-creation meetings, the group decided to create a profile on the Instagram and Facebook social networks, to put in an appearance on the networks, even if the posting was centralized by one person. It was decided that a university researcher, who had received a scholarship to support the ecosystem, would be responsible for updating the profile. The group would continue to collaborate in determining the guidelines for posting,

Fig. 2 Sustainable fashion ecosystem's manifesto

in contributions inserted in the digital collaboration tools used by the group (Slack and WhatsApp). After two months of creating the Instagram account, we already had 1000 organic followers.

The second device was the setup of the laboratory in a physical space, in the quarto distrito (fourth district) of Porto Alegre, in the collaborative space named Vila Flores, using funds from recurring collective fund-raising and exchanges based on the abundant resources we had to offer: knowledge and action in sustainable fashion (Fig. 3).

At the same time, as activist activities for the dissemination of sustainability culture took place, the group found that it is important to have a physical space in order to be able to enhance exchanges. We chose a space in the city that was consistent with the established principles and values and, through the collaboration of the different participants, the ecosystem was physically set up at a collaborative cultural space called Vila Flores. The costs for setup and maintenance of the physical space were shared among the participants through donations and exchanges. For the maintenance of the space, a recurring financing strategy and the offering of services with reduced costs for supporters were devised. The collaboration and financial support of people who believed in the cause were important to understand whether the path should be followed. The absence of a leader or single financial supporter was an important driver for the space to be open to experimentation through the horizontal collaboration of all participants. In previous studies [13], we identified that the centralization of an investor made the continuity of the collaborative space unfeasible. The physical space intention was to serve as the basic structure for the development of sustainable fashion projects. The main objectives of the space include: (1) creating modelling

Fig. 3 First configuration of laboratory's phyical space

and sewing courses to qualify the workforce of low-income women; (2) developing projects related to sustainable fashion with the surrounding community, (3) being open to local sustainable fashion brands; (4) supporting the creation of new brands; (5) establishing a partnership with Banco de Tecidos [Fabric Bank], as a way to encourage the re-circulation of materials in good condition for use. As a result, the expectation was to give rise to female empowerment, income generation, valuing of the surrounding community, valuing and promotion of local businesses and raising of awareness for sustainability culture.

The sustainable fashion ecosystem proposes a self-managed system with the objective of sharing actions, knowledge and practices of (small, medium or large-sized) fashion businesses, which, specifically because of its aggregative nature and dialogue, takes all the actors involved to another practical level—that of creation: of mutual, on-going and changing collaboration, which is adapted to the time and space of each project and each actor. Ecosystem exchanges can lead to new opportunities for each participant of the ecosystem in order to have sufficient means (and to receive assistance from other actors) to evolve, technically and creatively, through new opportunities that give rise to a new ecological mentality, the development of work methodologies and the application of more effective procedural methods that also include co-design practices and concepts.

Over this period, we have collaborated with different initiatives to spread the culture of sustainability through fashion: Vila Flores, Lojas Renner Institute, Fashion Revolution Movement, Virada Sustentável, Colóquio de Moda em Porto Alegre (Fashion Colloquium in Porto Alegre), Business Professional Women of Porto Alegre and the Biennial of Contemporary Textile Art. Each case of collaboration was a project co-created and led by one of the ecosystem actors (Fig. 4).

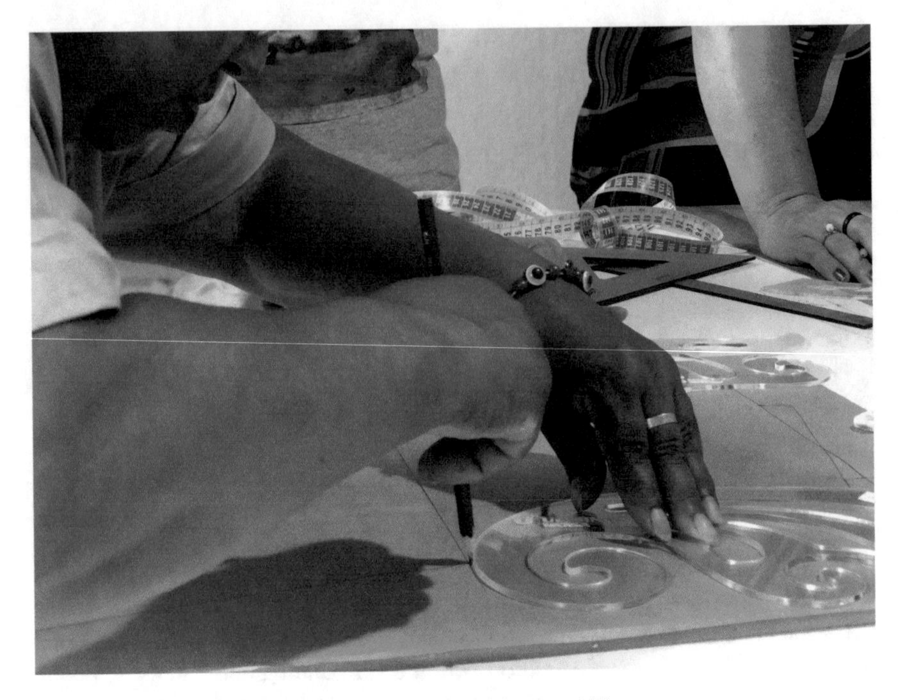

Fig. 4 Fashion classes for people in situation of social vulnerability

Throughout 2018, there were lectures, training courses on modelling for a seamstress cooperative, collaborations with Vila Flores Association in exchange for a discount on space, workshops on sewing and product development using waste textile, and workshops on photography, communication online and product exhibition at fairs. We reached an audience of over 600 people at events promoted by different organizations. In the courses, we trained 20 women who live in vulnerable situations. We supported the creation of a female entrepreneur's business, with assistance in the development of visual identity and business design. We reused textile waste from a clothing manufacturer for product development, which was retailed in order to help maintain the space (Fig. 5). We publicized partner brands, with products displayed in the physical space for visitors who circulated at Vila Flores during events. We experimented with different arrangements of the physical space, in order to favour the development of activities. The whole process was developed collaboratively, with different levels of intensity of participation by those involved.

The project for putting in place a sustainable fashion ecosystem is now two years in the making. Throughout this period, we have enhanced local brands by organizing the M.A.R. fair (fashion; art, revolution), by holding courses and lectures that are free or involve a conscientious contribution, by empowering women, and by creating a sewing group that is re-inventing waste into textile initiatives connected to the ecosystem.

Fig. 5 Product developed in collaboration to support the lab space

5 Final Remarks

The project shows us that strategic design processes can trigger design coalitions and stimulate the formation of a collaborative organization capable of promoting new values towards sustainability. The open and dynamic co-design processes enabled the development of the manifesto that represented the value shared by the participants of the design coalition. The manifesto's dissemination strategies enabled other actors to join the collaborative organization (sustainable fashion ecosystem). A collaborative organization needs to be open and characterized by the freedom to connect to (and disconnect from) the life project that guides the collective's actions. This openness and freedom led to several participants passing through the ecosystem during the first year, leaving their contributions to the scenario, vision and manifesto and taking with them learning from the collective construction experience. 2018 was a particularly difficult year for Brazil. Several initiatives ceased to exist and some participants reduced their collaboration. We understand that collaboration in times of economic crisis diminishes because people turn to activities that ensure their survival. On the other hand, we also understand that in political and environmental crises, collaboration strengthens once again into activism. People who are really connected to the organization's purpose, as expressed by the manifesto, joined in various activities to fight against threats to the desired way of life.

The results of the first year show that, in the words of one participant, "collaboration is a utopia". However, it is a utopia that deserves to be pursued. Collaboration means "giving up on individualities to create a system of bonds with other individuals who share the same dream" [8]. The sense of collectivity can be an intrinsic value of the individual or worked on along the process, which is a cultural innovation for this society marked by individualism. Collaboration among different people requires an openness to the other's point of view, an understanding of the co-dependency relationship that is established for actions to take place and a sense of responsibility so that the ideas generated by the group come to life. In this process, openness to the possible and the possibilities of experimentation caused the collaborative organization to change every month, in order to learn the best format to follow. This way of operating through dialogic strategies thereby testing the possibilities, learning from the responses of the environment resulting from the action and giving rise to new movements based on these learnings was very positive.

We understand that collaboration requires negotiation with others about when and how to do something together, requires openness and goodwill and is based on relational values of empathy, trust and friendship. The projects that were implemented and yielded good results were developed by a group of people who had this set of values and this pre-disposition. With collaboration as the foundation, we were able to experiment with ways of living closer to *Buen Vivir*, in which most of the projects were developed by a group of women who sought, through sustainable fashion, ways to break with the patriarchal system of oppression of bodies and the objectification of women.

So far, we understand that taking action through projects is what connects people and moves them towards action; and that a possible model for action might be a continuous design process, capable of identifying the signs of socio-cultural change and raising provocative questions capable of giving rise to reflection and action in society, through mutant projects, connected to local and global movements to pierce the bubbles created by the algorithms that surround our lives.

References

1. Santos B, (2013) Para além do pensamento abissal: das linhas globais a uma ecologia dos saberes. In: Santos, B, Meneses, M (eds). Epistemologias do Sul. Cortez, epub. Loc.362– 1243
2. Boff, L (2012) Sustentabilidade o que é – o que não é. 1st edn, Petropolis, Vozes
3. Illich I (1973) A convivencialidade, 1st edn. Sociedade Industrial Gráfica, Lisboa
4. Franzato, C. et al. (2015) Inovação Cultural e Social: design estratégico e ecossistemas criativos. In: FREIRE, K. (ed.). Design Estratégico para a Inovação Cultural e Social. 1st edn, Kazuá, São Paulo, p. 157–182
5. Levy P (2014) A inteligência coletiva: por uma antropologia do ciberespaço. Loyola, São Paulo
6. Manzini E, Cullars J (1992) Prometheus of the everyday: the ecology of the artificial and the designer's responsibility. Design Issues, Cambridge 9(1):05–20

7. Manzini, E (1999) Strategic design for sustainability: towards a new mix of products and services. In: Proceedings First International Symposium on Environmentally Conscious Design and Inverse Manufacturing, Tokyo, Japan, p. 434–437, https://doi.org/10.1109/ecodim.1999.747651
8. Manzini E (2015) Design when everybody designs, 1st edn. The MIT Press, Cambridge
9. Mauri F (1996) Progettare progettando strategia, 1st edn. Masson S.p.A, Milano
10. Morin E (1981) Para onde vai o mundo?, 2nd edn. Editora Vozes, Petropolis
11. Morin E (2015) Ensinar a viver, Manifesto para mudar a educação, 1st edn. Editora Sulina, Porto Alegre
12. Mariotti H (2013) Complexidade e Sustentabilidade: o que se pode e o que não se pode fazer. Atlas, São Paulo
13. Morais, G (2018) O poder da colaboração: cenários para a atuação dos coworkings como pontos focais no desenvolvimento de novos modelos para a moda sustentável. Master thesis. Universidade do Vale do Rio dos Sinos. Available at: http://www.repositorio.jesuita.org.br/handle/UNISINOS/7142. June, 15 th 2020
14. den Ouden E (2012) Innovation Design: Creating value for people, organizations and society. Springer, New York
15. Solón P (2019) Bem Viver. In: Solón P (ed) Alternativas sistêmicas: bem viver, decrescimento, comuns, ecofeminismo, direitos da Mãe Terra, desglobalização, 1st edn. Elefante Editora, São Paulo
16. Ribeiro, R (2016) O design estratégico como catalizador do desenvolvimento da arquitetura organizacional de laboratórios de inovação social. Master thesis. Universidade do Vale do Rio dos Sinos. Available at: http://www.repositorio.jesuita.org.br/handle/UNISINOS/5712. June, 15 th 2020
17. Wood, R (1999). The future of strategy: the role of the new sciences. In: LISSACK, Michael; GUNZ, Hugh (Ed.). Managing complexity in organizations: a view in many directions,1st edn. Westport, Quorum booksp, p 118–164
18. Zurlo, F (2010) Design strategico. In: XXI Secolo: Gli spazi e le arti. Roma: Enciclopedia Treccani. 2010. v. 4. Available at: http://www.treccani.it/enciclopedia/design-strategico_%28XXI-Secolo%29/. Accessed June, 15 th 2020

Pursuing a Circular and Sustainable Textile Industry in Latin America

Sebastian Garcia Jarpa and Anthony Halog

Abstract The textile industry profiles of key countries in Latin America and their developments have been analysed, in terms of sustainability. Different aspects related to social consciousness, environmental, and labour aspects have been considered. Countries like Mexico, Brazil, Colombia, Uruguay, Chile, and Argentina are presented as examples to describe the sustainable development in the Latin American textile sector. There was a decrease in Latin American textile markets due to the introduction of products from Asian countries. On the one hand, this scenario has put much pressure on Latin American countries to improve their textile products quality, a way to give a better value for their local consumers. On the other hand, labour costs in Latin America are increasing due to its economic development, which has been a big challenge for Latin American textile industries to compete in the international market. New technologies are also playing a major role to compete with products from Asia. Some examples are Mexico with its smart textileñ Colombia has enhanced its production speed as well as introduced high-quality customized products from local producers. In most Latin American countries, a sustainability agenda has pursued by the private sector, including artisanal textiles on their markets and recycled textile materials into their markets. These cases can be seen in Argentina, Chile, Colombia, Brazil, Mexico, and Uruguay. Countries with strong governmental support (Brazil, Argentina, Uruguay, and, Peru) have a solid raw sustainable material export system, the most important ones are organic cotton, wool, and alpaca. A special case is Chile with its pulp cellulose production which will start its production in 2020, exporting to Asian countries, principally. Uruguay and Brazil are being considered as "slow fashion" capitals in the region; Uruguay, for its rich green chain of biological fibre production management, its eco-friendly materials like ceramics, hemp, silk, organic cotton, and recycled materials, and it is artisanal way to process high-quality wool; and Brazil, with its second-hand market, and new sustainable materials, such as

S. G. Jarpa (✉)
Chemical Engineer, Universidad Tecnica Federico Santa Maria, Valparaíso, Chile

A. Halog
Faculty Member in Industrial Ecology & Circular Economy, University of Queensland, Brisbane, Australia
e-mail: a.halog@uq.edu.au

M. Á. Gardetti and R. P. Larios-Francia (eds.), *Sustainable Fashion and Textiles in Latin America*, Textile Science and Clothing Technology,
https://doi.org/10.1007/978-981-16-1850-5_6

natural dyes cotton, linen, and PET, based on natural plants. The volume of waste textiles is expected to increase over the years if every Latin American country does not have programmes to remediate this problem, particularly due to the importation of products. In conclusion, Latin America is therefore a potential growth market in the textile industry, due to its domestic demand, economic growth, and purchasing power. These countries have a strong possibility to develop sustainable textile industries, but a big challenge, because being sustainable requires a high initial investment cost as well as a strong policy and strategy to achieve it.

Keywords Latinamerica · Textile market · Sustainability · Sustainable materials · Circular economy · Recycled materials · Sustainable market

The textile industry profiles of key countries in Latin America and their developments in terms of sustainability have been analysed. Different aspects related to social consciousness, environmental, and labour aspects have been considered in this work. Countries like Mexico, Brazil, Colombia, Uruguay, Chile, and Argentina are presented as examples to describe the sustainable development agenda in the Latin American textile sector.

In this review, we found out that there was a decrease in Latin American textile markets due to the introduction of products from Asian countries. This scenario has put much pressure on Latin American countries to increasingly improve the quality of their products and has become more sustainable to give a better value for their local consumers. However, the labour costs in Latin America are also increasing due to its economic development, which has been a big challenge for the Latin American textile industry to compete in the international market.

New technologies are also playing a major role to compete with products from Asia. Examples like in Mexico with its smart textile Colombia has enhanced its production speed as well as introduced high-quality customized products from local producers to recover its strong textile industry presence before it was overtaken by Asian textile suppliers.

In most Latin American countries, the private sector with the development of artisanal textiles has pursued sustainability agenda actively with the transformation of recycled textile materials into new products. These cases can be seen in Argentina, Chile, Colombia, Brazil, Mexico, and Uruguay.

Countries with strong governmental support are related to the export of raw sustainable materials like organic cotton, wool, and alpaca. These examples are found in Brazil, Argentina, Uruguay, and Peru. A special case is Chile with its pulp cellulose production (which caters to the Chinese textile market) which will start its production in 2020. Additionally, Andean countries have an excellent opportunity to be more sustainable with camelid raw materials from local communities that could bring more economic benefits and ecological development.

Uruguay and Brazil are being considered as "slow fashion" capitals in the region; Uruguay, for the management of its rich green chain of biological fibre production, the use of eco-friendly materials (ceramics, hemp, silk, organic cotton, and recycle materials like tyres), and its artisanal way to process wool with high quality; and Brazil, with the buy of second-hand market, and the use of new sustainable materials, such as natural dyes cotton, linen, and PET, based on natural plants, with the aim to reduce costs.

The volume of waste textiles is expected to increase over the years if every Latin American country does not have programmes to remediate this problem, particularly due to the importation of products from famous brands that give clients a status symbol that differentiates them from others.

Latin American countries have a strong possibility to develop sustainable textile industries, but also feel pressured to implement new ideas over the years. This is because being sustainable requires a high initial investment cost as well as a strong policy and strategy to achieve it. Latin America is therefore a potential growth market in textile industry, due to its domestic demand, economic growth, and purchasing power.

1 Introduction to Sustainability

Sustainability pursues processes and operations that can be profitable with environment-friendly, natural raw materials and suppliers, and with fair working conditions. This is related to the term "Circular Economy" which has the objective to close the loop of a cycle with a group of sustainability-related actions, involving, for example, the avoidance of chemicals that are not environment-friendly, the reuse of materials or products, the reduction or recycling of waste, and other cleaner production initiatives.

Sustainability must improve with social, economic, and environmental projects in conjunction with the United Nations Sustainable Development Goals [1], an intergovernmental agreement in sustainable development. Challenges related to apparel and textile value chains include the support of sustainable agriculture and land management, producing sustainable raw materials, use of innovative manufacturing practices, standard working conditions, and implementing a circular economy model. To achieve these sustainability objectives, the certification from the Global Organic Textile Standard (GOTS) will become an important consideration to determine the veracity of processes and their improvements with respect to good practices as well as agricultural, social, and responsible management [2] (Fig. 1).

2 The Textile Industry in Latin America

Figure 2 shows the import and export of every country, showing the importance of Mexico and Brazil in the region being the biggest markets of textile industry in Latin America. Countries with the highest growth in the industry are Colombia and Peru [4]. The net balance of the region is negative for every country,

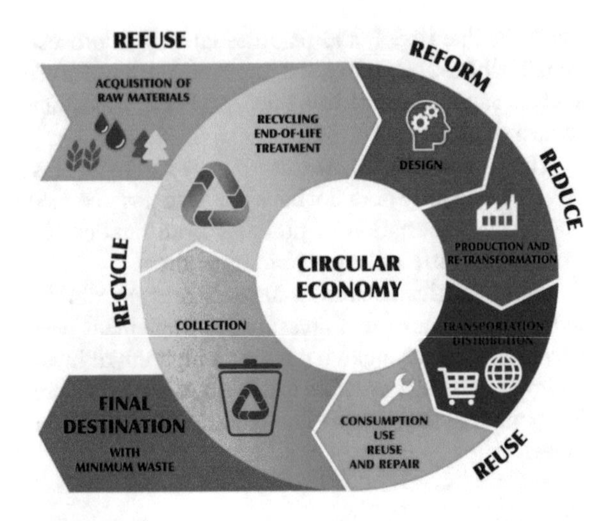

Fig. 1 Fundamentals of Circular Economy [3]

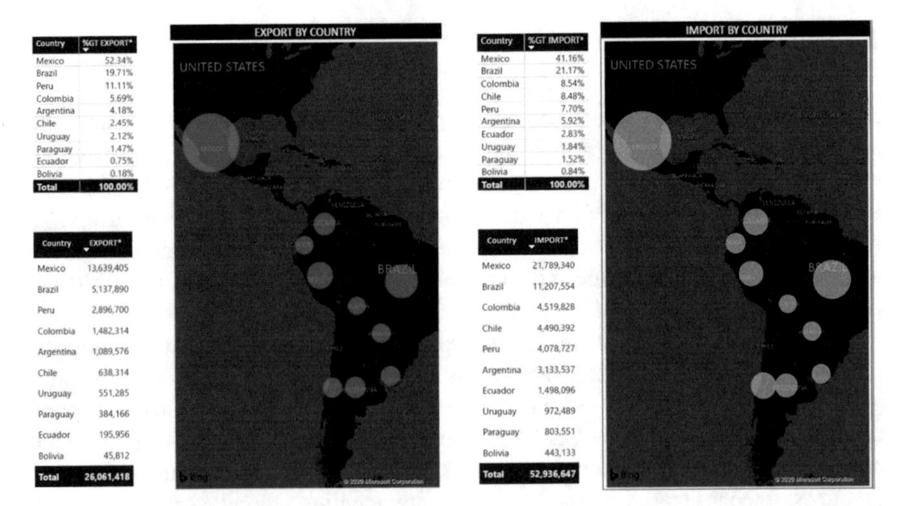

Country	%GT EXPORT*
Mexico	52.34%
Brazil	19.71%
Peru	11.11%
Colombia	5.69%
Argentina	4.18%
Chile	2.45%
Uruguay	2.12%
Paraguay	1.47%
Ecuador	0.75%
Bolivia	0.18%
Total	100.00%

Country	EXPORT*
Mexico	13,639,405
Brazil	5,137,890
Peru	2,896,700
Colombia	1,482,314
Argentina	1,089,576
Chile	638,314
Uruguay	551,285
Paraguay	384,166
Ecuador	195,956
Bolivia	45,812
Total	26,061,418

Country	%GT IMPORT*
Mexico	41.16%
Brazil	21.17%
Colombia	8.54%
Chile	8.48%
Peru	7.70%
Argentina	5.92%
Ecuador	2.83%
Uruguay	1.84%
Paraguay	1.52%
Bolivia	0.84%
Total	100.00%

Country	IMPORT*
Mexico	21,789,340
Brazil	11,207,554
Colombia	4,519,828
Chile	4,490,392
Peru	4,078,727
Argentina	3,133,537
Ecuador	1,498,096
Uruguay	972,489
Paraguay	803,551
Bolivia	443,133
Total	52,936,647

Fig. 2 Map of import and export of clothes per country. Brazil and Mexico lead the commercial business with the highest investment of import and incomes from export [6]

showing a need to import textile products to satisfy their national and international demands [5].

There was a decrease in Latin American textile markets due to the introduction of products from Asian countries. This scenario has put much pressure on Latin American countries to improve the quality of their products and to be more sustainable by giving a better added value to their products.

Latin America is a potential growth market in textile industry, due to its domestic demand, economic growth, and purchasing power [7]. Many countries are exporters of raw materials that are important for the textile industry, such as cotton or leather. However, labour costs in Latin America are increasing due to its economic development, which has been a big challege for the Latin American textile industry in international markets.

3 Environmental Impacts of Textile Industry

Environmentally, textile industry has the second most contaminated process in the world, consuming 35% of all pesticides [2]. The high level of contamination in chemicals and the toxicity while using synthetic dyes to process textile with colours (i.e. 8000 chemicals are needed to give raw materials its different colours) has been a subject of worry. Related to water consumption, the textile industry is one of the industries that consumes a high volume of water. For instance, 20.000 litres of water is necessary to produce 1 kg of cotton [2] (Table 1).

In waste generation, there is a big volume of unworn garments, which is approximately 40% of the total production [7]. The production of unworn clothes is due to 90% of consumers do not have money to buy them, and 87% of them are burned or disposed of in landfills [9]. The textile industry loses 500,000 million dollars due to non-recycling and reusing of waste textile products, whereby only 3% is recycled.

Related to recycling, the linear production system has generated big waste volumes and a higher worker requirement. In Latin America, consumers prefer to wear foreign clothes with famous brands to give the clients a status symbol that differentiates them from others, [8]. Therefore, the volume of waste textiles will increase over the next years if every country does not have programs to remediate this problem.

75 million people are working in the fashion industry and most of them are young women [7]. Furthermore, 11 over 100 children in the world work in the textile industry [8].

Latin America has a big potential of producing raw materials with a high fibre quality, like organic cotton and wool. This is primarily due to its big land and free space to farm mixed with a rich culture and textile traditions that make this region of the planet interesting to pursue sustainable production. One of the important policies

Table 1 Environmental impacts of the fashion and textile industry [8]

Chemical products	The textile industry contaminates 20% of global water production
Water scarcity	The cotton life cycle requires 2700 litres of water
Greenhouse effect gases	The textile industry is responsible for 10% of the global CO_2 emissions
Waste management	Wastes from the textile industry are around 5% of total wastes
Resources: soil and energy	58% of textile fibres produced in the world are from petroleum
Biodiversity	In India, the loss of cotton seeds is due to transgenic cotton contamination

in the future will be to emphasize the local economic development and give them a resilient spirit, due to the climate change impacts, that affected cotton and wool farming.

4 Cotton in Textile Industry

The most important raw material in the textile industry is cotton where its farming has been maintained in around 1.5 million hectares, and the fibre production has averaged 1.8 million tons with 350 million workers, considering the whole production chain [2]. Cotton farming represents 30% of the consumed fibres in the textile industry worldwide focused principally on India, China, United States, Pakistan, and Brazil. These countries contribute to 80% of the world's production, which are managed according to international standards (Fig. 3).

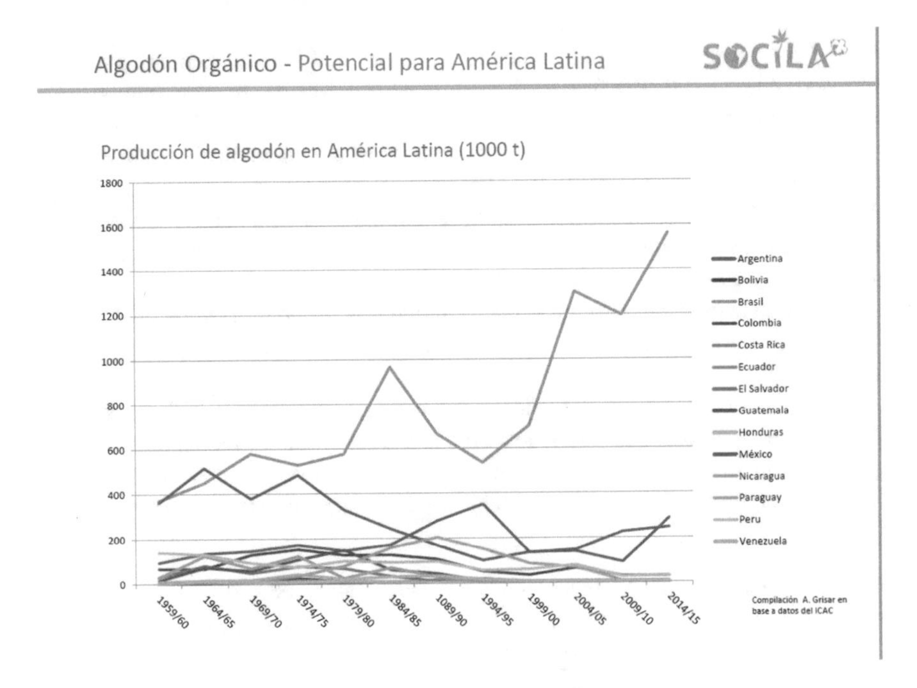

Fig. 3 Cotton production in Latin American countries per 1000 [ton] [10]

5 Organic Cotton in Latin America

One of the sustainable raw materials in the textile industry is organic cotton, which is an environment-friendly fibre produced from crops free of synthetic chemicals [11]. Latin America is the third region of organic cotton production behind Oceania and Europe. Its production had reached 6.7 million ha in 2017, representing 13% of the world production [12] (Table 2).

Latin America has an advantage against other suppliers because of its short distance to USA (low costs of transportation), where organic cotton demand is growing faster than in other countries. Other advantages are the ease of communication, and shorter lead times than other organic cotton exporters, due to short distances from the USA, compared, for example, to Asian countries [14].

On the other hand, poor conditions of farmers that are not allowed to develop new techniques on their own is a problem that Latin American countries must face. Farmers also face pressure from important clients to be more productive [14].

There are several projects to increase production and implement strategies to get sustainable production related to fair working conditions, natural production, and the use of technology.

Important projects are developing in Argentina with the use of biodynamic farming and in Brazil with the production in arid regions, and there is an increase of certifications for organic cotton. Also, companies like C&A are participating in sustainable projects, with the plan to make farmers more resilient to climate change impacts in organic cotton farming areas and to ensure a good quality of this sustainable raw material [12]. Research institutes like Embrapa Cotton in Brazil are developing important projects to enhance the sustainability of this product, such as the development of natural-coloured cotton, genetic cotton improvement, the use of technologies to increase the efficiency of cotton extraction, and others [12].

The most important producers of organic cotton in Latin America are Brazil and Peru. Brazil has a surface of 680 ha producing 17 MT with 232 farmers, while Peru has 337 ha with 312 MT of production with 141 organic farmers [14] (Table 3).

The production of organic cotton in Latin America is related to big companies in the textile and fashion industries that are promoting a sustainable brand in their stores to encourage and support environmental responsibility and fair trade. A summary of important companies in the region is detailed in Table 4.

Table 2 Key regions of organic cotton production (2018) [13]

Region	Percentage of organic cotton production (%)	Organic cotton land (million ha)
Oceania	45	22.8
Europe	25	12.7
Latin America	13	6.7

Table 3 Production of organic cotton in Brazil and Peru in 2017 [13]

Item for organic cotton	Brazil	Peru
Organic farmers	232	141
Organic certified land (ha)	680	337
Organic cotton fibre (MT)	17	312
Organic In-conversion land (ha)	16	85

Table 4 Top companies which use organic cotton as raw material for their products and promote sustainability in the textile industry [13]

Company	Sources
C&A	Mexico
Inditex	Brazil–Argentina
Outerknown	Peru–Mexico
Patagonia	Colombia–El Salvador–Nicaragua

6 Certifications and Standards in Latin America

A sustainable, certified product must comply with a list of requirements that are evaluated by certified organizations, like the Global Organic Textile Standard (GOTS). All requirements are based on the International Labour Organization (ILO) norms. The scope of GOTS is improving lives and organic products, which dictate the requirements along with the textile production to achieve ecological conditions and dignified jobs. They have two definitions of organic products: "*Labelled products*" which must have a minimum of 95% of organic fibres in their products, and "*Made with organic material*" which must contain 70% of certified organic fibres (see Table 5) [4].

According to GOTS, they have a list of environmental and social requirements before a product is considered sustainable. These requirements are described as follows:

Environmental requirements:

- Chemical supplies for a sustainable product must provide all the information related to toxicity, biodegradability, and elimination. Toxic heavy metals, formaldehyde, aromatic solvents, functional nanoparticles, genetically modified organisms (GMO), azo-derived dyes, and enzymes are prohibited. Furthermore, oils for textiles must not contain heavy metals.

Table 5 Label types of sustainable products, depending on the decomposition of their textiles [4]

Grade	Description
Grade 1	More than 95% certified organic textiles
Grade 2	More than 70% certified organic textiles, but 10% maximum of synthetic fibres

- It is forbidden to discharge carcinogenic substances like aromatic solvents, phthalates, and polyvinyl chloride (PVC). Companies with wet processes must have a complete record of their chemical uses, energy, water consumption, and effluents treatment, including local waste disposal. Wastes must be treated before discharge.
- Workers must adhere to an environmental policy that includes objectives and procedures to minimize effluents and discharges. Wastes must be clearly analysed with their limit values for complying with a non-desired waste or contamination risk based on GOTS norms, done by verifiers taking additional samples according to ISO 17025.
- PVC is not permitted in packaging processes. Every paper and board used in the process must be recycled or certified, according to the norms of Forest Stewardship Council (FSC) or the Programme for the Endorsement of Forest Certification (PEFC) for the chain of custody of forests.
- Raw materials, intermediate products, end products, and accessories must comply with strict limits on non-desirable wastes.

Social requirements:

- The place of work must have good labour conditions related to safety and hygiene. Employment must have a minimum wage for living, free of discrimination and violent abuse, and fair working hours under the legal regulations.

7 Organizations that Support Sustainable Development Programs in Textile Industry

Fairtrade is an organization that takes care of the sustainable development processes, establishing a fair price, ensuring good work conditions, and fair trade with agricultural workers. For example, Fairtrade has helped to increase the production of 2 million tons of cotton, compromised with a sustainable cotton plantation [2]. With the help of increasing sales with Fairtrade programme (established in Brazil and Guyana in Latin America), textile companies have developed new programmes to respect the environment and encourage water optimization in cotton farming [2].

Better Cotton Initiative (BCI) has a holistic focus on sustainable cotton production and covers the three pillars of sustainability: environment, social, and economic aspects. Members from BCI produce 5% of the total world consumption of cotton. They want to continue attracting members to comply with the following objectives:

- Finding better economic profitability for agricultural workers of cotton.
- Reducing effects of water and pesticide usage to improve the human health and environment of workers and to make a better farming with soil biodiversity.
- Promoting decent work in agriculture communities and among workers of cotton production.

- Facilitating an exchange of knowledge with the world on more sustainable cotton production.
- Increasing the traceability along the cotton supply chain.

Although all the organizations named above act in a global way, there are some organizations that act regionally, like the Responsible Brazilian Cotton. Their objectives are like BCI, increasing gradually social good practices and environmental and economic benefits in the cotton unity production. With the help of the Brazilian Association of Cotton Suppliers (ABRAPA) and the Government, they have established some rules to give accreditation for sustainable production:

- Labour regulations.
- Compliance with safety standards.
- Prohibition of child and forced labour.
- Prohibition of discrimination.
- Freedom of union and support of collective negotiations.
- Legal protection and environmental conservation.
- Agricultural good practices application in Brazilian cotton production.

8 Sustainable Textile Industry Per Country

Sustainability in Latin America is influenced by designers, large companies, artists, inventors, the retail market, fashion stores, and others. The development differs in every country and thus the development of their textile industry as well. The main objective in the Latin American textile industry is to minimize the environmental impact of the actual textile production and to change the consumer behaviour (i.e. fast consumption with less importance of recycling and quality of the products) nowadays [7].

In this section, each Latin American country is analysed separately, and examples of companies are given to demonstrate how sustainability has been established in their respective country market.

8.1 Mexico

Mexico has a mixed textile industry, where there is strong competition between the national textile products and imported products [15].

Important textile companies called "maquilas" are responsible to make new textile products with raw materials from the USA. Strategically, they are in the north frontier of Mexico with the USA. Maquilas are important because of their low-salary workforce and governmental support [15].

The textile industry involves 72,000 companies and 470,000 employees in the whole country [16] and constitutes 1.3% of the gross domestic product (GDP) [16].

Mexico imports raw materials from the USA and China principally to make textile products to export back to the USA (60% of the total), which are made mainly from cotton and denim. The second important market of Mexican textile products is the United States with 22%, and 18% is exported outside of the American continent [17].

A big risk of its industry is the lack of competitiveness with Asian textile products, particularly from China and Vietnam that export textile products to the USA, becoming a strong competitor for Mexican textile products [15].

Mexican companies have limited financial capability and are dependent on importing the materials and supplies. Their lack of technology minimizes the chance to compete with other textile products coming from overseas.

The other problem in Mexico is the contraband textile with low quality and price. These products come from the USA and Asia. The contraband has favoured poor working conditions, with informal and illegal structures [18].

Related to recycling, textile wastes are 1.4% of the total waste generated in the country, where only 0.5% is recycled, that is around 3.152 tons, according to Natural Resources and Environment Secretariat (SEMARNAT) [17]. The Mexican Government has implemented new programs to increase recycling among its citizens; for example, the *"Tu Ola"* programme, a reward system based on points to exchange with new products when people follow eco-initiatives promoted by the government [16]. Other initiatives come from the private sector, where they collect textile waste and transfer this material to recycled companies in the country. The aim is to reuse materials and people receive a reward based on green points that are convertible to food products [19].

Some companies are implementing a sustainable system in their production that involves the three aspects of sustainability. Results have a good response from the people with the image of the company, and the growth of the company economically [20]. SME companies, for example, have challenges in their finance, formalization, administrative capacity, and technology, which limit them to compete in an efficient way [21].

In cotton farming, there has been an introduction of innovative technologies, such as adoption of genetically modified seeds and narrow planting system. These new ideas have decreased costs in phytosanitary controls, regulated better water consumption, and have increased fibre production performance [2]. Mexico also has an organic cotton industry that is managed by small family farmers. This production is encouraged by C&A company, which inspects the whole value chain [17].

Mexico is the seventh-largest exporter of denim worldwide and the main supplier to the USA [20]. According to INEGI, 40% of denim produced in Mexico is exported to Latin American countries, like Peru, Chile, and Colombia, and 60% is sent to the USA [21].

Two problems in denim production that make a non-sustainable process are the high consumption of water and toxic chemicals used to make the colour in denim (heavy metals).

Sustainable denim is produced by the company Triarchy made of eucalyptus fibres, a product 100% renewable. For its production, they consume 80% less water than traditional denim. There is not enough information about sustainable denim in

Fig. 4 Designs with sustainable products from Amor & Rosas [23]

Mexico, but it seems that the impact of a denim sustainable production would be extremely high, considering the high export volume of this product, giving Mexico an important niche market.

Examples of sustainable companies are Amor & Rosas, which uses eco-textiles, hemp, and recycled cotton to produce high-quality textile products. Their products are masks, bags, and different styles of clothes for women and men. They mix the Mexican culture with sustainable products to promote local artisans as a trend in the fashion sector [22].

Someone/somewhere also has sustainable textile products inspired by the Oaxacan craft, using recycled cotton yarns. Oni Original company uses recycled material to produce shoes, with PET yarns and recycled skin bags. Their incomes are reinvested for environment-related projects (Fig. 4).

8.2 Colombia

This country is famous for the export of garments that are located mainly in Medellin. 49% of the garments used in Colombia are imported to cover its high demand [24].

Sustainability initiatives in Colombia can be found in three ways:

1. **Direct working with communities**, where practices and materials from artisans are used to make a production industrially. The aim is to combine the identity of folk cultures with the textile industry. These processes are well regulated by the government, which is worried about working conditions and the preservation of manufacturing processes. Moreover, they protect the product prices, to make them profitable and more competitive in the industry [6]. Examples are *Artesanias de Colombia, Wayuu Products, and Nasa Community*. According to *Artesanias de Colombia* (2019), there are 31,000 artisans in 29 regions of the

country, where 71% of them are women. 73% of them work manually on their products, and 22% have simple machines to support the production [24].

2. **Companies working with ancestral techniques linked with industrial processes**, where artisans and industrial companies work together to produce sustainable textile products. Examples are *Pre-Columbian* and *Indigenous Loom.*
3. **Circular systems or neo-craft**. They avoid the production of waste by using it as a raw material to produce new textile products. Companies examples are *Croquis* company.

There are also projects that support sustainable production like *Pro Colombia*, an organization that promotes tourism, foreign investment, and exporst of non-mining products to improve the image of the country. There are other companies that have considered implementing technology in the textile industry, specifically a 3D fashion design, to make more efficient and sustainable fashion collections [25]. This project is being implemented for companies like *CLO, Audaces*, and *MorganTecnica*, dedicated to the development of Industry 4.0 in Colombia.

Colombia wants to be the fashion centre in Latin America. The fashion sector in Colombia has increased notoriously in Latin America, with special environmental care and ethical work practices, promoting the slow and eco-fashion involved in small and medium-sized industries. Additionally, the Colombian government has participated in this process by making social and sustainable policies [7].

Paloma & Angostura works with the hiring and reintegration of violence victims, implementing a code of behaviour based on the World Labour Organization (WLO). They also use biodegradable and natural fibres and evaluate the life cycle of their products [26].

Fokus Green is a company that cares about environmental impact reduction and water consumption. Their products are free of pesticides and artificial dyes, and they use cotton and recycled PET bottles. They avoid 30 million litres of water per year and they recover 50,000 plastic bottles from the ocean annually [27].

Bareke has created new labour opportunities for indigenous people and artisans of the region. They offer training for sustainable production to produce bags made of natural fibres. They plant seeds of palm trees and use natural dyes for their products (mud or banana peels as raw material, for example) (Fig. 5).

Little & Ramonas, famous for its slow fashion products, uses only origin-certified suppliers and recycled raw materials with eco-friendly management of leather. Their products are famous because they can be worn any season of the year [29].

True Love & Poems makes their business virtually, promoting a circular economy system, local production, and a responsible business (Fig. 6).

Fig. 5 Presentation of Bareke on its webpage [28]

Fig. 6 Presentation and products of True Love & Poems [30]

8.3 Uruguay

Uruguay is famous to produce wool, a sustainable raw material. They export 39.3 million kilograms to 40 markets in the world [7], being the third exporting country in the world of combed wool behind China and Czech Republic (2018) [8].

Uruguay is considered a slow country, having many national companies with textile products made from local brands with natural resources. Slow fashion has a low-scale production system with a high standard of innovation, unseasonal clothes design, no-waste generation processes, and organic or natural raw materials. Slow fashion pursues less environmental impact, fair manual labour, and manufacturing that improves quality and reduces quantity.

Fig. 7 Promotion of Ana Livni products, following the slow fashion via Instagram [31]

Uruguay has a system of animal welfare—special care of animals with sustainability, protecting them from hunger, fear, distress, discomfort, pain, and diseases [31].

Ana Livni, one of the Uruguayan designers with a sustainable business, uses fair prices for their products and utilizes ethical work practices [7]. Ana Livni started the slow fashion 15 years ago (Fig. 7).

Calmo promotes responsible consumption and production. They work with renewable raw materials with sustainable processes. Their products are timeless and with their own labels.

Madame Hibou has noble materials like Peruvian organic cotton. They care about the value chain of their products and the working conditions to produce them. They also have a production of Green Denim with a 100% sustainable process. This brand is being compromised by slow fashion.

Mola is an event which organizes conferences, exhibitions, workshops, and promotes sustainability programmes [7]. *Mola* integrates sustainability, innovation, and technology with Latin American products.

8.4 Argentina

Argentina produces fine wool called Lana Merino, different from traditional wool. Lana Merino is softer, with higher quality and adapts better to the skin. Their weaves have more fibres due to small air bubbles that help to keep the body heat. Argentina ranks third in Lana Merino supplier. Clients of this kind of wool are with a high purchasing power and value more than the high wool quality, besides the products are environment-friendly, animal care, and fair working conditions [33].

The Sustainable Textile Centre applies a holistic and multidimensional vision of fashion, helping the sustainable sector of textile and fashion industry. They want to be an important research and academic centre with international recognition [33].

Wool farming is especially important in Argentina, giving them an economic and social edge. This raw material is produced mainly in Chaco, Santiago del Estero, and Santa Fe, where 80% of the producers are small farming centres. Argentina produced 42,400 tons in 2017/2018, and 43,750 tons in 2018/2019 [35] of national wool, with sustainable programs promoted by the Ministry of Agriculture, Livestock and Fisheries (PROLANA), with the aim of care for animals with conscious and fair working conditions [35]. The major destinations of wool export are China, Germany, and Italy, with 6,186 tons, 4,998 tons, and 2,167 tons of wool, respectively [35].

The cotton industry (the second important raw material in the textile industry) considers Argentina as the second consumer and producer of wool in the Latin American region, employing 26,500 workers in the whole value chain [36]. 75% of its production is in Chaco and Santiago del Estero regions and they are exported. Argentina satisfies the national demand of 140,000 tons, and the other 40,000 tons are exported to Indonesia, Vietnam, Colombia, and India [2].

The challenge for farmers is a plague that affects cotton farming. To fight this problem, the National Service of Agri-Food Health and Quality has a program to eradicate the cotton weevil. The program is focused on phytosanitary to reduce the damage of the plague with diverse actions like management, planification, monitoring, technology, and training [37].

Related to fashion, clients in Argentina follow European and American styles, rather than artisan styles. In this line, Argentinian designers have their stores with luxury clothes and sustainable [38]. One of these places is *Galería Patio del Liceo*, a shopping mall of mini stores of sustainable designers.

An example of sustainable fashion is *Animana*, a company famous for their textile and materials made from Alpaca, Guanaco, and Llama, mixed with a French style of their clothes. *Manto Abrigos* is famous for its organic cotton dresses, with the use of llama wool coats, and other products like leather clutches and flat-brimmed hats [39]. Maydi has hand-knitted products like sweaters, tops, bottoms, dresses, and bags made with natural fibres under fair working conditions, using an environment-friendly method to make textile products. These small sustainable locals are increasing in the country due to the preferences of clients for clothes to show respect to the environment and animals.

About recycling, there are some private initiatives that cover some percentage of waste textiles to produce other new products.

There is a potential project from the company *Vitnik*, called *Retazos*, which will take waste textile material from the big industry to make new materials like designed furniture, toys, and other decorative products [40]. *Ventures*, like *Reinventando*, of the designer Lucila Dellacasa, transforms unworn shirts into night suits.

Recycling is also present in some sustainable stores in Argentina like *Facon*, which have an upcycled effect, with denim and leather jackets, knits, leather purses, and other products [40]. *Contenido Neto*, another company that recycles PET bottles to produce strands and fashion accessories [6]. Their interest in sustainability is with the research of nylon socks biodegradation, which lasts 2000 years. With this information, they have decided to make a mix of waste of nylon socks and natural fibres, making a new crafted material to sell [7]. The small sustainable fashion in Argentina has one important disadvantage, which is its difficulty to scale up to an industrial size using traditional techniques.

Instead of these initiatives, in Argentina, there is no special program to manage the recovery of textiles [40]. The only governmental institution found, is the Environmental Control Agency (APRA) recovering non-woven textiles.

8.5 Chile

Nowadays, Chile is not a famous producer of raw materials in the textile industry. The fashion market in Chile is 70–90% imported [41].

Most of the sustainable ideas come from the private sectors which have management and programs for recycling of worn clothes to get new materials, with the aim of showing awareness about the contamination of the textile industry and also to be more conscious of the impact of the fast fashion.

Rembre Mill is a company aligned with recycling. They collect worn clothes and waste textiles to create eco-filling to make pillows and box bags. The textile in good conditions is donated to foundations for reuse in the market [41].

Modulab produces jewellery, containers, and handbags from university wastes [7]. They also have a link with important brands in the country like *Patagonia* and *Natura,* with the support of the government leading sustainable projects. *Alma* and *Hibrida* are other examples, famous of jewellery collections made of recycling materials and for their special awareness of recycling and footprint of consumption, with the production of pendants, bracelets, chokers, organic volumes, and bangles [7].

Retail companies and famous brands are protagonists with initiatives and campaigns, giving discounts to clients to buy new clothes if they recycle worn clothes, like *Paris* store. According to this company, they have recycled 1,000 tons of waste textiles per year. Waste textiles are processed by *I:CO*, a German company located in Berlin. They classify the textile and then transform these into new products or reduce them into their primary materials. Furthermore, they can reuse textiles and introduce them again to the second-hand market.

Another project started in Chile is the use of forest products as a new raw material, which is exported to China, principally. This is the case of *Arauco*, a forestry products company that will start in 2020 to produce textile pulp to produce viscose and rayon fabric. This is a sustainable fibre that diversifies the market of the company. They will produce 550 tons per year, of which 70% will be exported to China as a substitute of polyester, helping the Chinese textile industry to follow a circular economy, with zero-waste generation and to cover the needs with bio-sustainable products [42].

Sustainability in Chile is promoted by local designers and small textile companies that make a mix of artisan and fashion textile products. *Raiz Diseño* is an example of local sustainability production in Chile, in which they show the benefits of sustainable products and how they can be sold to consumers [5]. Institutions like The Institute of Agricultural Development (INDAP) and ONA Foundation are focused to support the craft sector in the competitive area of textile industry [7]. Another company is *Volver a Tejer* with their Chilean artisan products [7].

Challenges in Chile are related to making government policies to implement recycling, to have a bigger impact, to teach people about the impact of fast fashion in the environment and the waste volume, and also about fair working conditions.

8.6 Ecuador

Ecuadorians are influenced by tendencies of fashion that encourage clients to have responsible consumption. There are a few clients that claim about the origin of the products and do a responsible consumption.

Enkador is an Ecuadorian company with a sustainable model of production to make new products from bottles and polyester yarns to produce clothes and carpets, pillow fillings, and others. The company keeps a production volume of 6,000 kilos per year, which exports 47% of the total production to Colombia, Venezuela, Chile, Peru, Central America, and the rest for its own national consumption.

In the traditional model, there are a few links between industry and academia. One proposal is to link these two in the fashion industry by four ways: making proposals to reuse the wasted clothes to be applied in new clothes or textile objects; use waste clothes as foundry fill; create new fibres from wasted clothes; and use computer-aided design (CAD)/computer-aided manufacturing (CAD) systems to decrease the waste in industries.

8.7 Brazil

Brazil is one of the biggest textile manufacturers in Latin America, being the fifth behind China, India, USA, and Pakistan, with 2.4% of total global production. Furthermore, Brazil is in fourth place in the volume of garment production and organizes the fifth largest fashion week in the world [6]. The textile industry in Brazil represents more than 5% of the gross national product (GNP), highlighting their importance in the country [43].

Brazil has 29,000 textile companies that offer 1.5 million direct jobs and 8 million indirect jobs, where 75% are women [7].

Brazilian must import a big percentage of synthetic fibre, compared to its national consumption. 60% of the polyester is imported and the remaining 40% is bought from Brazilian companies [47].

Arab countries are the most important countries where Brazil exports its textile and apparel products [44]. The Brazilian government has a special program called *TexBrazil*, with the aim to support the export of Brazilian products, making agreements with international companies mainly in fashion weeks and fashion events.

Related to working conditions, all Brazilian industries must pay the minimum wage, due to government policy [45]. Workers are mostly women from families based on agriculture activities [7].

Brazil has a big potential for biological fibre production, such as rubber, cotton, bamboo, coconut, jute, sisal, flax, hemp, etc., and can be improved by using technological development.

Brazil has a strong cotton production (the fourth largest exporter of cotton in the world) which allows it to be a self-sufficient country [7]. Its importance is due to good weather to produce cotton during the off-season in the Northern Hemisphere, which allows Brazil to export cotton throughout the whole year.

The organic cotton in Brazil is managed by the Responsible Brazilian Cotton Program, certified by Better Cotton Initiative (BCI). The Brazilian organic cotton is naturally coloured, without using dyes, and without using contaminated processes as bleaching and dyeing. Embrapa Cotton is an organic cotton production company that exports organic-coloured cotton to countries like France and Germany.

The Brazilian Leather Certification of Sustainability (CSCB) is a multidisciplinary team that searches for sustainable processes to change the Brazilian leather production. CSCB has a transparent interaction with customers and efficient deliveries [45]. The certification is given by the National Institute of Metrology, Quality and Technology (INMETRO).

Rubber is another sustainable raw material in Brazil that is produced in the Amazon region. Its extraction has international support, and the process is regulated by the Worldwide Fund for Nature (WWF) and other academic projects.

In Brazil, there are controls in place at every part of the sustainability initiative to introduce Brazilian textile products in markets, which are focused mainly on United States, Japan, and the European Union. These companies are also aware of social and environmental standards:

1. **Social standards**, related to working conditions, legal employment, discrimination rules, and all related to workers.
2. **Environmental standards**, related to water scarcity, waste and effluents management, greenhouse gas emissions, energy-saving, and dangerous substances to people and the environment.
3. **Product safety and ethical integrity**.

For ethical consumption, sustainable companies have created the sustainable supply chain (SSC) and the eradication of slave labour in Brazil, with the aim of changing consumer behaviour, encouraging products with less socio-environmental impact in their production.

One of these companies is the Sustainable Fashion Lab that works in three key sustainable goals in the fashion industry:

1. Education: With the use of professional training courses for the industry, which adds themes like gender, diversity, fair working conditions, and sustainable issues.
2. Culture and consumerism: Making sure that every market has fair working conditions and sustainable processes.
3. Product life cycle: Giving information to stakeholders about the effects in the environment of the products, related to their toxicity, water use, and CO2 emissions.

An example of sustainable companies is *Pesquisa*, highlighted by its coloured primitive species program, which investigates the improvement of natural colour of cotton.

Another example of sustainable textile industry is *Justa Trama*, a brand with an agroecological productive of the whole cotton value chain, recognized for a good programme for workers and responsible consumption. The company is an example of a sustainable development model. The company does not use intermediaries, which allows for better commercialization and remuneration.

Other company associated with sustainability activities in Brazil is Ethical Fashion Brazil, which is a company that has the aim to improve a better fashion system with values such as sustainability, fair treatment of workers, scientific and critical thinking as a key to improving fashion, and courage to face the reality in the fashion industry and try to change it [46].

Apex-Brazil is a company that helps 25,000 Brazilian sustainable businesses to improve competitiveness in the international market and to create links with larger companies to include their sustainable products in global value chains. Apex is also concerned with good sustainability practices. Apex is the fourth largest producer of apparel and textile production worldwide, the second largest employer in manufacturing industry in Brazil, the fourth largest producer of knitwear, and the sixth largest producer of denim internationally [47].

Additionally, there are some companies that are involved in slow fashion, namely as follows [7]:

1. *Flavia Aranha*: Use of biodegradable materials and suppliers that are free of chemicals (dyes and printing) and raw materials from artisan processes.
2. *Agustina Comas*: Use of recyclable materials like scraps and pieces of clothing, giving them a new cycle or reuse.
3. *Insecta Shoes*: Making shoes with second-hand clothing materials, with a B+ certification.
4. *Contextura*: Company that augments their creations with slow design principles, testing sustainable methods studied in academia.
5. *Marcia Ganem*: Use of polyamide fibres to produce new clothing.
6. *Atelie Vivo*: Their policy is to reduce consumption with a do-it-yourself culture (DIY).
7. *EcoEra Award*: Searching of companies with conscious practices in their production chain.

Related to fashion, Brazil has one of the most important Fashion Weeks in the world, which showcases beachwear, fitness, jeanswear, and lingerie segments called *Inspiramais*, a fest where more than 190 exhibitors show their sustainable textile products to clients around the world [7].

The challenges of Brazil are focused on two aspects: increasing the ecological production of cotton, getting financial resources, and developing the dying process and the fixed colour process in the natural cotton.

Challenges of sustainability in the fashion industry in Brazil are managed by sustainable Fashion Lab, a multisectorial platform of 40 leaders, with the collaboration of the Brazilian Textile and Apparel Industry Association (ABIT), The Brazilian Association of Textile Retail (ABVTEX), and the International Labour Organization (ILO), have created a strategy to achieve sustainability goals in the country by 2027, related to fair and sustainable fashion industry. Sustainable Fashion Lab bases their sustainable development on, firstly, creating possible futures for the Brazilian fashion industry, and secondly, identifying key actions to advance efficiently with sustainability. There are four objectives:

1. Understanding the risk and opportunities of the industry.
2. Identifying strategies to achieve the main sustainable challenges.
3. Innovative ideas to help in transforming the process.
4. Establishing good contacts with ecosystem stakeholders.

The future of sustainability in the textile industry are:

1. Certifications for sustainable products and ensuring their standards.
2. Industry 4.0, introducing automation to textile industries, which can be saved until 70% of water in production manufacturing, and contributing to the transformation of circular economy processes.
3. Conscious consumption, higher quality of products with greater durability.

In recycling, according to the Brazilian Support Service for Micro and Small Enterprises (SEBRAE 2014), Brazil produces 170 tons of textile waste annually, of which 80% is sent to dumps and landfills. 20 recycling companies exist in Brazil, which most of them prefer to import recycled textile industries more than using national textile waste, because of the poor management of this material in the country [48].

There are three major problems of recycling in Brazil:

- Dirt and mixture waste textile products with a high cost of separation.
- No fiscal or tax incentives to make a business of recycling products.
- Logistic costs.

A movement created to raise consumer awareness of the true cost of fashion in the whole value chain of textile production is Fashion Revolution with participation via online or physical presence. In Brazil, the movement has created promotions of ethical fashion and has spread in 45 events, 29 cities, and 31 universities in 2016. One important focus of the movement is the students involved in debates about ethical fashion and bad working conditions of the traditional value chain of textile production. An example is Rio Grande do Sul state, known as the sustainable fashion production state.

8.8 Peru

Peru is famous for its global alpaca production, which contributes 80% of the total production [49]. Alpaca is recognized by its less need of water for washing and its properties: it is hypoallergenic, breathable, lightweight, and soft. Also, Alpaca has a variety of natural colours, and is free of dyes.

Peru is the principal exporter of alpaca fibre and *tanguis* cotton in the region [5], produced for small families which operate farms in Northern Peru. The textile industry in Peru has vertical integrity in the whole value chain [50]. They produce mainly high-quality raw materials such as alpaca and cotton.

The Peruvian cotton is called scientifically "*Gossypium barbadense*" and it has two varieties:

- **Pima**: recognized by its extra-long staple length, softness, quality, durability, resistance to pilling., shine, and hand feel.
- **Tanguis**: good cotton for blending with natural or synthetic fibres to create advanced materials [50].

Peru is facing a decrease in cotton production because of the decline of its quality, not investing in a genetic improvement of the species. Also, there are products more productive than cotton, like crop, rice, and sugarcane, another cause for the decline of cotton production [2].

Peru is managing a sustainable production of its textiles with specific projects in the whole value chain, optimizing alpaca breeding and improving its genetic resource management with the aim to advance in quality [50].

According to Designs company, Peru has made a sustainable positive impact, saving 48,450,000 litres of agricultural water annually, keeping out 20.750 [kg] of carbon dioxide to the air, and avoiding the use of 185 [kg] of deadly toxic pesticides annually.

The government has made some rules to protect the value chain of Alpaca production, promoting companies the use of alpaca in their production, highlighting the concept of sustainability in the whole value chain: eco-friendly, care and well-being for Alpacas and their farmers, good practices in alpaca shearing, and its management [50]. International organizations like Alpaca AIA and the Association Civil Alpaca del Peru-ASCALPE in alliance with the Ministry of Agriculture and Irrigation (MINAGRI), have specified technical norms, control points, and minimum criteria to comply with alpaca breeding good practices. They also promote the export and national sales in volume and value, contributing to the positioning of alpaca fibre and its derived products nationally and internationally with the brand called "*Alpaca del Peru*". The strategic importance of protection and promotion of the alpaca fibre value chain has the aim to assure the sustainability of thousands of families involved in this activity, friendly with the environment and prioritize the protection of the workers and animal well-being.

Companies are assuming the sustainable production challenge. One of them is Indigenous Designs, products made by Peruvian artisans that operate under fair trade alignments and training programs for artists communities. They work strongly in social sustainability, taking care of fair living wage, safe working conditions, spread of prosperity among local artisans. The company helps farmers to reach organic certifications, ensuring a higher price, and avoiding the use of pesticides and herbicides. They invest in promoting naturally coloured alpaca fibre without the need for dyes. One example is Indigenous Designs that has a lot of certifications to produce organic cotton and export it. All of them are listed as follows:

- USDA certified organic cotton since 1994.
- OEKO-TEX certification, independent testing that ensures 100% of non-toxic and safe use of materials.
- Coloured alpaca: it creates a demand for coloured alpaca wool, activating the genetic diversity of Peruvian alpaca populations.
- Safe dyes: Ensures non-harming dyes certified by the Oeko-Tex Standard 100, using a dyeing process that eliminates harmful chemicals and waste.
- Coloured cotton: use of cotton with few pesticides and water, a more sustainable choice.
- Pure collection: using only nature fibres, with security to people involved in the production and respecting animals.

Another company that is following sustainable production is Anpi Organic, famous for its baby clothes promotion with 100% certified organic cotton, taking care of the health of babies, the environment, and promoting a fair-trade business. Broaches used are free of nickel and they use recycled paper and recyclable bags for packaging.

Related to fashion, Flashmode is a fundamental event of Peru, where they promote ethical fashion. The aim of this show is to access an exigent market in Europe, North America, or Japan, where the environmental aspect has high importance nowadays. The last collection was ethical fashion, where they promote minimum environmental impact in textile production, use of organic and natural textiles, reduced use of pesticides, limited use of chemical dyes products, and use of recycled materials.

Peru has an interest in promoting sustainable production in the textile industry, but there are no deep statistics to analyse it. The major challenge they have in the future is the less formal labour practices as the most important area to improve.

9 Conclusion

In this review, it is found out that there was a decrease in the Latin American textile markets due to the introduction of products from Asian countries. This scenario has put much pressure on Latin American countries to increasingly improve the quality of their products and become more sustainable to give a better value for their local consumers. However, the labour costs in Latin America are also increasing due to its economic development, which has been a big challenge to the Latin American textile industry to compete with the international market.

In most Latin American countries, the private sector with the development of artisanal textiles has pursued sustainability agenda actively with the transformation of recycled textile materials into new products. These cases can be seen in Argentina, Chile, Colombia, Brazil, Mexico, and Uruguay.

Countries with strong governmental support are related to the export of raw sustainable materials like organic cotton, wool, and alpaca. These examples are found in Brazil (cotton), Argentina (wool), Uruguay (wool), and Peru (alpaca), respectively. These programs improve the working conditions of farmers, invest in new projects to improve the quality of species and to promote internationally their products, and be more competitive against other exporter countries. There is a big potential to be more productive, due to space and nature found in Patagonia to make a sustainable plan with cared animals.

New technologies are also playing a major role to compete with products from Asia. Examples like in Mexico with its smart textile and Colombia has enhanced its production speed as well as have introduced high-quality customized products from local producers to recover their strong textile industry presence before it was overtaken by Asian textile suppliers.

Uruguay and Brazil are being considered as "slow fashion" capitals in the region. Uruguay, for the management of its rich green chain of biological-fibre production, the use of eco-friendly materials and its artisanal way to process wool with high quality; and Brazil, with the buy of second-hand market, and the use of new sustainable materials such as natural dyes cotton (organic cotton), linen, and PET based on natural plants, with the aim to reduce costs.

Mexican companies have limited financial capacity and are dependent on importing the materials and supplies, especially from the USA. They have a lack of technology, which do not have the possibility to compete with other textile products from overseas. Another problem in Mexico is the contraband of low quality and price of textiles, from the USA and Asia which has favoured poor working conditions, with informal and illegal structures. The potential to make a sustainable textile market in Mexico is big, considering that 80% of their products are exported to the USA.

The waste textile volume is expected to increase in the next years if every country does not have programs to remediate this problem, because of the tendency to import products from famous brands that gives clients a status symbol that differentiates them from others.

Latin America is a potential growth market in textile industry, due to its domestic demand, economic growth, and purchasing power. Countries in this region have a strong possibility to develop sustainable textile industries, but also feel pressure to implement new ideas over the next few years. This is because being sustainable requires a high initial investment cost as well as a strong policy and strategy to achieve it.

References

1. Balanay R, Halog A (2019) Tools for circular economy: review and some potential applications for the Philippine textile industry. Circ Econ Text Apparel Text Inst Book Ser. 49–75. https://doi.org/10.1016/B978-0-08-102630-4.00003-0. Accessed 10 May 2020
2. FAO (2018) Estudio Nichos de Mercado del Algodón. http://www.fao.org. Accessed 15 June 2020
3. Fernandes, P (2020) Circular Economy as a way of increasing efficiency in organizations, APCER. http://www.portoprotocol.com. Accessed 10 June 2020
4. Como se incorpora la sustentabilidad en la industria textil en Latinoamerica (2018) CEPAL. http://www.conferencias.cepal.org. Accessed 29 May 2020
5. Mordorintelligence (2019) Latin American textile industry – growth, trends, and forecast (2020–2025). https://www.mordorintelligence.com/. Accessed 10 Feb 2020
6. World Integrated Trade Solution (2020) Textiles and clothing export and import by region. http://www.wits.worldbank.org/. Accessed 15 May 2020
7. Gwilt A, Payne A et al (2019) Global perspective on sustainable fashion. Part 1: Latin America, pp 1–43
8. Silvia Z (2017) Towards a sustainable and ecological fashion. DAYA, Diseño, Arte y Arquitectura. Nr. 2, December 2016 'June 2017, pp 61–73. ISSN 2550-6609
9. Inexmoda (2018) Economía Circular. http://salaprensainexmoda.com. Accessed 10 May 2020
10. SOCILA (2016) Algodón Orgánico Potencial en América Latina. http://www.socila.eu. Accessed by 13 Jun 2020
11. Uddin F (2019) Introductory chapter: textile manufacturing processes. Intech open. http://dx.doi.org/10.5772/intechopen.87968. Accessed 10 Feb 2020
12. Embrapa (2020) Embrapa cotton. http://www.embrapa.br. Accessed 2 June 2020
13. Textile Exchange (2017) Organic cotton market report. http://textileexchange.org. Accessed 2 June 2020
14. Romero P (2000) Sustainability and public management reform. Am Rev Public Adm. 30(4):389399
15. Garcia A, Diaz O (2019) La evolución del sector textil en la región centro-occidente de Mexico: "Del taller de costura al tianguis". núcleo Básico de Revistas científicas argentinas (Caicyt-Conicet). Nr. 32 Summer 2019
16. Schlomski I (2018) Mexico: cotton production recovers. Bremen cotton report, Issue 37/38. Accessed 28 Sept 28
17. Comercio Exterior (2019) Logística retail: la industria textil en México. https://ibercondor.mx/blog/. Accessed 10 May 2020
18. Eyhorn F, Ratter S et al (2005) Organic cotton crop guide, Research Institute of Organic Agriculture (FIBL)
19. Deschamps T, Carnie B et al (2016) Public consciousness and willingness to embrace ethical consumption of textile products in Mexico. Text Cloth Sustain 2:6. https://doi.org/10.1186/s40689-016-0017-2
20. Dussel E (2017) Efectos del TPP en la economía de México: impacto general y en las cadenas de valor de autopartes-automotriz, hilo-textil-confección y calzado. Centro de Estudios Internacionales Gilberto Bosques, México
21. Maldonado G (2016) Hablemos mezclilla: Mexico, uno de los mayores productores de jeans en el mundo. Fashion United. https://www.fashionunited.mx/noticias/retail. Accessed 15 May 2020
22. Soto R (2019) 5 marcas mexicanas de moda que apuestan por reducir su impacto ambiental. The Happening. https://www.thehappening.com/. Accessed by 15 May 2020
23. Amor y rosas (2020) www.amorandrosas.com.mx . Accessed by 31 May 2020
24. Artesanías de Colombia (2019) Panorama Artesanal Ilustrado. 1st Edition, June 2020. Bogota, Colombia
25. ProColombia (2019) Los beneficios de la tecnología 4D para la moda colombiana. https://procolombia.co. Accessed 5 June 2020

26. Contreebute (2019) Cinco marcas de moda sostenible en Colombia. https://www.contreebute.com. Accessed by 5 June 2020
27. Fokus green (2020) Fokusgreen official webpage. https://fokusgreen.com/. Accessed 5 June 2020
28. Bareke (2020) Bareke official webpage http://www.en.bareke.com/. Accesed by 5 June 2020
29. Little & Ramonas (2020) Instagram profile of Little & Ramonas. http://www.instagram.com/littleramonas/. Accessed 5 June 2020
30. True Love and Poems (2020) Webpage of true love and poems. http://trueloveandpoems.com. Accessed 5 June 2020
31. IWLO (2019) Wool sheep welfare. Accessed 5 May 2020
32. Livni A (2020) Ana Livni, Moda lenta Uruguay. http://Instagram.com/ana_livni/. Accessed 5 May 2020
33. Centro Textil Sustentable (2020) Centro textil sustentable. http://www.cts.org.ar. Accessed 5 May 2020
34. Argentine wool federation (2019) Argentine wool statistics. EL 728 (06/2019)
35. Senasa (2020) el Encarpado, una herramienta para prevenir la dispersion del picudo del algodonero. http://www.senasa.gob.ar. Accessed 5 May 2020
36. Gate away to South America (2018) Patagonian Lamb: a worldwide reputation of quality. http://www.gatewaytosouthamerica-newsblog.com/. Accessed 5 May 2020
37. Gonzalo E (2019) Wool: global and national markets, perspectives, and possibilities. Laboratorio de Lanas Rawson-EEA Chubut INTA. Accessed 5 May 2020
38. Risso N (2019) Industria textil: cuales son las perspectivas de un sector debilitado, El Cronista (in press)
39. Animana (2020) Webpage of animana. http://www.animanaonline.com.ar. Accessed 2 June 2020
40. Maurello M (2019) Sustentabilidad. Donde empiezan los residuos textiles y como terminar con ellos, La Nación. https://www.lanacion.com.ar/moda-y-belleza/. Accessed by 2 June 2020
41. Valenzuela L (2019) La ropa tambien contamina: Con iniciativas aisladas Chile avanza en el tratamiento de residuos textiles de la industria de la moda, País Circular. https://www.paiscircular.cl/industria/. Accessed 5 May 2020
42. Mundo Marítimo (2019) Arauco desde 2020 producira por primera vez pulpa textil en Chile para exportar al mercado asiático. https://www.mundomaritimo.cl/noticias/. Accessed 3 May 2020
43. Al Mahuz A (2018) A brief history of Brazilian Textile Industry. Available via Textile Today. https://www.textiletoday.com.bd/brief-history-brazilian-textileindustry/. Accessed by 22 May 2020
44. Legal Team Brazil (2018) Business opportunities- introduction to the Brazilian textile industry, available via Bizlatinhub. https://www.bizlatinhub.com/opportunities-brazil-textile-market/. Accessed 22 May 2020
45. Bassi M, Galleli B et al (2015) Brazil's fashion and clothing industry: sustainability, competitiveness and differentiation. Int J Environ Sustain Dev 24:280–295
46. Ethical Fashion Brazil (2019) Agency and platform for sustainable fashion businesses. http://ethicalfashionbrazil.com/ethical-fashion-brazil/. Accessed by 20 May 2020
47. Apex (2020) Apex Brazil. http://www.portal.apexbrasil.com.br. Accessed 20 May 2020
48. Amaral M, Zonatti W et al (2018) Industrial textile recycling and reuse in Brazil: case study and considerations concerning the circular economy, Gest. Prod, Sao Carlos. https://doi.org/10.1590/0104-530X3305.
49. Peru Moda (2020) Peru Moda textile industry. http://www.perumoda.com. Accessed 21 May 2020
50. Tinoco O, Raez L et al (2009) Perspectivas de la moda sostenible en el Perú. Revista de la Facultad de Ingenieria Industrial 12(2):68–72

Sustainable Latin American Aesthetics

Diana Patricia Gómez García

Abstract Aesthetic is a key factor in the construction of sustainability, especially when related to the fashion field. People used to interact with fashion mainly in a sensuous way, not only regarding how a garment looks and feels but also in the way that fashion becomes an identity-building tool. For sustainable fashion to be successful, it has to, besides the obvious ethic aspect, display an aesthetic content capable of thrilling the user by the integration of its own contextual factors that enable its identification. In the case of Latin America, adoption of a sustainable fashion characterized by minimalism, which has been a recognizable aesthetic representation of this fashion movement, has not been easily accepted due to the lack of particularly self-aesthetic referents within this construction. Hence, it is necessary to understand the Latin American context, including its endogenous aesthetics and its relation to sustainability. In order to achieve an intertwining between the sustainable design principles, certain design strategies are proposed that promote the adoption of sustainability from an autochthonous minimalism. This article is based on a wide literature review and it is organized as follows: first, the description of local contexts followed by the definition of key concepts such as aesthetic, fashion and minimalism; then, the depiction of the Latin American aesthetics characteristic features aiming to establish a relationship with the aforementioned concepts to conclude proposing the conceptualization of Latin minimalism as a strategic route to design sustainable aesthetics within the continent.

Keywords Latin American aesthetics · Sustainable fashion · Minimalism · Latin minimalism · Sustainable aesthetics

D. P. Gómez García (✉)
Bogotá, Colombia

M. Á. Gardetti and R. P. Larios-Francia (eds.), *Sustainable Fashion and Textiles in Latin America*, Textile Science and Clothing Technology,
https://doi.org/10.1007/978-981-16-1850-5_7

1 Contextual Framework

Within the continent, the tensions around the territory are a transversal factor that determines its configuration. Notwithstanding, Latin America can be defined by Canclini's proposal as "a geographical, political and ideological space that also has a deep symbolic burden" [1]. A space that derives between economic and social realities allows us to understand the structures from which the region was and is built in order to identify the identities that have been gestated within, which have been largely determined by a colonial past and continuous miscegenation, and that ultimately gives Latin America an eclectic view of itself.

The discovery of America and its subsequent conquests is a turning point that tags the locals first as natives and later as colonized peoples. During the pre-Hispanic era, native people organized themselves autonomously into complex socio-political and cultural structures that were erected and evolved for years, giving character to the representations of both the material aspects (from works such as textiles, basketry or gold-smithing) and the spiritual side (from their common world-views) of their culture. It is pertinent to understand the pre-Hispanic materialities and world-views in relation to the construction of the local sustainable identity, where the former points to the tangible and the latter to the narrative. This cosmogonic narrative derives from an animistic conception of the world; therefore, these world-views were based on the worship and respect of nature as an agent of creation and as a sensitive and guiding being who deserves to be cared for and heard as the only way to ensure existence due to nature being able to influence the destiny of peoples. "Making the man-nature relationship therefore almost indissoluble" [2].

Later, the conquest periods were marked first, by domination and, subsequently, by the miscegenation of both individuals and knowledge, which led us as a continent to a configuration where the divine and the profane, the ancestral with the modern, and the black and the indigenous with the white were mixed; thus, erecting a hybrid territory, hybrid but colonized.

> Latin America is a melting pot of dilutions – not always peaceful, built upon exterminations and enslavement –but its imaginaries and images, necessary in the manufactures of a collective identity, always refer to the hybridization of three components: all that range of mixture that favours asymmetrical and enchanted encounters between Spaniards, indigenous and African [3].

It is important to highlight the long-term effect of colonization on the construction of the Latin American gaze and by extension, its identity and its connection with the surroundings. Since the colonization, the history of the indigenous people was erased in order to be told from the gaze of the settler, eliminating the voice of the native inhabitants. In colonial fiction, the indigenous needs to be undermined for the colonialist to establish its own narratives, which implies the appropriation and annulment of the other. Ultimately, their conquest comes from an act of "despiting the other" [2]. Consequently, a moment is created in which we begin to be alienated from our identity as owners of our cultural legacy and condemn it to inferiority for not belonging to the cultural and moral constructions of the conqueror, placing the native legacy in a vacuum. Therefore, the place where we constantly have to be educated, modernized and civilized is under the instruction of the white, which leads us to constantly disown what is ours with the intention of overcoming a type of man "who is part of what is called the West and at the same time excluded", [2] evidencing the "tropicalist discourse" [4] that does not end up recognizing us as Western neither modern.

Regarding the territory, what the tropicalist discourse evokes is the inexorable relationship created between Latin America and the imaginary of the tropics, framing it in a space that does not acknowledge the other geographies that, by their diversity, have been decisive in their physical and idiosyncratic construction. Conversely, the highly rugged geography allowed the constituent of peoples to develop independently and plurally from what their territory offered and permitted them. Therefore, not a single identity was formed, but multiple identities in the said geography are worth differentiating in order to explain the aesthetic constructions that represent the literal or the iconographic.

Besides historical, it is also important to understand the contemporary Latin American social construction that, due to the different socio-political events derived from ideological clashes that perpetuate the different acts of violence exerted in the territory, make it a broadly unequal society. These acts of violence, in addition to deepening ideological and class ruptures, also shape urban structures as arrival centres for a human diaspora displaced from their territories towards the peripheries of the cities. The result is a continent that quickly became eminently urban, where 80% of its population is concentrated in cities, which in turn depend on the extensive and unpopulated rural area.

> "The cities of Latin America and the Caribbean went through an unprecedented development process: a rapid and run-down urbanization in a scenario characterized by high levels of poverty and misery. Immigrants from the countryside came to live on the periphery, in often environmentally vulnerable places... Latin America's cities have largely been shaped by poverty, inequity and segregation" [5].

As a consequence, the socio-economic scope reveals that Latin America has 30.2% of its total population living in poverty and 10.2% in extreme poverty (CODS América latina y el caribe; universidad de los andes n.d.). In contrast, Latin America also has a traditional, rich and minoritarian class that persistently accumulates the majority of land and economic resources.[1] While in the midst of these extremes, we find the growing middle class that strives to not back down and projected itself towards wealth through, among other things, constant and cumulative consumption. A society where, as Bauman says, "status is measured not by the ability to buy but by the speed of throwing and replacing" [6] sheds light on the fragile relationship between this voracious consumption and the resources it feeds on. "The rising global middle class will increase the rate at which resources are being consumed, affecting the (fashion) industry as a whole" [7].

---

[1] According to the world bank in countries such as Brazil 58% of the income belongs to 20% of the population.

Therefore, class configuration is an element that requires special attention because of its deep rooting within the collective imaginaries and how decisive its results are for the construction of contemporary. In this class configuration we can even talk about a class interbreeding, as Canclini mentions, "Despite attempts to give elite culture a modern profile, secluding the indigenous and colonial in popular sectors, an interbreeding interclassity has generated hybrid formations in all social strata" [2]. From that hybridization of classes, features are recombined to generate their own aesthetic constructions. In Latin America, although not exclusive, in terms of class, the majority represent the popular and economically marginalized crowd [8], so its expressions dominate and energize the panorama through the appropriation of what belongs to the privileged minority so that this appropriation that merged with the historical component could be the place from which Latin American fashion builds sustainable language and, as Barbero argues, it becomes not only a tool to "cover up class differences" [9] but to overcome them.

In short, Latin America is modelled within a hybrid construction arising from a meeting of cultures that find a way to adapt in order to survive. Cultures which in turn have been crossed by multiple violence acts and which are found in dense urban configurations that are representative of modern longings: security, opportunities, consumption, from which the class gaps are evident and where in turn ways to blur them are found. It is a space with defining geographies and natural relationships, but that expands beyond these, to build upon the symbolic, by redirecting both symbolic and real representations that Latin America begins to be recognized and re-established by its inhabitants through postcolonial narratives, while pursuing a quest to belong/overcome the traditional Western project. Thus, the understanding of these characteristics is fundamental for the construction of a profuse aesthetic discourse, which recognizes and integrates sustainability in fashion and in the different contemporary spheres from which it feeds.

2 Conceptual Framework

In this section, we will define the concepts of aesthetics, fashion and their relationship with sustainability as follows, starting with defining aesthetics from the traditional and etymological view and then complementing it from a contemporary concept of everyday life aesthetics. The following definition will allow us to approach the concept of fashion as a system of symbols that builds identities as proposed by semiology; then aesthetics and fashion will combine within sustainability and Latin America. And later, the aforementioned will be framed in the concept of minimalism as a determinant feature in Saito's proposal and as formulation baseline of aesthetic strategies.

2.1 Aesthetics

Traditionally, our aesthetic notion is entrenched in the dichotomous and Western gaze adopted mainly in art, where our possibilities oscillate between the beautiful and the sublime as the only ways of understanding the phenomena that occur to us. This approximation promotes an approach to experience objectively without cognitive biases and therefore rests entirely on the perception of the senses as the only way to assume reality, being completely phenomenological. The said construction is based on its etymological definition where:

> The term 'aesthetics' comes from the Greek aisthetikos, meaning to perceive, or 'relating to perception by the senses' ('perception' from the Latin percipere, to 'seize, understand'), aesthetics is understanding that arises through the senses (from the Latin verb sentire, to 'feel') when bodies come into contact with another. And so, aesthetic judgment is located in the world of things, not concepts, and aesthetic meaning is the result of physical experience rather than the 'reading' or 'decoding' of the abstract, symbolic or metaphorical. In short, aesthetics does not produce meaning in the manner language does [10].

However, while it is important to understand aesthetics from the sensory, this approach does not allow break through the material barrier and avoids the personal relativism of the observer. This notion of aesthetics is insufficient in fashion since in fashion one is no longer a spectator but becomes a user and gives the garments the meanings that communicate identity. This makes it necessary to perceive and formulate beyond the physical and objective and include the symbolic as "we could say that fashion becomes a language whose basic element is the sign, is therefore a non-verbal system of communication" [11]. Therefore, we propose to complement the traditional aesthetic concept with a view from the everyday life aesthetics [12], which together with a sensory factor recognizes a subjective load that complements phenomenological reading with semiotics, surpassing the object by extending to the experience of use that surrounds it.

The everyday life aesthetics allow a symbolic reading by extending its scope. Firstly, to a multiplicity of aesthetics, removing the uniqueness of the term and encompassing scenarios that have traditionally been underestimated. Secondly, towards inclusion "in its study (of) interactions with other people, everyday activities and the objects that we use everyday life as clothing" [12]. And finally, by abandoning the binary of the beautiful and the sublime giving way to more familiar interpretations of aesthetic experiences cross-cutting to all:

> such as pretty, cute, messy, gaudy, tasteful, dirty, lively, monotonous, to name only a few. These items and qualities are characterized by their ubiquitous presence in the daily life of people, regardless of their identity, occupation, lifestyle, economic status, social class, cultural background, and familiarity with art [12].

In addition, the everyday life aesthetics also raises theoretical issues as an indeterminate identity of the object of aesthetic experience; changes and modifications everyday objects go through; absence of any clear authorship behind everyday objects; bodily engagements with objects and activities and their pragmatic outcome [13]. Some of those issues are extremely pertinent around the fashion understanding and its construction within Latin America and sustainability. As a final idea on this concept, we propose to address the everyday life aesthetics as a way to transgress the design summarized by Forsey as "functional, immanent, mass-produced, and mute [14]". A proposal that the fashion industry has been able to profit from and which represents everything we should overcome if we intend to face new scenarios around fashion within sustainability extends aesthetic values beyond the superficial.

2.2 Fashion

Regarding fashion, we will establish the differentiation between functional dress and fashion from a semiotic approach to develop the ideas of how it is managed in Latin America. Firstly, when we talk about dressing it responds to the need for shelter. What Maslow would place at the base of its pyramid as it is a primary necessity that appeals to the protection and survival of our bodies is a necessity that, although it is not totally stripped of a particular aesthetic or codes mediated by the climate, gender or religion, mainly responds to a utilitarian need.

However, from very early on, the protective function of dressing migrated to an emotional need derived from the awareness of our appearance [15], making evident the idea of fashion as we know it. Fashion satisfies axiological and identitarian needs; therefore, it is configured as a system of symbols and contents that allows us to load with different readings objects that are perceived through the senses with a semiotic content. As Eco explains, "Semiotics has to do with anything that can be conceived as a sign. Sign is all that can be understood as a significant substitute for something" [15]. Thanks to that interpretive content, a garment becomes fashion. Therefore, it becomes a construction that allows us to communicate beyond words since "the symbolic value of the dress [...] in a harmonious interaction with the other modalities of non-verbal communication, forms a visual language well-articulated by the multiple psycho sociological and cultural implications" [16]. Then, through fashion, multiplicity of expressions are modulated that allow each individual to build his identity and its own particular language; it should be noted that "The union of the signs of dress ends up generating a language, but there is not a single language of clothing but many, just as there are many different languages, dialects and accents and that each individual has its own repertoire of words" [17].

But like any language it has to be validated within the collective. As Iglesias mentions "This nonverbal syntax will be a cover letter of ourselves, provided that between the sender and the receiver there is the same code of understanding and, therefore, the same cultural basis" [15]. This gives importance to the fashion context for the correct decoding of its signs, which, if appropriate, would lead to acceptance, both of representation and the individual, and the potential replication by the recipients. In addition, it will influence its ability to be sustainable to the extent that it may or may not be read properly as "If a product is removed from the context and the time in which it is considered fashionable, it ceases to make all sense, even if this does not affect the quality of the product itself" [15] so that the context will be the one that provides that language with the necessary qualities for a correct interpretation of its meaning:

> Objects have no meaning in themselves, rather they are prompts for a field of possible meanings that are dependent on context. Meaning often implies something fixed, but in this instance, let's understand meaning as that which arises as the result of an object's exposure to a specific circumstance. That is, objects provided certain outcomes rather than contain certain meanings, and each interaction presents the possibility for a range of outcomes to arise that are not wholly predictable. These interactions accumulate over time; thus, the meaning of an object is ever evolving [10].

In relation to fashion as a symbol, Latin America has been able to understand these dynamics of representation and construction of meanings through clothing from the peripheral. On the one hand, this has endowed it with an added value that ends up shaping its own identities and therefore a strong contextual language. On the other hand, in Latin America we struggle with the idea of belonging to "hegemonic fashion" [9], which relates to the dynamics of fashion that poses it as a differentiating and integrative at the same time. The key in this relationship is to assume marginality as its own language that becomes striking for the centres and thus re-signify the periphery.

2.3 Fashion and Aesthetics

It is important to combine the concepts of fashion and aesthetics to understand the dynamics between them, their relationship with consumption and sustainability. When we talk about fashion nowadays, we cannot detach it from consumption which is its main engine of growth. "With projections through 2023 that show annual growth of approximately 5%. By 2030, the global apparel and footwear industry is expected to grow to 102 million tons in volume and USD 3.3 trillion in value" [18]. This overproduction and hyper-consumerism have led fashion to be disposable, both in monetary terms,[2] as in social and environmental ones.

To get to this point of growth and waste, fashion has been continuously supported by an obsolescent aesthetic where its improvement lies in the cosmetic, rather than the functional [20]. In this way we are caught up in a constant replacement frenzy for the new and improved, of perishable satisfaction, ephemeral beauties and identities that are dismantled at convenience because they are in themselves made also for consumption [6] and therefore will never be satisfactory and lasting. Because of these continuous replacement dynamics, it is necessary to establish a counterweight from the same place that has been used to promote it; that is, from aesthetics, but making an effort to propel them from more holistic constructions, which lead to a connection

[2] According to the Ellen Mchurart foundation, "Globally, customers missed on USD 460 billion of each value year by throwing away clothes that they could continue to wear" [19].

with the subject beyond the material and give rise to positive, pleasurable and consequently more lasting relationships, and thus leading to long experiences and identities. Hence, as Saito mentions "The aesthetics of sustainable consumerism, therefore, is not simply a matter of promoting specific sensory features of the object regarding beautyness. It is more importantly a paradigm change about our relationship with the material world" [20].

This paradigm shift needs, in order to be effective, to include an additional conceptual load that contemplates the moral and political ramifications [20] in which the experiential object ceases to have aesthetic immunity to its underlying processes and its entire cycle is visible to be considered either aesthetically positive or negative. The new paradigm also requires assuming a sentimental content as a fundamental element for an aesthetic extension, as Saito proposes:

> "I believe that what may be dismissed as sheer sentimental attachment to an object in the objectivity-driven-discourse does have an important role to play in the sustainability-driven consumer aesthetics". Saito continues saying "for the purpose of cultivating longevity of an object in the hands of consumers, I would include associational and historical value as a part of sustainability aesthetics" [20].

2.4 *Sustainable Design and Fashion*

Just as we can't deny consumption in fashion, we can't deny its origin either. Fashionable objects, and by extension their experiences, are not natural but, on the contrary, they are designed with the aim of serving human functions. Hence, as designers[3] we have the capacity and responsibility to direct these processes from the aesthetics that we have previously raised since they are the ones that have as "the main objective to put in the mainstream ecological, social and environmental values." [21] and therefore, serve as a tool to understand the characterization of sustainability within Latin America in order to design it. Punctually, within green aesthetics, Yuriko Saito proposes within what she narrows as the everyday aesthetics, principles for contemporary design, that "offers the possibility to combine environmental values with aesthetic experiences" [22]. "These principles that cover the aesthetics, the functional and the contextual are minimalism, durability and longevity, fittingness, appropriateness and site specificity, past present contrast, preservability of natural processes, health and fostering and caring and sensitive attitude" [12].

As the intention is to look for an aesthetic configuration, we will focus on minimalism as a starting point because, on the one hand, it has been the archetype that finds the least correspondence within the Latin; and, on the other hand, it is the most aesthetically recognizable principle within Saito's proposal. In addition, minimalism finds a rather strong objective representation within what sustainability proposes in fashion; and since the 90 s it saw a boom in what Fletcher points out as aesthetics of

[3] "Anyone who shapes matter, processes and energy to need perceived needs can be understood as a designer" (St. Pierre 2019).

sustainability [23] which has been implanted almost dogmatically among those who seek to approach a sustainable way of characterizing their products.

The approach through minimalism can have a double reading within the realm of contemporary fashion. On one side, we find fast fashion that aims to be universal and operational in any context (cultural or social), so it needs to operate without the need to make further readings and thus achieve ubiquity, massification. On the flip side, we find sustainable fashion that uses neutrality to transform objects/garments into simple and easy-to-adopt because they do not require special knowledge to understand them and therefore to use them for a long time avoiding an early disposal. For both cases we could use the following example: a black turtle neck, which is validated within fast fashion or within sustainable fashion, it will depend on how we approach minimalism, whether from the aesthetized or the conceptual side.

Hence the fact that the configuration of minimalism today has more to do with aestheticization than with a real aesthetic, so it is necessary to emphasize this difference to avoid confusion when it comes to raising strategies. The first does not promote sustainability values; instead, it only seeks to make products on its surface attractive to speed up or perpetuate the consumption of objects designed to be obsolete from mutism and eternalize the oppression. "Aestheticization of everyday life is often used to cultivate consumers for the market, audience for political spectacles, and accomplices in perpetuating oppression and injustice" [13]. On the contrary, formulating design proposals from true aesthetics that respond to sustainability aims to create a nutritious experience that generates in the interlocutor an attachment that will make viable many of the other principles mentioned by Saito, which are necessary for a holistic design strategy.

Once this differentiation has been established, we can address the minimalism itself without the fear of entering into a contradiction between the objective of promoting a contextual and multidimensional aesthetic language and the apparent emptiness of the archetypal minimalism.

2.5 Minimalism

Minimalism largely represents the aesthetic and conceptual ideal of sustainability but seems excluded within the Latin, which makes it relevant to break down what minimalism refers to and how this serves as a basis for articulating sustainability in the Latin American context.

Minimalism corresponds to the familiar, to the beautiful "that complies with the aesthetic ground principles concerning to symmetry, comfort, order, predictability, demarcation, shape and balance" [24]. Therefore, as a principle it is relevant within sustainability because it tends to be easily decodable, as Harper mentions in a first approximation:

> from the aesthetic analysis of sustainability so that there is a durable expression objects have to be easily decoded, therefore they have to be neutral and minimalist so that they can encompass a wider range and seek a universality, meaning they are based on basic and simple characteristics so that there is a flexibility of adoption to the context [24].

On a first basis, to achieve universality minimalism uses reduction. Either from the modernist design that dictates to avoid the ornament in pursuit of supreme functionality that silences the noise of excessive forms and contexts in order to become adaptable and therefore lasting. Or from the artistic approach, where this minimalism in addition to a formal simplicity also calls for a simplification of the content; where the pieces do not carry any abstract message that needs a cultural baggage, and where it is only necessary to be in the same space of the work that only exists in the presence of the observer [25]. In this manner, minimalism becomes enunciative of sustainability by creating "time and placeless expressions" [24], reducing in addition the ornament and content, materiality. Here we might ask ourselves, why if what we try to do is to make an aesthetic reading of the contents of the objects from new perspectives, we propose minimalism being that this appeals to the silence of those same contents?

The answer is found through three steps. Firstly, expanding the scope of what is understood as minimalism, leaving behind the stereotypes that have shaped through the years depicted as "a kind of enlightened simplicity, a moral message with particularly austere visual style", erected in monochromatic solid colours, organic textures, clean, modernist reduction, emptiness and silence [25]. Secondly, going beyond traditional moralist modesty, the efficient functionalism of modernist productive systems, aestheticization governed by postmodernist means [26] and contemporary elitized precarization, where little is a sign of detachment (disposable) and freedom (vacuum). Thirdly, detaching minimalism from being sustainable per se, because a simple appearance does not mean that its underlying processes are simple as well [25].

Understanding that minimalism is narrative and not empty, let us approach it as a paradigm shift from the formal and symbolic based on our need to live the world from the appreciation of details that can be ephemeral in form but durable as an experience. In this way, minimalism is not about a "perfect cleanliness or a specific style. It's about seeking unmediated experiences, giving up control instead of imposing it, paying attention to what's around you and accepting ambiguity, understanding that opposites can be part of the same whole" [25], where instead of a universal story there is a blank space to build a narrative of its own. Finally, since it is from Saito's design proposal that we extract minimalism, we must understand it from his vision, which refers to a reduction of materiality through processes "such as reduce, reuse and recycle from the discipline of sustainable design" [12].

With this expanded vision of minimalism, we can investigate the construction of Latin American aesthetics and how this minimalism is welcomed or rejected within that context in order to configure a sustainable language, and then propose design strategies that integrate those components effectively.

3 Latin American Aesthetics

So far the Latin construct is settled in a symbolic and real space, and now we have to establish the aesthetic identities that represent it. As we mentioned, minimalism tends to be easily decodable and therefore it is beautiful. From there, if minimalism is beautiful and Latin American at first glance is not minimalist, we would have to go into understanding the Latin aesthetic experiences, how do they look and search for the answer to where they belong?

The first thing to say is that Latin American aesthetic experiences belong to plurality. Taking into account the typologies proposed by Rodriguez-Plaza, there are three large groups evident: the magical-religious that responds to the ritual, the political-revolutionary that refers to the challenge of power and the erotic-love, all crossed by the ludic that appeals to joy [27]. Such categories are not mutually exclusive, rather they are intertwined through the expressions of their peoples. To elucidate this aesthetics, we have to refer again to the context emphasizing the synergies that their constituent elements—social, economic, cultural, racial and geographical—produce when they recombine. Because these are the conditions that allow what Juan Acha calls "aesthetic objectivity" [8] or the ability to perceive and appreciate the beauty and, ultimately, reproduce it from design, arts, crafts and, by extension, fashion.

3.1 The Corporeal

Talking about fashion, aesthetics happens inescapably through the body, because it is through the senses that aesthetic perceptions are experienced, and it is the body that determines on which fashion it is built. Being that fashion is habitable we have to literally be inside it and personify it; it is a 1:1 scale and therefore the body is the module of creation. Whether or not the body responds to the aesthetic postulates of fashion marks a line between the bodies that belong and those that do not, thus building the bodily imaginaries of physical beauty where identity is widely constituted from one's own and collective appearance and where, as Saito mentions, oppressions are also exerted against the excluded:

> Aesthetics is also implicated as a primary instrument of justifying the societal oppression of disabled, sexed, gendered, or racialized bodies. Those individuals whose bodies do not meet the aesthetic standard suffer from an unfounded perception that their physical appearance correlates with their competence, intelligence, and moral character, and are subjected to many forms of injustice and discrimination [13].

Likewise, to talk about the aesthetic in Latin America is to speak of the body, one that was unknown, appropriate and subdued and that is often still turned into exotic. However, the body also converts itself into a vehicle that crosses the limits of words and mediates the Latin expressions from the corporeal. A body inherited from the races has been mixed here and that finds, for example, in the dance a way of existing in the world, of naming something that has not been said elsewhere [27]. The corporeal also refers to a culture where experience revolves around the sensitive—what is felt—and where "language, sound, movement, singing is linked to things and both produce not a definition, but an action, a presence" [27]. Therefore, the corporeal in Latin America results in existing within spaces that have been expropriated and where bodies have constantly been targeted, disappeared and denied; in Latin America the body is a constantly moving existence.

In this constant development, the idea of the corporeal in Latin America arises from three visions. Firstly, what it should be, which corresponds to the imaginary body hegemonic, young, thin and white and on which longings and frustrations are built. Secondly, the being, which refers to the diverse and invisible but lived—and clothed—reality. And thirdly, the opinion, which responds to the constructions that others have made around the Latin body, where the body turns out to be co-produced to fulfil those fictions. The conjunction of these three visions entails the ideal that is distributed and should correspond with any vision that comes from these latitudes, perpetuating the idea of the exotic, curvilinear, provocative and willing that oozes Latin exuberance and it is on that body that aesthetic constructions and own desires are projected.

3.2 The Urban

This being a continent whose population concentrations are crowded in cities, the aforementioned corporeal figures are located within the urban, where they respond to recurrent informal dynamics. The urban is established as the common area in which cultural and social intersections develop and determine the majority aesthetic experiences. Therefore, it is a relevant place to address the Latin American, as mentioned by Rodriguez-Plaza:

> A Latin American critical-aesthetic theorization means, in my opinion, to assume the widespread cultural expressions and experiences deployed and arranged in the open, anonymous and public realm of the city as an environment culture. Scope built from the theatricalities, iconoclasts, sonorities and narratives of daily life. Which from the formalization of sensitivities, produce total aesthetic facts; that is, structuring networks of significant perceptions and operations. Resemantizations about materialities, signs and signatures and where the popular and the massive are often intertwined. This because there is where the large human groups rejoice and suffer an extended and extensive substrate of Latin American urban culture… A city where modernity and tradition invigorate time, simultaneously, conflicting and creative [8].

When we talk about the urban, it is key to understand that Latin American cities have been erected as a reflection of that aforementioned desire to belong to the West but that it continues being built from the periphery of this longed-for centre. However, this has provided a place for each periphery to gravitate over new centres, which allow it to deconstruct and appropriate closer narratives through heterogeneous and fragmentary expressions, but above all popular. Narratives that have shaped the local imaginary by translating the everyday life effectively into culture through "the telenovela, the dance, the street painting, the joke or the plastic that adorns the patron" [8]. Being "the other" [2] or the popular which is not validated, what dominates and sustains the self-construction of the city and identity and the path to overcoming hegemonic Westernism from the juxtaposition of times and stories. Within these urban constructions, the popular is postmodern, aestheticized, massive, consumable and cumulative and it seeks a kind of life "poured into the order/disorder plot of cities and their multiple construction." [8], a construction that is "crossed by the race, the class and the genre where its differentiation is marked but also belonging, therefore the interaction" [27].

The Latin American city that has been steadily imagined as a place of deficiencies, marginalities and subtraction is, however, a space of handmade cultural profusion. And, with the body, work and leisure and therefore is the propitious environment for—popular—fashion to develop. As fashion is a tangible object to represent and materialize the mass and daily urban experiences, the city serves as a stage for its display and commodification.

3.3 The Popular

It is therefore clear that the urban is built largely from and by the popular, taking care to distinguish it from "the folkloric, the past or the rural and thus avoid seeing it as the opposite pole of mass processes" [9] and instead equate it with the widespread. Therefore, giving agency to the popular is to be configured as opposed to the hegemony of aesthetic constructions from "relatively own cultural resources" [9]. Being the popular is consumption of masses that takes, reinterprets and problematizes the material culture of clothing produced by hegemonic fashion, while at the same time producing its own references that must be considered as fashion for the fundamental premises of the concept: expiration, aesthetic reference, mass consumption, production chains, circulation, appropriation and disposal [28].

These reinterpretations, by not belonging to the hegemonic, have readings that must also be given outside the beautiful or good taste. As Saito proposes, Salazar relates the popular with "the pretty" which "does not belong to the world of art, but either to the vernacular creation" [28], thus remarking its everyday character and the possibilities that this gives it to self-construct. Nevertheless, where, in addition, when we distance ourselves from good taste, we find a way to overcome class bias by recognizing other nominations of aesthetic constructions that have bidirectional mobility within the society and where the popular appropriates what belongs to the upper class while including a popular repertoire in its speech.

> "Thanks to the wide circulation of objects/clothes on the market… The gradations that objects suffer when they move from one social group to another are demonstrable. These gradations, which can be reproductions, adaptations and creations give account of the life of objects in popular sectors as aesthetics of similarities and distances from hegemony" [28].

Lastly, among the popular, novelty is essential; we seek to build the instant of the present that begins again and again after it is consumed, but that is rebuilt from the recycling of its own and appropriated symbols.

3.4 *The Copied*

This might seem like an exclusively popular phenomenon; however, it also traverses traditional fashion systems. Within the popular, as mentioned earlier, the aesthetic values of hegemonic fashion are emulated and reinterpreted, to compose the values within their own visual language, which results in almost reliable imitations of the appearance and even the inspirational brand, which in popular jargon would correspond to the "triple x copy", or the reproduction of those aesthetic values but contextualized within the massive, for example the use of "fusible plastic glassware" in reference to Swarovsky.

On the contrary, copy in the field of traditional fashion enters a less objective plane, where what is sought to imitate are at the same time aesthetic values, industrial systems and universal languages. In Latin America, this copy acquires great relevance since the 80's where there is an Americanization preceded or parallel to a Europeanization, thanks to the different approximations to these markets that bring with it the phenomenon of fashion, surpassing the local constructions that lay mainly in textile production but not of styles [29]. This openness transformed the idea of dressing by moving it to the scene of the trend and therefore to the commercial context. A change that also transformed the industrial landscape by turning much of Latin America into sweatshops of these foreign fashions, blurring for a long time the local creation. "Gestating fashion more from imitation rather than from reflection" [30] as Rosales mentions for the same commercial character that it entailed, thus producing elements of consumption but not own recognizable aesthetics.

However, over the last few years this dynamic has begun to reverse and Latin American fashion has gone from copying to being recognized and imitated. Latin American fashion has taken a first step towards translating its context and culture, away from folklore, to the material means that are its own—the garments and the accessories. It has also begun to shape an aesthetic idea that, although still under construction, already shows its zeitgeist and manages to connect beyond the trend by referring to narratives that surpass objects just as proposed by Saito within its aesthetic discourse.

3.5 *The Cumulative*

Within the Latin context, perhaps the most visible and recurring expression in all representations is the cumulative approach. The historical approach is the most apt way to meet this fact to understand the latter practices around the construction of own visual languages.

From this approach, we have the colonization that does not know our cultural and social agency as individuals and that is the architect of miscegenation, being one of the most decisive factors of the sociocultural formation of Latin America. Thanks to this racial mixture, we have a hybrid cultural composition where "ethnic, regional and national identities are reconstituted" but also added to extra-American identities [31] and that through abundance they find a way to encompass all the representations that constitute them. Making Latin American an additive and cumulative composition, which although is well influenced by foreign impositions, finds a way to capture its identity in these alien ways and thus appropriate them within the symbolic. It is therefore an "identity system by membership; to grow and create with each other, not against each other or the other" [31] as the colony did. In this way, it composes a baroque aesthetic recharged with signs and representations where the endogenous and imported are mixed and accumulative in its signs and in its "renationalizations" "resemantizations" [8].

Equating the cumulative with the Baroque, then Latin America is also read from the extravagance of its forms and its contents fed by an equally abundant environment "of rich chromatic and botanical profuse vivacity" [3] which seeks representation. Hence, the baroque architecture, Creole portraits of the eighteenth century, is not only the colonial altars but also the contemporary everyday, like the cities built by layers and the fashion that recycles itself and recombines with the outside and the inside, the new and the old at the same time, where the edition is not subtractive but additive, striking.

It is only worth noting that the cumulative is fragmentary [31], indicating that in this accumulation their plural origins are recognized by claiming differences rather than amassing them homogeneously within the universal. Thus, giving a space for the coexistence of these origins not without first recognizing a hierarchy that is still problematic today and has to be reassessed.

3.6 The Caribbean

When we talk about the Caribbean as well as when we talk about Latin America, both concepts extend beyond a geographical location, covering mostly their symbols. Therefore, appealing to the tropicalist discourse, Latin America is circumscribed within the Caribbean narratives; of hot and exotic lands and bodies and of Macondian stories that have shaped the Latin imaginary and which have translated into fashion through what Rosales has described as "caribbean chic" [32]. A concept that she establishes as the possibility of an aesthetic identity is born from the readings of the Caribbean, and that this being a wonderful chaos mixes the magical of the references that compose it. The Caribbean, raises Rosales, is syncretic and eclectic, giving field to the beliefs, rituals and to the cumulative visual compositions that fit within this symbolic place.

This Caribbean imaginary poses practical challenges that determine its language, where the heat, the humidity and even the light reaches these territories, no longer symbolic but real, frame negotiations on functionality or colour, where novelty is ephemeral and salt and sun "prevent materiality from being lasting" [33, 34], but which finds in decay an own and unexpected beauty. From the Caribbean, both the syncretic and the practical are evident in dressing, as Rosales describes "White for freshness, pearls to evoke the maritime, fan to melt beauty and function in the green and floral humidity of the afternoon.... The inclement heat makes the dress not provoking in its excessive expression; therefore, triumphs white and linen and jewellery is sometimes impractical annoyance" [35]. Therefore, it shows more concretely by embossing the Caribbean chic in "fluidity, shoulders in the air, vibrant floral slogans, colorful ensembles, handcrafted accessories, colorful ornaments on the ears and on the head" [3]. And where the body it dresses is, as well, representative of that tropicalist dream.

Within Latin aesthetics, it is perhaps the Caribbean story that has had the most echoes within the continent, within the fashion and within foreign imaginaries. That is why it is relevant as it is a path that has opened the door to the conversation of local fashion and that can be followed by the rest of the aesthetics described here.

3.7 Identity Under Construction

Finally, it must be recognized that Latin American aesthetic identities are constantly under construction around a constant question Who are we? [27] and, which within materiality it is also questioned, what are we? An immutable question seeks the answer in the historical narrative and in the construction of the present, where we have gradually realized that this is a polyphonic continent and that it is from there from where we must continue to narrate those identities that are still being discovered. As Rodriguez-Plaza mentions "Any poetic effort that tries to deny contexts, I say not historical or social (which is already dangerous), but the imaginaries, is condemned to the deepest of failures" [27].

Regarding the construction of these identities in fashion, Valerie Steele mentions the importance of creating icons that help consolidate a collective imagination from what they represent [34]. When she talks about icons, she doesn't just refer to individuals but extends to ideals that represent the feeling and complexity of the people for later translation into the language of fashion. Along with this, other considerations for the construction of these identities are embracing the periphery with the urgency of overcoming central discourses, abandoning the neutrality that might be apparent, making statements of our narrative from critical places and feeling as legitimate all the productions that are built in order to empower them.

In conclusion, Latin America, precisely, is a space where a myriad of contents that are combined into a hybrid. But if the syncretic is cultural and religious, then the aesthetic is eclectic and baroque; therefore, if we add up all the characteristics of Latin American aesthetics, besides the plural, they belong and are experienced from the "pleasure of the unfamiliar, of the sublime, that which is: asymmetrical, chaotic, unpredictable, without limits, without form, unbalanced, distorted and uncomfortable" [24]. Although going further, Latin America belongs to every day that is named from the pretty, the cute, the cheesy that confront us to the conventions of what (hegemonic) fashion should be, but that still provides us with satisfying experiences. Then, it is necessary to learn to navigate these aesthetic experiences since Latin America has always been a challenging territory that has to be explored from within, from "the utopia, the memory and the instant" [36] around the places, the bodies and their experiences.

4 Minimalism and Latin American Aesthetics

When confronting minimalism with Latin American aesthetics, the marked formal disagreements that exist between them are evident, but we could see the disagreements are not decisive, instead they open the conversation to a relationship between them horizontally. These stresses could be summarized as shown in the following table and its further explanation:

The everyday life		
Minimalism	Latin America	Latin American aesthetics ID
Moderation function	Excess appearance	The cumulative and the corporeal
Anonymity silence	Brand noise	The copy and the urban
Aesthetic universal	Taste singular	The caribbean and the popular
Repetition value	Novelty price	The copy and the cumulative

MEASURE VERSUS EXCESS: Minimalism is often conceived from the scarce, from the necessary and functional; putting in relevance the rationality in everything that it encompasses, from the conception of the objects, their materiality, their use and even in their consumption. These conceptions originate from abundant societies that have been overwhelmed by an excess of ornaments, objects, and seek a depuration of their aesthetic, conceptual ideals and their ways of life. Minimalism is a path of liberation from material dependence. In Latin America, the opposite happens. This has been a continent that, although exuberant, has largely been shaped by the consecutive violence it has suffered, imposing prohibition and perpetuating scarcity, which helped build a language where abundance, its accumulation and its demonstration are important in building the aspirated life [37]. Being able to access this exuberance is the local path to liberation because it raises possibilities, belonging and possession of everything that has historically been denied. It is the ability of being seen through possessions. As Saito sums it up in terms of production and thus consumption: "Production of goods is accordingly geared towards satisfying consumers' aesthetic appetite or desire, whether for personal fulfilment or, perhaps more commonly, projection of a particular person through possessions [20].

THE ANONYMITY VESUS THE BRAND: Within the everyday life aesthetics, Saito shows an important difference between arts, where the work's author is relevant, and the everyday life, where the creator is predominantly anonymous. This differentiation is possibly analogous, within the Latin American, in the discourse mediated by the brand, where fashion would amount to art, putting in a relevant place the authorship of the pieces that circulate, that is, the brand itself. Hence, the appearance of phenomena such as the copying or emulation of brands and their aesthetics, [28] because of what they represent and their declaratory capabilities, where the possibility of branding is linked to the purchasing power, often matching the brand with wealth. In the opposite place, which would correspond to the everyday life, we find minimalism. Here an emulation is meaningless as it is an aesthetic prone to uniformity, where repetition and reproduction are intrinsic and where the apparent anonymity of its authorship is appealed because the characteristics that differentiate each brand, in the event that it exists, are not recognizable. Which results in an equalization between minimalism and the ordinary, crowded and anonymous; therefore, a place where there is no recognition or distinction that entails the discourse of the brand, but those minimalist characteristics find their echo in the popular where that anonymity becomes enunciative and the crowded is identity.

THE AESTHETICS VERSUS THE TASTE: We have previously stated that minimalism belongs to the familiar, to the easily decodable, and therefore, to the beautiful experienced from the proportional, symmetrical and demarcated, thus responding to universal aesthetic principles that derive from the common need of humans to, for example, understand and order their environment [24]. When we talk about minimalism, we propose to rationalize the environment, but not to particularize it, thus cancelling any possibility of the local or personal. This leads to a reduction of decoding errors as it is aesthetically neutral and adaptive by extension, although it overrides any cultural load. On the contrary, taste is a subjective construction that depends on the culture and lifestyles; therefore tastes are personal narratives that within the continent need to be claimed. Taste, as Salazar proposes, does not have to be built from the class encasing definition of Bourdieu. Rather it is possible to find other representations from "the pretty" as the possibility of communicating outside the universal construction from the historization of the objects, which gives the possibility of reinterpreting, of not being determined by the class, but above all of "connecting a personal biography with a biography of the objects" [9]. Making tastes a closer choice within the idea of the Latin context. But isn't minimalism also everyday life?

REPETITION VERSUS NOVELTY: Within what encompasses minimalism, repetition and constant rhythm operate in part as response to the inherent need for order of the human being; while they are silhouettes, palettes and materials that are constant and therefore recognizable, but mimetic to become part of the familiar [24]. Minimalism also appeals to be durable and stable, transcendent thanks to the adaptability of its intrinsic silence while maintaining its formal characteristics, avoiding the need to be replaced, highlighting its ability to be part of everyday life. When we establish this feature in Latin America, local readings about repetition are made by confronting: on the one hand its acceptance from the cumulative as a search for stability, of a pace that establishes its own order; but on the other, and more urgently, its rejection from the need to embrace novelty and assume the opportunities offered by the new social configurations where there is a possibility of differentiation, consumption and economic opportunities. As Salazar mentions: "As novelty has value for producers, sellers and consumers, the market is saturated with trends, of things that fulfil the mathematical function of being the best sellers". In addition, the novelty allows delight and generates continuous emotion when discovering what is not usual [28].

Having navigated the differences, both minimalism and Latin American are fed back from ambiguity and opposites that are created in everyday life. Speaking then of minimalism where "nothing combines, but everything goes together" [25]. As life itself, where humanity is allowed and whose main objective is to deconstruct the narratives and question how to live in the current world rather than uniform surfaces and silence the spaces, this being precisely the core of Latin American aesthetics, miscegenated, hybrid, additive and inhabited. Hence, it begins to be consistent to talk about building Latin minimalism.

4.1 Latin America and the Sustainable in Fashion

Regarding sustainable fashion, The State Of Fashion [38] sets that the new generations and the market in general is adopting more conscious practices and one of the most important purchasing drivers in recent months is sustainability. Punctually, according to WGSN "the propensity to buy from socially responsible brands is strongest in what has become fashion's manufacturing hot-bed of Asia-Pacific (64%), closely followed by Latin America (63%)" [7], showing the local interest in questioning traditional dynamics. However, despite the growing interest and the emergence of new voices throughout the fashion system that advocates for effective transformations, context is once again a determining factor for the adoption of these dynamics.

The pulse of fashion 2019 [39] notes that 75% of consumers placed sustainability as an extremely important issue, but only 7% make their purchases motivated by this factor as a determining criterion. Following through, when characterizing the type of consumer, only 16% of the total belongs to the group that fully adopts sustainability and it refers to people with, among other things, professional education and stable income. Therefore, when we contextualize these numbers within Latin America we face a reality, where there is a growing generation that opens up to conversations around social and environmental responsibility but falls within emerging economies where incomes are not stable and much less sufficient in most cases. Then, the sustainability proposed from the global north is reduced to a little-executable intent where niches that are "ready" to take it on are extremely scarce, which makes sustainability unique and exclusive.

In addition, Latin America experiences phenomena that are hardly flourishing here, such as overcoming poverty and violence, expanding its middle class, access to global content, and brands and objects commonly reserved for elites. It is important to recognize how this affects the adoption of sustainability within the imaginaries of a consumer who does not want to feel judged or limited but wants to enjoy those possibilities that often come with fast fashion.

However, sustainability in Latin America is not a strange issue; on the contrary, it has deep roots within what have been traditional practices. Sustainability accounts for the fundamentals of the man–nature relationship, thus showing that there is vernacular sustainability, which has been claimed from the traditional and the popular as reparation, the overlap, the no throwing away, the leveraging of the object and material resource, transformation and adaptation. Although assigned under other names such as the precarious or the recursive, it responds to the same concerns from the context that have been reached from the experience, not from the conceptualization. That is, they have been experienced rather than planned but are still systematic, so it must be our starting point when it comes to raising it within the local fashion system. In this way, Latin America will be able to enter into the conversation of sustainability in fashion from the endogenous, resulting in a construction that takes advantage of the traditional and the contemporary to respond to the needs that Latin American poses [40].

5 Design Strategies for the Latin American Sustainable Aesthetics

Given that the objective of sustainability aesthetics is to achieve a durable, adaptable and nourishing aesthetic expression that can be constantly rediscovered by the user, these must be settled from a perspective that is not punitive or prejudiced, since all approaches that have been taken, from what Kate Soper calls "anti consumerist aesthetic",[4] have turned out not to be effective due to its guilt-inducing approach. Instead, we are going to play with another approach proposed by Soper as "*the alternative hedonism*"[5] [20] in order to achieve a positive interaction looking for creating narratives that promote satisfying interactions beyond superficial normative beauty.

[4] The kind of aesthetic in which "the commodities once perceived as enticingly glamorous come gradually instead to be seen as cumbersome and ugly in virtue of their association with unsustainable resource use, noise, toxicity or their legacy of unrecyclable waste" (Soper 2008 in Saito 2018).

[5] "This involves identifying aesthetic qualities that can render sustainable good attractive or appealing, and in short fashionable" (Soper 2008 in Saito 2018).

It should be noted that the purpose of these strategies is not to be a checklist of formal features of how the things inscribed under a Latin American sustainable aesthetic should look like, but rather an approach to a translation of the Latin American at an aesthetic level that could be executable within sustainable systems. Therefore, it is a conceptual formulation route that can be appreciated and read universally but that should be developed, at a design level, from the particular needs of the context and the productive system that supports it, in order to find the narrative approaches that allow the construction of local icons from the concepts that constitute the identities—the corporal, the popular, the accumulative, the Caribbean—that help to build relevant, durable and pleasant experiences which are constantly rediscovered. In this way, it is possible to promote less consumption and extended use as the most appropriate way to achieve a sustainable future, avoiding the typical "kind of sustainable scenario for clothes, where they are old, torn and patched up, dirty and wrinkled, that in practice would correspond to environmental values but most certainly would not fulfil our need for beauty or social norms in appropriateness" [22].

Understanding the uniqueness of what is conceived as Latin American, we know that it responds to an extended and complex decoding aesthetic, often from the coincidence with popular. Therefore, it is necessary to make this decoding experience intentionally built, its complexity has to be predetermined and prevent it from being just a poor design experience [24]. Although I am not unaware of the development of adaptation and appropriation processes within the popular, it is necessary to establish design strategies that make them premeditated and, consequently a reproducible path, as Salazar says, what design does is synthesize and abstract to subtract the literality of the popular [41].

The formulation will be proposed within the axes of comfort and camouflage according to the semiotic decoding strategies proposed by Harper in order that the phenomenological (material) construction will be adaptable to particular situations and comes from the user himself. Both concepts arise within a spectrum where when we talk about comfort its opposite is discomfort and when we talk about camouflage its opposite is standing out [24]. In order to make the proposal easier to understand, we will start from the following graph to explain its configuration:

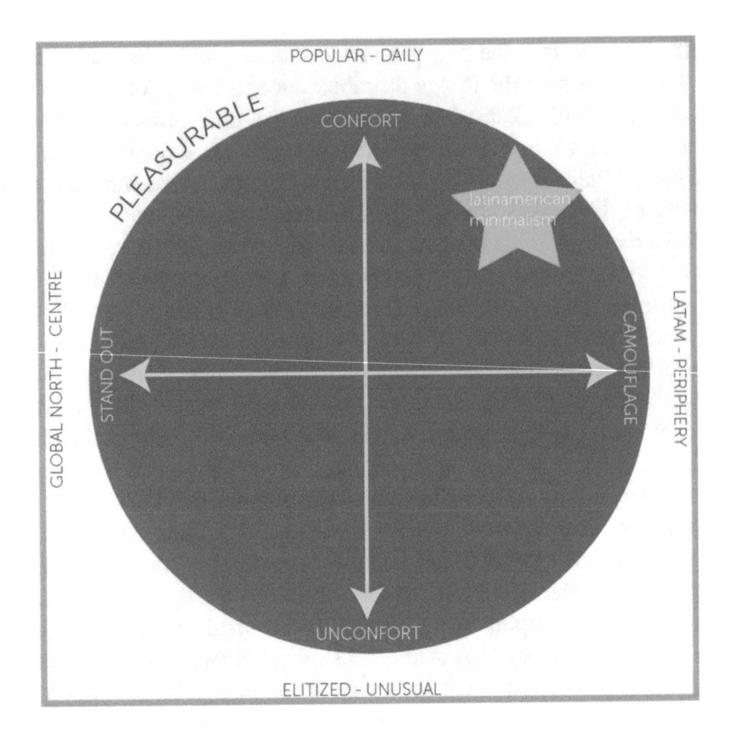

The features that determine these concepts come from Harper's strategy and have been confronted with Latin American aesthetics and context, hence, when reading the proposal, we must situate ourselves both inside and outside to understand how they are arranged.

5.1 An Outside Reading

Comfort: It is equivalent to the short decoding that has its foundation in *the familiar*, in what we recognize and provides us with security since they are all experiences where what is expected meets expectations, therefore there is no room for surprises [24]. In comfort, there is also the security of being able to satisfy our needs from the complicity generated by *the familiar*, mediated by narratives in our own codes that tell us an unambiguous story. In this case, this comfortable and safe place refers to the status quo of fashion since it has been the place from where discourses have been traditionally built and, therefore, we know its foreign language and its stylized and elitized codes.

Uncomfort: At the opposite extreme we find the *everyday* and *popular*, aesthetic constructions typical of Latin America that are uncomfortable for anyone who is outside the context. Latin uncomfort do not have to be understood from a negative connotation but as a long decoding experience provided by their fragmented and accumulative history. From discomfort, the aim is to question basic assumptions, experience chaos and the unexpected. It is the place to always be ready for "something happens" [24] and where each event serves as a catharsis. Within the Latin uncomfort we also find taboos, like those related to the body and the enunciative, where *the beauty* and *the tacky* are signs which need time to be assimilated.

5.2 An Inside Reading

Stand out: At this end, we place the western world—therefore the centre—and all the features that have been produced in it shaping the fashion language. Those features are remarkable within the Latin American context but by default not by excess, where a refined and silent aesthetic tends to stand out within the boiling semantic of the Latin. In Harper's terms, the characteristics of this concept are being exaggerated, wild, challenging, intriguing and unusual, which comes from the disapproval of the standard, those features also correspond to the Latin if the reading is conducted from the outside, hence, the positions of Latin in the spectrum can be interchangeable if the viewer stands in or out.

Camouflage: The intention of the camouflage is to blend in, in spatial and temporal terms, with the surroundings; therefore, it is located within the contemporary and is shaped by the tastes and habits of a given society. However, in order to avoid obsolescence, conjugations between durability and volatility have to be done and, in this way, be able to provide durable aesthetic nourishment promoting the belonging and cohesion of identities under construction. In this way, sustainability in Latin America has to find the correct camouflage within *the popular* and exuberant echoing its sociocultural needs; besides that, fashion will have to be built with the language of the periphery in order to create the appropriate clothing. As Goffman mentions, "the concept of 'appropriate clothing' depends entirely on the situation. Wearing clothing that is considered appropriate for a situation acts as a sign of involvement in it, and the person whose clothing does not meet these criteria may be more or less subtly excluded" [15].

All these experiences are combined within the pleasurable, which makes them relevant and, therefore, aesthetically nourishing, resulting in an extended experience. As a first conceptualization of this process, and in order to synthesize the contextual and aesthetic aforementioned, a Latin American minimalism emerges, an experience of long decoding but of rapid assimilation allowing, from a conceptual point of view, an expanded phenomenological approach.

5.3 Latin American Minimalism

Latin American minimalism refers to a multidimensional construction with an extended decoding, emerged from the discomfort of belonging to the popular and daily which is camouflaged in the context from the accumulative of the Latin periphery, setting as an alternative to hegemonic fashion. It uses the polyphony of their aesthetic identities to build a language "where nothing combines but everything goes together" [25] and, in this way, it changes the way we relate to the material world. To change this relationship, it is necessary to propose approaches from the Latin, translatable into materiality mediated by sustainable processes and resulting in actions or objects that can be read as "fashion acts".

> Uotila [42] argues that a consumer creates a "fashion act" when modifying fashion styles and rules to create an individual look according to their own inner values, emotions and identity. Hence highly unique and visually beautiful, interesting and even odd personal deviations from fashion styles can be considered "fashion acts", which may also be defined as aesthetic creativity [22].

With the semiotic approaches established and space where to build the materialities is settled, we will pose the sensuous of the Latin from questions of how the phenomenological characteristics of aesthetic concepts could be represented in fashion.

How to represent ***memory***? As we mentioned throughout the text, one of the fundamental tasks in the construction of a Latin identity is to be able to narrate, from this shore, the history and to claim the autochthonous, not from the folkloric but from the intrinsic. Memory serves as a counterweight to oblivion, silence and denial of which we have been heirs and architects. This approach makes aesthetics narrative and testimonial; therefore, they need to be exhibited and monumentalized, hence the idea of the altar as memory, pilgrimage as a process and catharsis as a result. Thus, the objects that make up these altars will be cherished for and their life will be prolonged.

Regarding fashion, and also in relation to the depiction of Latin American, the body is the most relevant communicator vector. From memory, we can turn the body into an altar and the garment into a monument. To achieve this from *the local*, we can refer to the embroideries and textile interventions that Colombian post-conflict victims have used as a narrative and reparative process and, from this example, build up languages of clothing that are capable of becoming coded monuments.

How to use ***the sentimental*** to construct reality? At this point the materialization of fashion must overcome the construction of reality from the conceptual to highlight the perceptual and the imaginative [27] and in this way give space, on the one hand, to cosmogonic narratives where stories have found a representation through embroidery, weaving, basketry, pottery or jewellery from the traditional and, on the other hand, to personal stories that capture the feelings of its users through personalization. From both perspectives, garments cease to be dispensable objects to become repositories of identities. For this, it is necessary that fashion objects reach the user with only half of the story told so that it is she who finishes configuring it from her experiences and from her own construction of beauty, prolonging the duration of the aesthetic experience thanks to the emotional bonding.

How do you embrace ***decay*** as a visual goal? In Saito's research, good ageing [43] is presented as an important factor for the prolongation of the aesthetic experience. From this idea, we could establish *the decay* that is evident in the Caribbean or the cities as a representative visual sign of the Latin to translate it into fashion experiences. it is worth to clarify that the decay that is intended to be achieved does not imply a neglection or abandonment, but rather the preservation from the recognition of the passage of time, evidencing it and noticing the narrative layers that it implies as a witness of experiences that comprise it. As an aesthetic strategy, decay has to be composed by a predetermined and designed part and by a spontaneous one, derived from private use where a convergence can be found between traditional practices and new technologies.

Just to name a few, within *decay* we could find depigmentation due to the sun or laundry washing, the wear and torn of selvedge or most exposed areas to contact, the peeling, the opacity and loss of trimmings, the rubbers elasticity loss, the wrinkles. All the aforementioned tearing could be approached by the user through reparation or even transformation, providing an additional story that strengthens the ties with these garments but that also gives possibilities for design and research. Examples of this research could be found in the use of endemic fibres suitable to vanish or tints that change with use. The evidence and cherish decay makes us not only to recognize the perishable nature of things in order to fully experience them but also to propose design paths where garments are self-consumed, fade away and where their ephemeral materiality leaves no traces.

How to create a ***utopia of the moment***? Latin America has been thought of as a western utopia where Europeans find the possibility of projecting themselves into this new lands and as a pantry where to satisfy their wishes; thought as a utopia of its natives from the symbiotic relationship with nature, a space of implausible stories of gold lagoons, mythical animals and heirs of the gods; thought as a utopia by the contemporary Latino as a place of potentialities and opportunities yet to be discovered, a Latino who broaden the vision of its culture from the exile and, as a spectator, has discovered how to build it from the outside to narrate its interior. Conjugated with the dreamy utopia, we need to live the instant and "be able to imagine an aesthetic that finds out how to invent performances that will not diminish our future nor make our past redundant" [36].

When construction is done on the instant, it is not with the intention of fleetingness or Bauman's pointillist time marked with ruptures and discontinuities, "*pulverized in a multitude of eternal instants*" [6] without possibilities beyond themselves. On the contrary, what is intended is to give time to each moment and fully enjoy it as a fragment of a story that has continuity and needs that moment to build itself continuously, not of utopias and fictions that fashion tells us but from the *everyday-life* realities built around it.

More than any other representation, fashion combines the utopia of the aspirational and the moment of novelty, both as consumer tools. Therefore, within this strategy, it is necessary to take them to the field of use from the experiences that compose it:

for example, how to bring home security to a garment?/, how to replicate, with each use (wear), the smell of childhood?/or how can the garments adopt the character of tribal tattoos where each drawing on the body represents a memorable moment?/.

All the previous questions are the departure point of the sustainable as a lasting experience and the narrative as an articulator of the Latin American constructions, in order to rethink the material relationship, where quantity is re-evaluated or reorganized so that all objects are part of the history, evoking again *the accumulative* of the Latino. From this proposal, the need to integrate the subject becomes evident, not only as a receiver of the proposed fashion but as a co-creator and maker, clarifying that it is not expected to return to a production society prior to consumption, as Bauman describes, but to an alternative one that breaks up the ongoing hierarchy, where the consumer only has *transactional* agency and the industries take all the other decisions around fashion.

5.4 The Prod(User) as a New User

Within Latin American aesthetic identities, the most important thing that has been evidenced is the need to decentralize discourses. In the same way, within sustainability, one of the main objectives is to decentralize both resources and means of production in order to empower each individual to build their reality on a path guided by design processes, but not hermetic to unique and finite outcomes. Therefore, it is necessary to trace routes of shared learning that is reproducible, adaptive and evolutionary, where fashion creation is not simply a *DIY* but a *Do it together* [44].

The moment the user crosses the consumption barrier and also becomes a producer, it will be the moment when the logic of novelty is broken, since the urge to *replace* will be overcome by the ability to continually *rediscover* aesthetic experiences from, precisely, a decentralized idea of *the creator* and, only this way, sustainability will be naturalized in all fashion dimensions.

6 Conclusions

This chapter aimed to sketch out an approximation to Latin American aesthetic representations, contrast them with global north sustainable aesthetic proposals and set up a dialogue between them, expecting to provide lasting and relevant experiences departing from the personal identification of the individual with its own cultural narratives, conjugate them in the fashion imaginary and, in that way configurate a proper Latin American sustainable aesthetic that merges the semiotic and the phenomenological approaches to aforementioned fashion. Throughout this route, minimalism as an aesthetic representation serves as the main axis to contrast both realities, to later, from its concept widening include the Latin American, understating it as a sublime experience. The outcomes of this search, just as the Latin American, are still under construction; however, they provide a relevant departing point to understand how Latin American vernacular features can be a robust alternative to the traditional fashion frenzy.

Understand Latin American from its own stories allow us to expose it from the polyphony and the periphery which perform as a reivindicative act from all the discourses that do not belong to the centre and shed light on the necessity to reinterpret and appropriate the hegemonic constructions through the popular, massive and everyday life aesthetics intending to usher different new insights to fashion. To the popular and daily belongs several practices that has been constantly underestimated due to its empiric background but that are remarkably sustainable, for that reason the Latin American experience as a whole, need to pass through design processes that enable the replicability of its extended decodification along a prederminated route and consequently, be able to incorporate within new fashion perspectives. Precisely from those Latin American diverse aesthetic narratives, the design strategies are framed addressing an approach to nourish the user relation with clothes.

Latin American minimalism emerges as an aesthetic synthesis departing from the narrative and the accumulative where the user stands out also as a creator. This construction allows implementation of the remaining concepts of Saito's theory, which was first addressed from minimalism, thanks to camouflage into the popular and autochthonous that are then easily executable. As a consequence, "durability, longevity, fittingness, appropriateness and site specificity, past present contrast, perceivability of natural processes, health and fostering and caring and sensitive attitude" [12] do not represent a sustainability imposition, instead a necessity for the aesthetic experience to be durable establishing and intertwining between those concepts and the memory and the instant as relevant paths to understand sustainable approaches in Latin America. Then it is imperative to highlight the importance of context and the blending with it to address proper transformations.

This is just the first tracking of what pretends to be a wider strategic proposition that, besides the conceptualization also be explored from the practice in the fashion field and that will work as precedent in the production and consumption systems transformation, founding aesthetic approximation as a key factor to embrace new paradigms in the continent.

References

1. Canclini NG (2014) Latinoamericanos buscando lugar en este siglo. Paidós SAICF, Buenos Aires
2. Fernández EN (2017) La mirada Latinoamericana. 30 de 0ctubre. https://www.redalyc.org/jatsRepo/5138/513853876002/html/index.html
3. Rosales V (2019) Notas sobre una cumbre de moda Latinoamericana. 17 de diciembre. https://vanessarosales.com/notas-sobre-una-cumbre-de-moda-latinoamericana/
4. Lasso M (2019) Erased: the untold story of the panama canal. Harvard University Press
5. Rodríguez Becerra M (2019) Las ciudades. En Nuestro planeta, nuestro futuro, de MANUEL RODRÍGUEZ BECERRA. Debate Penguin Random House, Bogota

6. Bauman Z (2007) vida de consumo. Fondo de cultura economica, Mexico
7. WGSN (2019) The vision 2030 how technology will drive a more sustainable creative industry. WGSN
8. Rodriguez-Plaza P (2005) Crítica, estética y mayorías latinoamericanas. Aisthesis [en linea] 38:99–122
9. Salazar E (2016) Estéticas en plural. La moda popular en Bogotá. Cuadernos de Antropología 26:51–68
10. Beshty W (2015) Lesson: notes for an introductory lecture. En akademie X: lessons in art + life. Phaidon, London, pp 14–27
11. Casablanca L, Pedro C (2014) La moda como lenguaje de una. 20 de Diciembre. http://www.aacadigital.com/contenido.php?idarticulo=1030
12. Saito Y (2007) Everyday aesthetic*s*. Oxford University
13. Saito Y (2018) Aesthetics of the everyday. 18 de November. <https://plato.stanford.edu/archives/win2019/entries/aesthetics-of-everyday/>
14. Forsey J (2013) The aesthetics of design. Oxford University, Oxford
15. Iglesias J (2015) El papel de las marcas de moda en la construcción de la identidad personal. Tesis doctoral. Univesitat Ramon Llul, Noviembre, Barcelona, Cataluña
16. Squicciarino N (2012) El vestido habla. Ediciones Catedra, Madrid
17. Lurie A (1994) El lenguaje de la moda. Una interpretación de las formas de vestir. Ediciones Paidós Ibérica, S.A, Barcelona
18. Global fashion agenda; Boston consulting group(2017) The pusle of fashion industry. https://www.globalfashionagenda.com/wp-content/uploads/2017/05/Pulse-of-the-Fashion-Industry_2017.pdf
19. McArthur FE (2017) A new textiles economy: redesigning fashion's future. http://www.ellenmacarthurfoundation.org/publications
20. Saito Y (2018) Consumer aesthetics and environmental ethics: problems and possibilities. J Aesthet Art CritIsm 76:429–439
21. Zafarmand SJ, Sugiyama K, Watanabe M (2003) Aesthetic and sustainability: the aesthetic attributes promoting product sustainability. J Sustain Prod Des 3:173–186
22. Niinimäki K (2014) Green aesthetics in clothing normative beauty in commodities. Artifact III(*3*):3.1–3.13
23. Fletcher K (2014) Sustainable fashion and textile. Design journeys. Earthscan form Routledge, London
24. Harper KH (2018) Aesthetic sustainability product design and sustainable usage. Earthscan from routledge, New York
25. Chayka K (2020) The longing for less. living with minimalism. Bloomsbury Publishing, New York
26. Gong (2015) Aestheticization of daily life and post-modernism design. School of Design & Art, Beijing Institute of Graphic Communication, China, Beijing
27. Rodriguez-Plaza P (2000) Experiencia estetica e identidad en America latina. Aisthesis 33:35–59
28. Salazar E (2016) Muy bonito todo. Investigar imágenes y objetos contemporáneos desde la cultura mediática y la cultura popular. Campos 4:199–217
29. Salazar E (2014) De los textiles a las apariencias. Los tránsitos de la moda en Colombia entre 1970 Y 1999. Tesis para optar al título de Magíster en Estudios Culturales. Universidad de los Andes, Bogotá, Cundinamarca
30. Rosales V (2014) La moda en Colombia es caribbean chic. 30 de julio. https://vanessarosales.com/la-moda-en-colombia-es-caribbean-chic/
31. Canclini NG (1995) Narrar la multiculturalidad. Revista de Crítica Literaria Latinoamericana 9–20
32. Rosales V (2015) Un caso por lo caribbean chic. 9 de abril. https://vanessarosales.com/un-caso-por-lo-caribbean-chic/
33. Rosales V (2014) Caribbean chic. 3 de noviembre. https://vanessarosales.com/caribbean-chic/

34. Rosales V (2014) Caribbean chic un estilo varios iconos. 30 de septiembre. https://vanessarosales.com/caribbean-chic-un-estilo-varios-iconos/
35. Rosales V (2015) Invencion de una idea llamada caribbean chic. 23 de Enero. https://vanessarosales.com/invencion-de-una-idea-llamada-caribbean-chic/
36. Canclini NG (2004) Aesthetic moments of Latin Americanism. Nat Hist Rev 89:13–24
37. Ramón, Ana López, entrevista de Diana Patricia Gómez (2020) Notas sobre la sostenibilidad y la pobreza (5 de Mayo)
38. BOF; Mckinsey (2019). The state of fashion 2020
39. Lehmann M, Arici G, Boger S, Martinez-Pardo C, Krueger F, Schneider M, Carrière-Pradal B, Schou D (2019) The pulse of fashion update 2019. Publisher Global Fashion Agenda, Boston Consulting Group, and Sustainable Apparel Coalition
40. Rubio, Laura Beltrán, entrevista de Diana Patricia Gómez (2020) Notas sobre sostenibilidad en la historia (17 de Abril)
41. Salazar, Edward, entrevista de Diana Patricia Gómez (2020) Notas sobre estéticas en Latinoamerica (22 de abril)
42. Niinimäki K (2013) Sustainable fashion: new approaches. Aalto University, Helsinki
43. Saito Y (2017) The role of imperfection in everyday aesthetics. Contemp Aesthet (J Arch) 15, Article 15. https://digitalcommons.risd.edu/liberalarts_contempaesthetics/vol15/iss1/15
44. Niinimäki K (2013) Sustainable fashion: new approaches. Aalto University, Helsinki

Bordado De Valors. Design-Oriented Actions to Support Paraguayan Crafts for Local Female Self-Determination

Giovanni Maria Conti

Abstract This paper presents the design actions and the results obtained in the project "Development of fashion products", a joint operation carried out by the Design Department, Milano Fashion Institute consortium and the IILA, Instituto Italo-Latinoamericano, based in Rome; aimed at creating a training course for trainers and local development activities, respectively, in Paraguay, to understand, support and develop actions related to micro-entrepreneurship and women's self-determination. The aim of the project "Development of fashion products" was initially to create a training course for trainers from Latin American institution, especially for trainers from Instituto Paraguayo de Artesanía de Asunción, to strengthen the training activities already underway in the institute. The course held in Milan in2weeks through three macro themes: Fashion Design Management, Technologies and Merceology for Fashion, Design and Product Development—the latter in the form of a design workshop on research methodologies, organization and management of a fashion collection—with the aim of bringing participants to the modus operandi that today characterizes the development of a fashion product within the most prestigious made in Italy brands. The second phase of the project is focused instead on two international actions, in two municipalities of Paraguay, Pilar and Yataity, with the aim of achieving a co-design workshop with two groups of women artisans specialized in embroidery, to transmit to them the concepts of micro-entrepreneurship, female self-determination and improvement of female workers in their communities. Through this field experience, we can observe that creativity in design combined with the manual work and the ability of the artisans to know how to manipulate materials represent the expression of a society wish, able to understand the changes; today, the object of market and consumption is not only the simple possession of a specific product but also it is the experience, the "story" that the customer can live inside the object, according to values of the manufacture that create or add value to the existence.

G. M. Conti (✉)
Design Department, Polytechnic of Milan, Milan, Italy
e-mail: giovanni.conti@polimi.it

M. Á. Gardetti and R. P. Larios-Francia (eds.), *Sustainable Fashion and Textiles in Latin America*, Textile Science and Clothing Technology,
https://doi.org/10.1007/978-981-16-1850-5_8

Keywords Craftsmanship · Design · Co-design · Feminine self-determination · Manual work · Craft

1 How Design to Improve the "Know-How"

Designer's skill to build connections and be the mediator between producers and consumers is not a new feature of the profession. By nature, designers find themself having to communicate with actors of different levels but may be involved in a project with a single company, assuming then the role of "facilitator". In this context, the designer must make use of his/her ability to interpret the various languages and build a relation and to push for innovation in the relative system.[1]

At the international level, unprecedented opportunities to build connections, networks and supply chains are available now. This phenomenon has certainly been facilitated and accelerated by the development of communication and information technologies that allow the exchange of design data at a rate that cancels physical distances. Still, the fact is that the "world of design" reports are not attributable to individual actors but enlarged to the community of scholars who represent the scientific debate and design verification at the international level. Today, the contamination of the design languages, cross-fertilization between disciplines methodologically related and/or different represent unprecedented mergers and design produce specific local know-how.[2]

2 From Micro to a Macro Reality

"The guru of the 'global village' concept, Marshall McLuhan, predicted in 1966 that in the future, the role of the craftsman will not be more important than ever before".

"Four decades later, there are some interesting signs sustaining this forecast: the growing awareness by the public and private sectors as well as regional agencies for International Cooperation of the dual role of crafts in their blending of traditional and modern skills, creativity, economics and in their social-cultural impact on sustainable development, and so have increased the public's preference for eco-friendly, handmade, quality products and the greater recognition of the very qualities we take for granted in crafts—qualities of timelessness and permanence, the adaptability of

[1] According to Zurlo, among the different roles that have been attributed to design, "*design has (also) the possibility to build stable relationships with different skills and exploit them if necessary. This system of relationships, external to the company and linked to the designer's system of relationships, favors cooperative processes of creating new knowledge*". In Zurlo F., (1999) Un modello di lettura per il Design Strategico. La relazione tra Design e strategia nell'impresa contemporanea, Final dissertation of Ph.D. in Industrial Design, XI ciclo, Politecnico di Milano, p. 162.

[2] Conti G. M., Dell'Acqua Bellavitis A. (2006), *Cross Fertilization: the path for Innovative Fashion and Design*, Proceedings of D2B The 1st International Design Management Symposium, Shanghai 2006.

artisans to their materials and to changing needs, and above all, the spiritual dimension of crafts. These are all favorable trends, nevertheless, counter-balanced by some disturbing contradictions. In today's "global village", the artisan is, paradoxically, more and more disconnected from consumers' needs and tastes. With the expansion of markets and the spectacular growth of tourism, the traditional direct, personal contact between makers and users has been disrupted. Can the artisan take any longer, as in the past, the combined roles of a designer, producer and marketer? In this context, there is an increasing demand for well-applied design, much of which comes from the local cultures and from the imagination and creative skills of artisans".[3]

As evidenced by Indrasen Vencatachellum's comments, there are interesting signs that show a greater interest in the peculiar character of micro and small enterprises' productions, with widespread difficulties for small producers to connect to international networks of production and distribution. Many international institutions, including the Design Department and Milano Fashion Institute—MFI have found how difficult it is for small producers not to have tools capable of helping them in the changes in the socio-economic world and above all the inability to relate to other actors to improve their growth and development of their products. According to Indrasen Vencatachellum, in order to face the continuous change, it is necessary to break with the traditional ways of working of craftsmanship and make the artisans dialogue more with the designers whose task will be to understand how to alternate the "ethnic" aspect of the product based mainly on the characteristics local, to a "narrative of the product" in which the value of the product is linked to its quality of the materials that take on new meaning through the symbolic, emotional or identity relationship that the consumer is able to experience. Based on these premises, the course "Fashion product management" was designed for four trainers from Instituto Paraguayo de Artesanía de Asunción with the aim of presenting participants with the main themes that characterize the various stages of the process of developing a collection, even on a small scale, within the textile fashion contemporary sector.[4] In particular, the course (given at Milano Fashion Institute, Milan, the Interuniversity Consortium formed in 1997 from the Polytechnic of Milan, Bocconi University and Catholic University of the Sacred Heart) consisted of three consecutive modules:

- Fashion Design Management, with the aim of presenting the main business models of the companies that operate within the chain of the Italian contemporary fashion industry, the key factors and the main operating business models and the main evolutionary scenarios that characterize the sector; and relevant to manage the relationship between management processes and creative processes, with particular emphasis on different approaches to the management of creativity, from the stylistic identity to the different product strategies.

[3] Indrasen Vencatachellum, Foreword, in "Designers meet artisans. A practical guide", craft revival trust, Grass root Publications, New Delhi, India, 2006.

[4] In Conti [8] Towards a Cross Cultural Society; from ethnicity to design, "narrative" heritage drives innovation. Mapuche Weavers and Italian Designers co-create for fostering diversity. Projecting Design 2012, International Conference Santiago, Chile, 2012. Pag. 181.

- Fashion Technologies and Merceology, whose objective is to offer an overview of the main production technologies for the fashion product as well as a solid knowledge of the materials and their uses. The technological developments and innovation in recent years have made available a series of fabrics commonly used for clothing. In this specific case, the module was designed to provide the knowledge of the participants as a "trader of basic products", from fiber to yarn and from textile to woven, also proposing to offer knowledge of leather articles and accessory materials.
- Product Design and Development, the module to study the various stages of the development process of a fashion collection, identifying both the activities related to the organization and management of creative processes and the relationship with the planning of the architecture of the collection according to the links of companies and market. From the definition of the elements that characterize the stylistic identity of the company and the DNA of the company, it has been demonstrated, according to the different business models in the system, the different design processes adopted, the factors involved, the types of fashion products developed. Each phase of the development process of a fashion collection is deepened: the development of research and the definition of the collection scenario, the definition of the architecture of a collection and the development and engineering of the collection.

The teaching activities, in addition to offering lectures, also involved several tours; the first in the Vitale Barberis Canonico to Pratrivero (Biella), a vertical business reality in the creation of high-quality fabrics, in order to show how the quality and "properly done" Italian part of the raw material and the way in which the Industrial textile sector has been operating for more than 150 years in the form of quality control, attention to industrial and production processes, strong participation in environmental issues and sustainability. The other visits are organized within the LP fashion studio (the largest research center for leather and leather goods), the Design Museum of the Milan Triennial and in the showrooms of Armani Casa and Bottega Veneta Home, with the objective to show how the Italian textile sector is transversal to the Made in Italy product categories so that the fabric can be used both in clothing and in interior design and contract.

The course has exalted the points of view of the product-fashion in its transversal character, which shows the complexity and richness that characterize the Italian industrial system that, in the textile-confection sector, is very vast and irregular.

With regard to the way of learning content, since we dealt with professionals, teachers, experts in textiles and craftsmen, the purpose was not to give purely theoretical lectures but to seek a continuous exchange to understand how the content and the "way of doing" of Italy could be read, understood and, in a second moment, adapted to the territorial realities of Paraguay, instead of saying "how it is done" the objective has been to stand out as an international training entity, such as Milano Fashion Institute, which could accompany, through the training of trainers, the countries that approach Italy to understand the business dynamics and design.

3 Actions: From Ethnic to *Narrative* Product

Before starting the critical analysis of projects, it is useful to have an overview of some suggestions given by past experiences in this field. There are two main examples that we can refer to in planning the transfer of design knowledge in "developing countries" or "peripheral countries".

For Gui Bonsiepe[5] in particular in South America, Design knowledge transfer projects should keep into consideration that:

> *1_ the socialization of strategic production tools (technologies) must be sustained by the elaboration of innovative projects characterized by a high value of use;*
>
> *2_ the import of external design ideas (design from other cultures) should be reduced to minimum [...];*
>
> *3_ the influence of foreign consumption models [...] have formed and deformed the (local) consumers conscience. New products should be introduced slowly and accompanied by information in order to let the (local) consumer to form an authentic conscience of their (local) needs.*
>
> *4_ it is important to socialize the design process in order to let workers (local community) participate directly as producers of (their) material environment [...]. (italic words are the authors writing)*

According to the Ahmedabad Declaration (1979) in India, Design knowledge transfer projects should aim:

> *1_ to understand the values of a society and to define a good quality of life inside its parameters;*
>
> *2_ to look for local solutions for local needs using local materials and competences and applying advanced technologies;*
>
> *3_ to build new values, to satisfy primary needs and to preserve the plurality of cultural identities.*

The typology of technologies that should be used and transferred to communities within these projects has also been an important topic of a discussion lead by the pioneers of Design knowledge transfer: V. Papanek's belief in the use of "autochthonous technologies"[6]; G. Bonsiepe's theory on "intermediate technologies versus appropriate technologies"[7]; K. Schumacher's attempt to apply "intermediate technologies" and develop "vernacular solutions".[8] However, the aim of this paper is not to investigate this specific matter.

These two examples do not aim to give and an exhaustive and omni-comprehensive picture of the past actions in the field of design knowledge transfer projects, but

[5] Text translated from the author taken from Bonsiepe Gui, *Paesi in* via *di sviluppo: la coscienza del design e la condizione periferica* in Storia del Disegno industriale 1919–1990, il dominio del design, Electra, Milano, 1991, pp. 252–269.

[6] Papanek Victor, *Design for the real world*, Thames and Hodson, London, 1972.

[7] Bonsiepe Gui, *Paesi in* via *di sviluppo: la coscienza del design e la condizione periferica* in Storia del Disegno industriale 1919–1990, il dominio del design, Electra, Milano, 1991, pp. 252–269.

[8] Idem.

merely aim to underline the importance that Design knowledge transfer projects have had in the past and to stimulate an international discussion on the meaning of these projects in today's new economic and geopolitical paradigms.

Summarizing, we can affirm that design fits into the processes of globalization with tools and methods developed in the discipline, which allow:

- to understand the **cultural contexts**;
- to understand the **technologies** (i.e. understand what the specific skills are);
- identify the **consumption contexts** (in terms of new markets, new consumption scenarios).

This approach relies on the belief that this project can create more knowledge and innovation. Specifically, the project developed with the Paraguayan community has moved on two levels:

_ **Promote greater access** by local artisan and industrial associations to international relations to create a virtuous circle between the exchange of knowledge and mutual recognition while keeping the strategic aspects of production "at home".

_ **Promote projects** between "hybrid cultures" capable of generating sustainable design scenarios for all those involved.

4 Paraguay. Places Need to Improve Local Products

A tacit phenomenon that goes beyond any study or possible schematization, but in which the ability of the designer to interpret the needs of the interlocutors and transform them into tangible responses, is combined with the ability to interpret multiple cultures, capture values, signs, traditions and customs to translate them into innovative solutions[9] for the consumer; this has been the basis of the most complex project implemented in Paraguay in the cities of Pilar and Yataity.

Why in Paraguay? First of all, because in that context there is a lot of small micro craft's company where work-only women whose skills in manual embroidery are of the highest level. Second, because, in a society still strongly focused on "machismo", women as person, their rights, their needs, are still not respected and, sometimes, the same job becomes a "servility forced" because it is the only income's source for the family. These premises are, therefore, necessary to better understand the responsibility of the actions of both training and workshops: design as a tool not only for training but for "awareness of a work". The manual work becomes the fulcrum of knowledge and the "respectability" of women itself and the products represent their skills.

In this case, not only was the training status of the trainers who had attended the course in Milan monitored and verified but also a real project of territorial and social valorization was carried out.

[9] In Conti G. M., Galli F., Pino B., op. cit. Pag. 182.

Fig. 1 Start of the design workshop in Pilar and first moodboards, Paraguay

At the headquarters of the City Hall in Pilar, a design workshop was created that, after the modification of three wedding dresses (received as donations from the IILA and used as a basis for the development of new design ideas), integrated elements of the culture Paraguayan in the form of embroidery, new volumes, different portability (Fig. 1).

The workshop allowed the 55 artisans involved to participate while respecting their work hours. After a practical demonstration on the actions to be carried out, the management and redesign of the bridal garments were in charge of the artisans, monitoring the design activities and the transformation phases of the clothing (Figs. 2 and 3).

The objective of this workshop was to elaborate together with the artisans involved, the project of the renewed garment to enhance their skills of clothing, embroidery and manuals in general, so that they could understand the "value", understood not only from an economic point of view but also from their skill. The designer by nature finds himself having to communicate with people with different knowledge both when he is involved in a project with a single company and when he assumes the role of "facilitator" for the production fields. In this context, the designer must use the hermeneutical skills to interpret the different languages to build a shareable

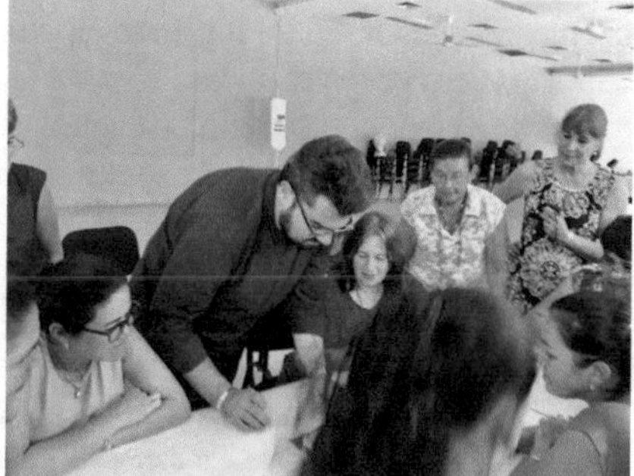

Fig. 2 The design workshop in Pilar, Paraguay

Fig. 3 Reviwe's chatting and first results in Pilar, Paraguay

third party and to drive the reference system towards innovation.[10] It is then when the designer becomes a "mediator of knowledge",[11] a production orchestra director, an attentive observer of the forms and work processes, in order to enhance the full potential of production, based on what is done and that is not only dedicated to creating always new forms.

5 Some Final Considerations

Italian design is recognized as the engine of innovation for small and medium-sized Italian companies and "Made in Italy" represents the success of companies in the production of products with a high level of qualitative excellence in materials and finishes. Furthermore, it is a "meta-brand" identified mainly in the furniture, textile and clothing sectors capable of describing the Italian production system, describing lifestyles, quality, ethics, creativity. The designer is a professional capable of identifying and enhancing all the endogenous and exogenous conditions by fusing them to create innovative products.

So the designer becomes a "knowledge broker",[12] an orchestra conductor, a careful observer of the ways and processes involved, so as to exploit the full production

---

[10] In Conti G. M., Galli F., Pino B., op. cit. Pag. 182, Zurlo F. (1999) Un modello di lettura per il Design Strategico. La relazione tra Design e strategia nell'impresa contemporanea, Dissertazione di Dottorato di Ricerca in Disegno industriale, XI ciclo, Politecnico di Milano, p. 162.

[11] In Conti G. M., Galli F., Pino B., op. cit. Pag. 182, Auricchio [1] Internationalization of design research and education centers. Promotion of international design networks, Dissertazione Dottorato di Ricerca in Disegno industriale, XX ciclo, Politecnico di Milano, p. 13. Per approfondimento si veda anche [4] *Milano distretto del design*, Il sole 24Ore, Milano.

[12] Auricchio [1] Internationalization of design research and education centers. Promotion of international design networks, Dissertazione Dottorato di Ricerca in Disegno industriale, XX ciclo,

potential, based primarily on what can be done and not only involved in creating new forms.

The experience with all the participants of the course and with the Paraguayan artisans, in their territory and with their instruments and their wisdom in the "know how", was above all a human adventure in which the designers have tried to improve the normal skills of the artisans beyond repetitive experiences, proposing to express a creative freedom in what is usually processed in a "mechanical" way. This type of design model leads us to the reminiscence of the "theory of the de-equilibrating system"[13] by Ugo La Pietra in the 60 s, which is based on the possibility of intervening in the "rigidity of a system", in order to reveal the contradictions and/or open it to new expressive possibilities.

Especially in the workshops in Paraguay, it is possible to summarize the activity[14] in:

_ **understanding**. From the encounter with the artisans, to the participation in the phases of reworking, dismantling, re-assembling existing garments;

_ **action**. The designer carried out some first seminary activities and, later, interaction with the craftsmen so that, implicitly, the two working methodologies could meet and confront each other;

_**consolidation**. Those are the activities developed between artisans and designers directly in the product.

We can talk about valuation strategies that provide a system of artisanal knowledge, with a formulation-oriented approach, so that objects become icons and witness the origins of all cultures. The strength of these objects is expressed through the ability to project future influences into a present that fascinates with the intensity of its history.

This type of design action has determined cross-fertilization between the design methodologies that have led the craftsmen involved to redefine the goal of manufacturing and the designers to reflect on the methodological experience and on the relationship between design and craftsmanship. On this occasion, we want to communicate through designing the possibility of going out of one's "way of doing" to learn another or to integrate one's knowledge with other skills.

Another objective was to communicate to the participants of the territory, both the commercial value and the strategic value of the recovery and the improvement of the textiles as a plus identity. In this case, we can say that design represents the interface between tradition and modernity, at the moment in which it reconfigures the codes of tradition to make them coherent with the languages of contemporaneity.

Politecnico di Milano, p. 13. Per approfondimento si veda anche [4] Milano distretto del design, Il sole 24Ore, Milano.

[13] La Pietra [13] *L'artigianato per l'industria*, Artigianato tra arte e design, Edizioni Imago International, Milano, p.18.

[14] In Conti G. M., Galli F., Pino B., op. cit. Pag. 183.

References

1. Auricchio V (2008) Internationalization of design research and education centers. Promotion of international design networks, Tesi di Dottorato, Scuola di Dottorato, Politecnico di Milano
2. Balaram S (1998) Thinking design. National Institute of Design, Ahmedabad
3. Bauman Z (1998) Globalization. The human consequences. Columbia University Press, New York
4. Bertola P, Sangiorgi D, Simonelli G (a cura di) (2002) *Milano distretto del design*, Il Sole 24 ore, Milano
5. Colombo P (2009) Mestieri d'arte e Made in Italy. Marsilio, Venezia
6. Colombo P (2013) Artefici di bellezza. Marsilio, Venezia
7. Conti G, Vacca F (2008) Traditional textile on fashion design. New path for experience. In: Proceedings changing the change design, visions, proposals and tools, Allemandi, Torino
8. Conti GM, Galli F, Pino B (2012) Towards a cross cultural society; from ethnicity to design, "narrative" heritage drives innovation. In: Mapuche Weavers and Italian designers co-create for fostering diversity. Projecting design 2012, international conference Santiago, Chile
9. Conti GM, Gaddi R, Motta M (2017) Hand stories of italy. The Burano's jacket. How fashion designers can use the territory to develop a collection. In: Proceedings of the IInd international conference on environmental design, Torino
10. Conti GM, Panagiotidou MA (2020) Social innovation in fashion design: can design provide opportunities of inclusion to refugees in greece? In: Advances in industrial design, proceedings of the AHFE 2020 virtual conferences on design for inclusion, affective and pleasurable design, interdisciplinary practice in industrial design, kansei engineering, and human factors for apparel and textile engineering
11. Fagnoni R (2018) Da ex a next. Design e territorio: una relazione circolare basata sulle tracce. MD J N. 5, Anno III, 2018
12. Germark G (2008) Uomo al centro del progetto Design per un nuovo umanesimo. Umberto Allemandi & C, Torino
13. La Pietra U (2007) L'artigianato per l'industria, Artigianatotra arte e design. Edizioni Imago International, Milano
14. Manzini E (2004) Design as a tool for environmental and social sustainability. In: Design issues in Europe today, the Bureau of European design. associations - BEDA white book, ed. By Stuart Macdonald, The Publishers, NL
15. Paris I (2006) Oggetti cuciti. L'abbigliamento pronto in Italia dal primo dopoguerra agli anni settanta. Franco Angeli, Milano
16. Piscitelli D (2018) Professione designer, basterà una norma tecnica per riconoscerla? Il Giornaledell'Architettura. https://ilgiornaledellarchitettura.com/web/2018/05/02/professione-designer-bastera-una-norma-tecnica-per-riconoscerla/. Accessed 26 Feb 2020
17. Schön DA (1993) Il professionista riflessivo. Per una nuova epistemologia della pratica professionale. EdizioniDedalo, Bari
18. Serlenga L (a cura di) (2007) I dettaglicambiano la moda, EdizioniTessiliVari, s.l., Milano

Public Policy and Legislation in Sustainable Fashion

Analia Pastran, Evangelina Colli, and Hazriq M. Nor

Abstract The fashion industry is notoriously one of the most polluting and damaging sectors of the worldwide economy, not only is the harm evident in the natural world but also within societies of people. In 2015, the United Nations General Assembly (UNGA) adopted the Sustainable Development Goals (SDGs) of the 2030 Agenda for Sustainable Development. In addition to this comprehensive set of goals, and with the addition of the Green New Deal and the New Urban Agenda, the international community (with emphasis on governments and the entrepreneurial ecosystem) is presented with several elements and guidelines in order to tackle the issues, problems and, thereafter, identify solutions to adhere to sustainability in fashion. A key constant is to ensure that the circular fashion economy is the mantra when dealing with this global dilemma. However, in order to enable all players to be involved, legislation and policies implemented on a governmental level should always factor the needs of the society and the environment. In addition, the fashion industry itself has to see themselves responsible if the change is to be carried out successfully, starting from production of raw materials to the reuse, recycle, repair and remake of garments and products, as well as the fundamental preservation of the environment and the societies that play their roles within the industry. Finally, fashion users have to play their role as a driving force of sustainable consumerism and sustainable fashion by demanding a higher level of empathy from the government and the fashion industry.

Keywords Legislation · Public policies · Sustainable fashion · Sustainable development goals · Circular fashion economy · Green new deal · New urban agenda · Climate change

A. Pastran (✉) · E. Colli · H. M. Nor
Smartly, Social Entrepreneurship on the SDGs, New York, USA
e-mail: apastran@insmartly.com

Smartly, Social Entrepreneurship on the SDGs, Buenos Aires, Argentina

M. Á. Gardetti and R. P. Larios-Francia (eds.), *Sustainable Fashion and Textiles in Latin America*, Textile Science and Clothing Technology,
https://doi.org/10.1007/978-981-16-1850-5_9

1 Introduction

According to estimates, the total combined consumption by all the human activity around the world is greater than the resources generated by the planet. In other words, every year, the Earth Overshoot Day is being reached earlier, and this overtakes the planet´s ability to generate those resources for that year. By November 1, 2000, all the resources of Mother Earth had been consumed. In 2015, it was by August 13, and last year, the Earth Overshoot Day was July 29.[1]

Scientists and international agencies have consistently reported on the root causes and consequences of climate change, thus alerting global leaders and the world about this phenomenon.

In 2015, the Sustainable Development Goals (SDGs) of the 2030 Agenda for Sustainable Development were adopted by the United Nations General Assembly (UNGA). It is an action plan in favor of people, the planet and prosperity, universal peace and partnership. The 2030 Agenda proposes 17 SDGs and 169 targets, covering the economic, social and environmental spheres. The global goals for a new world aspire to be more sustainable, inclusive and humane. They are a guide for action to transform the world in a way that no one is left behind.

The 2030 Agenda and the Paris Agreement on Climate Change were two important milestones achieved in 2015, which specified the actions that everyone, from all across the world, must accomplish in order to achieve a more equal and inclusive future, while enjoying economic growth, and strong protection of the environment.

In the context of climate change, the COVID-19 pandemic affects human life in the terrestrial ecosystem and impacts all communities, especially large cities or urban concentrations, with greater intensity, and which is estimated to leave millions of people poorer. We ask ourselves as to what are the essential agreements and actions that must be emphasized to avoid inequality, injustice and environmental insecurity. In this sense, the role of the state becomes a preponderant role. In this framework, in particular, the Green New Deal (GND), the SDGs and the New Urban Agenda (NUA) for the entrepreneurial ecosystem are vital, as is the rethinking of the type of city we need to defeat COVID-19.

- **The City we want: integrated—accessible—resilient**
- **The Humanity we need: responsible—conscious—active**
- **The Opportunity to undertake that we deserve: local—institutional—sustainable**

In terms of the role of the state that is needed at this moment, there is a special emphasis on the relevance of boosting the economy in a conscious and responsible way regarding the impact on the environment. In this sense, the Green New Deal calls on governments to reduce the carbon footprint, generate well-paying jobs, ensure pollution-free air, renewable energy, access to water and food as basic human rights, as well as ending all forms of degradation. The GND is inspired by the New Deal developed by President Franklin Delano Roosevelt to help the USA overcome the

[1] Global Footprint Network.

Great Depression after his election in 1932. It was an economic stimulus plan. In the twenty-first century, it was proposed to reorient another great economic stimulus, but in those sectors that are responsible for the preservation of the environment because **the Great Depression of this century is the Depression of the Environment.** Both the Chancellor of Germany, Angela Merkel, and the Secretary General of the United Nations, Antonio Guterres, addressed this issue during International Mother Earth Day 2020.

Power generates a space of interpersonal relationships in which the redistribution and appropriation of resources occur. Knowing and controlling these resources are important in order to act upon them in the world. But the question to be analyzed is the way in which each person or area manages their way of being in the world, the way in which they shape their lives with active commitment, and even more so when this commitment consists of integrating a system of social and economic organization, which will have an impact in social recognition and a sense of belonging for the people.

However, this active management of those resources depends, in parts, to a large extent on public policies, which are designed to promote the integration of the individual with a state that promotes access to opportunities for the population. In this article, we propose a model of responsible production and consumption that we need to promote in the context of climate change and COVID-19.

In the context of the COVID-19 pandemic, individual habitat has become a vital space like that of a defensive world in the life of the global population. Everyday life, family, work and localized life as well as free time are reconfigured to that of the area of individual experience, of micro-stories, where the links are created in relation to which political discourse constructs its new narrative for the "new normality" that comes.

History shows that the breakdown of the norms and dominant values has taken place as a consequence of new forms of production and the people's activities. The transformation of production relations has had an impact on the links between all the parties and has transformed the unspoken and written rules that regulate the social contract as well as the link between the actors. In reality, if one examines the present, it can be seen that a new order will be born. While we still do not know which it will be, here a possible postulate is expressed, **our proposal from and to sustainable fashion consists of legislating to institutionalize the processes that people and civil society organizations are constantly promoting. These must be accompanied by public policies that allow for further deepening of the processes in the medium and long term**.

Combining new ideas and intellectual audacity to strengthen alliances and promote sustainable development, and within it sustainable fashion, becomes essential. It is necessary to analyze the sector from the cultural system in which it coexists and find a group of guiding principles that allow for understanding, expressing and promoting this cultural-economic practice. In this sense, this article postulates recommendations as a conclusion.

Observing and relating sustainable culture from the perspective of the actors are essential to stimulate the world in which they create, create, coexist and produce. Another point is that whoever has something to say on a subject is.

> Always positioned from/somewhere, and that socialized space constitutes our being-in-the-world where history, geography and political economy merge [1]: 192.[2]

The place where we are standing in order to look at the world and to promote an initiative determines the vision that we have of the history in relation to this phenomenon, the prism with which it is currently viewed and the innovation or modification that is recommended.

As such, the questions occur spontaneously: where is the social group? What are the social relationships that define it? What is the social structure that characterizes it? What aspects or goods do both mediate the ways in which people produce sustainable fashion as for those who consume it? What are the ways in which people experience and act on their environment?

Efforts must be made to understand the scope of sustainable fashion. Not only do we take into consideration analysis on wages, markets, investments, taxes, raw materials, exports, economies, prices, credits, subsidies, etc.; but paradoxically enough we also discuss the relations of power, domination, manipulation and control, and much less about everything that describes and defines the social and cultural sphere where this production of sustainable fashion takes place as well as what definitions are set for the actors to be in the world.

A practice requires motivation. It is understood that until the moment prior to the pandemic, it has been produced and over-consumed in a context motivated by the interest of the material, and what is politically useful for its world, its historical context and its cultural matrix.

If interest-motivated actors are always actively engaged in profit, effort-motivated actors are seen as experimenting on the complexities of their situations and trying to solve the problems generated by those situations. We must, therefore, position ourselves as actors motivated by sustainable development, which generate social integration with preservation of the environment and economic growth.

[2] Wright, Pablo:. El espacio utópico de la antropología. Una visión desde la Cruz del Sur.

2 New Emerging Sector: Sustainable Fashion

Fashion has historically always been exclusively for the rich to show off their wealth and status. Bright colors, delicate fabric, exotic textiles and the latest in designs were used to determine who was ahead in the fashion race. The elite set the trend and that would eventually trickle down to the masses. In the nineteenth century, the Parisian courtiers determined what was considered up to date and in style. Tailors were sought after to the point that those who catered exclusively to the wealthy were regarded as celebrities. However, the Industrial Revolution brought in fast-moving technologies and the ability to increase production. Subsequently, mass production and consumer demands increased, which saw fashion accessible to everyone. It was not until the 60s when the age of environmental consciousness started to have an impact on the trend of fashion. The "*hippies*" generation, known to be rebellious toward the status quo, opted for natural fabrics and a return to a simpler way of life. This was later followed by the punk (in the 70s) and goth (in the 80s) movements that rejected the norm of fashion and instead preferred second-hand and vintage pieces of clothes, as opposed to new fashion. But this did not last long. By the 1990s, due to globalization and offshore manufacturing, the production of fashion became cheap and easily accessible to everyone. Materialism became the norm and the need to take care of the environment and the people who produced the materials was a mere second thought. It was not until the early twenty-first century that people started seeing the grave impacts that fashion had toward, not only on the environment but also on the economy and the exploited people manufacturing the clothes.

To achieve sustainable development, partnerships and concrete actions are required between governments, the private sectors and civil societies to reflect and define together a direction that promotes and revises the pre-established frameworks. In accordance with that, it is necessary to question if they are adequate and if they serve, in this case, the new emerging sector: sustainable fashion. In that regard, we recommend that the debate and institutionalization of sustainable fashion be done from the legislative level with the effective participation of the local actors.

As stated in SDG 17 Partnerships for the Goals:

> Goal 17.14 Enhance policy coherence for sustainable development.
> As well as SDG 16 Peace, Justice and Strong Institutions.
> Goal 16.b Promote and enforce non-discriminatory laws and policies for sustainable development.

To look at why sustainable fashion is a crucial sector, we have to see how disastrous it is to the environment. The textile sector is the second most polluting industry in the world, after the oil sector. According to the United Nations (UN), it produces 10% of carbon emissions in the world and 20% of wastewater. Although some companies have taken measures to mitigate the damage, they continue to generate serious impacts on rivers and oceans. To this is added that as a work environment, it presents a strong weakness in the regularization of the labor rights of people who are frequently trafficked by people, this is an acute symptom of this scourge.

According to the United Nations Environment Programme (UNEP), the fashion industry is valued at around $2.4 trillion and employs over 75 million people worldwide [4].[3] Of this, in Latin America, the fashion industry is estimated to be around $160 billion [5][4] (almost 10% of the worldwide market). Brazil and Mexico are its biggest markets, with Colombia and Peru having the top growing rates. On the flip side, the industry also loses around $500 billion of annual value because of lack of recycling as well as products being thrown into landfills before ever being sold. The industry itself uses around 25% of the world's chemical products and accounts for around 8–10% of global carbon emissions (that is more than all international flights and maritime shipping combined!). Part of these emissions comes from pumping water to irrigate crops like cotton, oil-based pesticides, machinery for harvesting and emissions from transport. The fashion industry is responsible for 24% of insecticides and 11% of pesticides.

Sustainable fashion is a growing movement that hopes to change the way actors in the fashion industry regard the industry and its system by placing a higher emphasis on environmental integration and social justice. The fundamental criteria of the sustainable fashion business model include the conservation of natural resources, the low ecological impact of the materials used—which must be capable of later joining the recycling chain, the reduction of the footprint carbon and respect for the economic and labor conditions of the workers who have participated from the raw material to the point of sale [6].[5]

In a gist, sustainable fashion places prominence on the lifecycle of products; including design, raw material production, manufacturing, transport, storage, marketing, sale and the final stages of use. However, this is not a locked-down

[3] UN Environment Programme: UN Alliance For Sustainable Fashion addresses damage of "fast fashion".

[4] Hecho X Nosotros: How do we incorporate sustainability to the Latin American Textile Industry.

[5] Sostenibilidad Para Todos: ¿Que Es La Moda Sostenible?

definition. As a matter of fact, there are many forms of sustainable fashion. Green Strategy's Dr. Anna Brismar has identified seven main forms of more sustainable fashion production and consumption, as seen in the figure given below [7].[6]

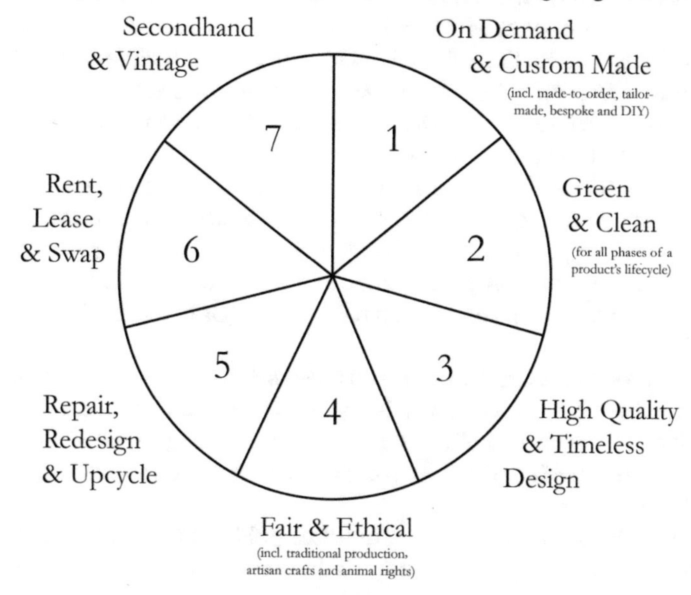

[6] Green Strategy: Seven Forms of Sustainable Fashion.

3 Public Policy and Legislation

From *Smartly Social Entrepreneurship on the SDGs*, we promote the sector to acquire the rank of public policy and legislation because we recognize the challenges that arise in the production of fashion clothing and the incipient sustainable work within it. The 2030 Agenda is the new social contract while the 17 SDGs are the guide of action to "*Legislate the Local thinking about the Global*" and thus contribute to local development.

Localizing the SDGs in public policies and legislation is a local challenge in the global opportunity of optimizing resources and making efforts for a sustainable local development in which the people have the possibility to live, study and produce in their birthplace. The New Urban Agenda guides us to promote the right to the city with people-centered development and is inspired by the SDGs to reach fairer, safer and more resilient cities.

We propose to generate a legal and fiscal framework that allows the promotion of the sustainable value chain in which sustainable fashion, as an emerging industry, as well as the generation of healthy and innovative employment, allows to provide a better quality of life (health) in the workplace as well as professional and personal development.

One area of work is sustainable fashion, in which public policy must promote the entire value chain that this innovation addresses, from the production of sustainable and renewable raw material necessary to carry it out until its sale and commercialization stage. Boosting the development of a sustainable and healthy textile industry both for the user of the garment and including the entire ecosystem that develops, processes and develops the sector is the key. We are interested here to incorporate the healthy concept for those who work with sustainable fashion, since it is observed that it is thought of almost exclusively as sustainable promotion to the consumer, and, in reality, when the raw material that is manipulated in the elaboration of a product is sustainable, it impacts in a way positive in the health of the worker who is in direct or indirect physical contact with said input. That is why we also cite the importance of addressing this aspect and its direct contribution to **SDG 3.9: By 2030, substantially reduce the number of deaths and illnesses from hazardous chemicals and air, water and soil pollution and contamination.**

Furthermore, **we observe that the consumption paradigm is changing where more and more people are looking for authenticity, transparency and social value in the products and services that they consume, leaving behind patterns of consumerism in favor of responsible consumption (SDG12).** Therefore, it is necessary to adapt the production paradigm to the needs of the current times. The implementation of the SDGs offers the opportunity for a new style of sustainable business development: conscious and responsible. For this reason, we have launched the Iberoamerican Platform of Sustainable Entrepreneurs as a space to empower those entrepreneurs who are taking into account the urgency of climate change and risk and are looking to produce sustainably starting from the stage of ideation of their business. A special chapter in this platform stems from sustainable fashion.

To achieve a concrete step toward this proposal, *Smartly* aims to generate indicators targeted toward the sector and thus innovate strategies to lobby and influence the local parliaments on legislation in order to strengthen and institutionalize sustainable entrepreneurship, both locally and institutionally, as indicated at the beginning of this article. To carry forth with this, we invite you to fill in the form, which can be found at https://www.insmartly.com/emprendedores-sostenibles.

In addition to the Platform of Sustainable Entrepreneurs, *Smartly* promotes the **Local Parliament Network on the SDGs**, a unique worldwide initiative that seeks to innovate from the legislative field to be able to provide solutions to the challenges that this particular sector presents and promotes partnerships of effective **collaboration between parliamentarians and actors of sustainable fashion so that it is incorporated from the legislation: sustainability and look in SDGs to fashion, as well as to public policies.** It should be noted that promoting sustainable fashion as an innovative and impactful industry that generates employment, production and sustainable consumption is the way to achieve the sustainable development that is so often postponed. Mission that we take as a shared responsibility and based on SDG 8 that refers to innovation and sustainability in two of its targets:

> 8.3 Promote development-oriented policies that support productive activities, decent job creation, entrepreneurship, creativity and innovation and encourage the formalization and growth of micro-, small- and medium-sized enterprises, including through access to financial services.
> 8.4 Improve progressively, through 2030, global resource efficiency in consumption and production and endeavor to decouple economic growth from environmental degradation, in accordance with the 10-year framework of programs on sustainable consumption and production, with developed countries taking the lead.

We all have the responsibility and, many times, the possibility of changing the state of things. When both aspects combine, embracing an agenda such as the SDGs and sustainable development drives the meaning of the action in a more noble way, and in a context of urgency of action to mitigate **climate change (SDG 13)**, where we cannot continue doing things as usual, promoting sustainable fashion becomes a necessary condition for good living.

4 Global Actions on Sustainable Fashion

In sustainable fashion, the material footprint is kept to a sustainable limit, where the raw materials meet the economy's final consumption demand. At the same time, the principle of a "circular fashion economy" (the continuous usage of clothes, textiles, and fibers)—as opposed to a "linear fashion economy"—are maintained at their highest value during use and re-entry into the economy to prevent them from becoming trash. Sustainable fashion puts into practice four main components of the materials: reuse, recycle, repair and remake.

The majority of the current fashion industry is still operating on the concept of a linear economy model. The transition toward embracing a circular fashion economy requires governments and consumers to prioritize brands that meet the criteria for sustainability. The market needs products that are designed for long life and for end-of life recyclability, well-functioning and convenient garment collection systems, sorting procedures that can efficiently serve both the reuse and recycling markets, recycling technologies that can recycle textiles into high-quality fabrics and other products, consumer readiness to buy used and recycled products, and brand commitment to uptake recycled fibers in new collections [8].[7]

[7] Niinimäki K. Sustainable Fashion in a Circular Economy.

Circular fashion economy has many advantages. By reducing the need for imported raw materials, fashion brands will put more emphasis on already existing as well as local products and materials, not only will this support the local market directly but the carbon footprint will be decreased tremendously. The damages caused toward the environment will be greatly reduced, especially with resource extraction minimized. By putting sustainability as a major focus instead of a side note, more eco-friendly industries, businesses and jobs will emerge, thus, creating brands with a better and more welcoming public image and reputation.

An international study on sustainable fashion was co-conducted in 2013 between Copenhagen Business School (CBS) and Growth from Knowledge (GfK). Among others, they concluded that:

- Due to the long, fragmented and complex supply chain, there exists a disconnection in the fashion supply chain between those who benefit from fashion and the many who pay the social and environmental costs.
- The existing "linear economy" created an unhealthy "throwaway" culture that breeds overconsumption and waste. Consumers are accustomed to cheap, poor-quality fashion that they can ultimately throw in the garbage after a few wears.
- "Fast fashion" is repeatedly criticized for creating a level of consumption that is neither socially or environmentally sustainable. There is a need to challenge the dominant "fast fashion" business model, which is based on large quantities of new, low-priced collections.
- In an existing "race to the bottom", the increasing demand for cheap fashion has a negative impact on social and environmental performances in the fashion supply chain.
- Fashion companies do not have sufficient knowledge and competence to address sustainability; likewise, the curricula at most design schools and universities do

little to bridge this competence gap. A number of fashion companies are introducing innovative materials, manufacturing technologies, management systems and business models that create value for both business and society.
- Sustainability should not be an isolated add-on project—for instance, within the communications department. Workshops, lectures, panels, etc., can be used to inspire employees across departments to work with company-specific sustainability challenges and opportunities.
- Multi-stakeholder partnerships are highlighted as an important precondition for bringing about changes toward sustainability in the fashion industry.
- Sustainable consumer behavior can be promoted through an increased focus on repair, customization, communication, campaigns, price mechanisms, labeling, competitions, etc.
- Policymakers can promote sustainable fashion by using a variety of "carrot and stick" mechanisms, including tax incentives, minimum product standards, mandatory reporting requirements and awareness-raising campaigns.

In general, policymakers can use a variety of tools to influence corporate sustainability policies and practices [9],[8] such as raising awareness, tax incentives, public procurement, etc. A blueprint of policy instruments to implement a circular fashion economy in Latin America should be identified, much of what has been undertaken in similar initiatives in Europe. These measures should include (but not limited to) financial incentives, policies that reward innovation, transparency and accountability framework, evaluating existing and future trade policies as well as to encourage actions voluntarily.

Policy and financial incentives such as procurement, extended producer responsibility, value-added tax (VAT) rewards, support for innovation and a tax shift to drive the market to embrace circular products and services. Fashion companies should consider these incentives as a way to discover new and innovative approaches while, at the same time, recover costs involved with implementing environmentally-friendly investments. They can research into innovative and local materials, manufacturing technologies, management systems, and business models that are valuable for both the business and society as a whole. Consumers can benefit from a return on their spending from sustainable brands. Tax cuts, refund policies on resource use and financial subsidies create a positive stimulation for industry players to embrace the circular fashion economy. On the flip side, penalties could also be enforced on those who do not embrace environmentally friendly products.

Facilitating trade and business ethics policies such as export of semi-finished products and sorted, reusable textile waste to producing countries, labor rights and gender equality, as well as avoiding negative social impacts in producing countries. Governments—with the help of organizations—could put into practice prioritizing sustainable fashion companies when it comes to international trade. Some countries are in the forefront of sustainability and have higher—and more rewarding—standards for fashion companies that wish to undertake business in their respective nations

[8] European Sustainable Business Federation.

or even in another country. One such case is the Dutch Agreement on Sustainable Garments and Textile, which includes industry associations, trade unions, NGOs and the National Government of the Netherlands. Those who have signed the agreement commit themselves to fighting discrimination, child labor and forced labor, support a living wage, health and safety standards for workers, and the right of independent trade unions to negotiate. In addition, they also aim to reduce the negative impact of their activities on the environment, to prevent animal abuse, to reduce the amount of water, energy and chemicals that they use and to produce less chemical waste and wastewater. The International Labor Organization (ILO) has also provided a forum to promote appropriate standards, policies and programs to protect the rights of workers. In addition, the Group of Seven (G7) has formulated due diligence standards for the textile industry to help improve working conditions and enhance workers' rights in the global textile supply chain.

In mid-2019, the French government entrusted Chairman and Chief Executive Officer of Kering, François-Henri Pinault, to "bring together the leading players in fashion and textile" and reduce the environmental impact of the industry. The "*Fashion Pact*"—which was presented to world leaders at the 45th G7 summit in Biarritz—tackles three major issues: climate, biodiversity and oceans. The fashion and textile companies and brands (many of whom are each other's competitors) signed this pact to contribute in their own way to making the industry more sustainable. The three science-based targets are: stopping global warming (zero greenhouse gas emissions by 2050 in order to keep global warming below 1.5 °C until 2100), restoring biodiversity (with a focus on restoring natural ecosystems and protecting species) and removing single-use plastics from our oceans as well as reducing the negative impact the industry has on them. The first meeting of the Fashion Pact was held in Paris in October 2019. As of then, the total number of signatories reached 56, involving about 250 brands.

Establishing and enforcing a common regulatory framework for knowledge, competence, transparency and accountability, circular design and improved end-of-waste status across the region. Policymakers can enforce that fashion companies meet minimum product standards, have mandatory reporting requirements, as well as carry out awareness-raising campaigns. Most, if not all, fashion companies do not have sufficient knowledge and competence to address sustainability; likewise, the curricula at most design schools and universities do little to bridge this competence gap. With the current trend towards sustainability shaking the market, more companies are working toward educating not only themselves but also their consumers. It is recommended that smaller organizations and government agencies work together to establish working groups to define sustainability priorities and metrics to uphold accountability.

Encouraging actions on a voluntary basis with the likes of covenants, commitments and standards in order to engage all stakeholders. If this fails, then a sound legislative measure should be enacted to ensure these measures are carried out legally. Companies can take the initiative by carrying out their own sustainable practices on a voluntary basis, whether locally or in their international locations.

An example of rewarding consumers toward their behavior in voluntarily helping can be found in the USA, where consumers that donate clothes to charity organizations, are eligible for tax deductions. This initiative to the public to donate unwanted clothes in order to receive benefits in return is widely embraced. Donated clothes are also resold to low-income communities in other countries in order to recycle textiles and reduce the amount of clothing that ends up in landfill.

The need for a robust and thorough approach toward sustainable fashion has not gone unnoticed by the United Nations. While separate methods have been identified in the past (mostly in tandem with the existing policies and initiatives), the UN Fashion Industry Charter for Climate Action was finally launched at the COP24 in Katowice, Poland, in December 2018 to identify ways in which the broader textile, clothing and fashion industry can move toward a holistic commitment to climate action. The Fashion Industry Charter for Climate Action was created to identify and amplify best practices, strengthen existing efforts, identify and address gaps, facilitate and strengthen the collaboration among relevant stakeholders, as well as to join resources and share tools in enabling the sector to achieve its climate targets. The industry charter specifies the following overarching areas of work to be further developed by specific Working Groups:

- Decarbonization pathway and GHG emission reductions
- Raw material
- Manufacturing/Energy
- Logistics
- Policy engagement
- Leveraging existing tools and initiatives
- Promoting broader climate action
- Brand/Retailer Owned or Operated Emissions

Like most everything aimed at being environmentally friendly and sustainable, the fashion business model should give significance to the need to conserve natural resources, the importance of a low ecological impact of the many materials used (new and reused), which has to be able to later join the recycling cycle, reduction of the carbon footprint, as well as empathy for the economic and labor conditions of the workers involved in production to sale of the goods. The latter includes the justification of a gender-equal opportunity in the fashion industry. In addition, the systematic sharing of knowledge will foster accountability and responsibility throughout the value chain. The fashion industry should also work in tandem with governments and non-governmental organizations in setting up policies, legislations and enforcements. All players in this field have a distinctive role to encourage that the market trend will move toward a sustainable and environmentally friendly economy.

> We propose a different and disruptive dynamic to the conventional fashion catwalk, in order not only to question the common spaces, hegemonic and exclusive that represent the "Fashion shows", but also understanding that fashion is part of our daily routine and not just an ephemeral performance.
>
> Analia Pastran, Executive Director of Smartly,

Social Entrepreneurship on the SDGs at the Urban Thinkers Campus: Vibrant and Inclusive Urban Life (Mexico) of the World Urban Campaign of UN Habitat.

5 Recommendations for Introducing Sustainable Fashion in Policies and Legislations

While we acknowledge that there is still a long road ahead of us in order to fully incorporate sustainable fashion as a policy and legislation in itself, there are step sand workable recommendations based on what has been mentioned earlier that could be implemented right now or in the near future to achieve this goal.

1. **Incorporate Sustainability and SDGs in the curricula of design schools and universities.** It is important that sustainability be taught in fashion and business schools as a transversal axis, which must be present from the moment of zero business ideation as well as in all its business strategy (sustainable canvas).
2. **Policy and financial incentives** such as procurement, extended producer responsibility, value-added tax (VAT) rewards, support for innovation and a tax shift to drive the market to embrace circular products and services. It should be noted that the entrepreneur who makes their products in a sustainable method has much more difficulties in obtaining raw materials and in developing the production process. For this reason, having tax incentives that protect the sector and make it competitive is the key.
3. Facilitating **trade and business ethics policies** such as export of semi-finished products and sorted, reusable textile waste to producing countries, labor rights

and gender equality, as well as avoiding negative social impacts in producing countries. **Governments—with the help of organizations—could put into practice prioritizing sustainable fashion companies when it comes to international trade.**

4. Establishing and enforcing a common regulatory **framework for knowledge, competence, transparency and accountability**, circular design and improved end-of-waste status across the region.
5. **It is recommended that smaller organizations and government agencies work together to establish working groups to define sustainability priorities and metrics to uphold accountability.** It is vital to generate a multi-sectorial work table or advisory committee with all the relevant actors to improve production processes, their rules and regulations so that the legislation is not outdated.
6. **Recycling and donating:** consumers are encouraged to donate clothes to charity organizations. To promote this act of recycling, those who donate are eligible for tax deductions. Donated clothes that are still in good condition are also resold to low-income communities, whether in their own societies or in other countries, in order to recycle textiles and reduce the amount of clothing that ends up in landfill.
7. **Encourage the production of raw materials**: preserving the ecosystem and the production of supplies in this sector becomes essential. For the production sectors to be identified, a registry must be carried out to protect the biological species while at the same time fulfilling the necessary stock to satisfy the production demand always guaranteeing the preservation of the environment.
8. **Parliamentary policies and actions to strengthen sustainable fashion and discourage green washing**. Institutionalize their existence in the local value chain in order to guarantee that the voices of the actors are heard without anyone being left behind.
9. **Awareness campaign on responsible consumption and production (SDG 12) in the context of climate change**: sensitizing citizens on responsible consumption and the importance of reusing clothing, as well as donating it, is important in order to generate the virtuous circle that allows for the value of responsible production toward the care for the environment and the people.
10. **Boosting clothing that promotes integration:** sustainable fashion must be increasingly accessible to people in their diversity of bodies and styles, allowing all those who want to contribute to the planet to do so.
11. Sustainable fashion should be understood as a **productive activity that creates clothes with decent jobs**, which begins as an undertaking to achieve economic development with a focus on people and which is guided by innovation using ancient knowledge and its implementation value.
12. **Sustainable fashion as a healthy concept for the planet and the actors, but above all for the worker.** Sustainable fashion must make a difference to the textile industry and give clear signals regarding its strategy to eradicate slave labor and abuse of people in addition to promoting environmental protection and becoming a way out of poverty.

13. It should be applied as a **field of scientific research to the economy of the productive development** of a nation and far beyond the design of the form of clothing: but as a survey of the production of raw material, its cycles of renewal, substitutes and its preservation and generation of seed.
14. **Sustainable fashion is important as a stimulus for rooting communities** for those who produce it as well as for those who consume it. The designer's philosophy could be promoted in the context of its productive corridor and inclusion of local cultural practices toward the use of the raw material and its final destination, the garment itself. Promoting and incorporating cultural integration, local commerce and indigenous knowledge should be seen as an innovation in fashion.
15. **Promote investment and research in a sustainable fashion** to encourage its development as a sustainable production model.

6 Reflections

Boosting and strengthening the sustainable fashion sector in the context of COVID-19 are presented as urgent and, since their importance, have accelerated awareness and adherence to the proposed style of production and consumption. We are going through an unprecedented historical moment that invites us to rethink the institutions, businesses and cities in which we want and need to live in. This has increased the already existing crisis of climate change and has highlighted the great challenges we face in order to achieve sustainable development goals.

Without a doubt, taking into consideration that the fashion industry is the second most polluting industry worldwide, it is an absolute priority to refocus and redesign it so that it is sustainable in its production and consumption SDG 12 as well as toward the preservation of the planet and its people. There are many who identify this as a true revolution in the fashion industry since it would imply redefining consumption and the permanent replacement that occurs in this industry (fast fashion). In that sense, young people are leading this change, as they are looking toward responsible consumption within the context of the environment and people. Furthermore, they are beginning to stop the consumption of brands that are considered premium and "fashionable" because these brands are in the antipodes of responsible production, which is, by using clandestine and low-cost workshops, inadequate labor wages and working conditions, soil contamination and water, to name just a few scourges that these brands carry out in various latitudes worldwide. This is also possible because there are governments that allow this type of situation, and it is there that we fight and propose legislation and public policy in a sustainable fashion. Because it is precisely in these regions where legislation is required to protect and promote the sustainable fashion sector and discourage Green washing.

References

1. Wright P (1995) El espacio utópico de la antropología. Una visión desde la Cruz del Sur. Cuadernos. Instituto Nacional de Antropología y Pensamiento Latinoamericano 16:191–20. https://content.bhybrid.com/publication/4c76415c/mobile/. Accessed 12 June 2020
2. Pastran A (2019) Social entrepreneurship on SDGs. In: CSB annual global micro-, small and medium-sized enterprises report, 76–79 p. https://icsb.org/wp-content/uploads/2019/09/REPORT-2019.pdf. Accessed 9 June 2020
3. Pastran A, Colli E (2020) SDG legislation to strengthen the entrepreneurial ecosystem. In: ICSB annual global micro-, small and medium-sized enterprises report, Buenos Aires
4. UN Environment Programme (2019) UN alliance for sustainable fashion addresses damage of 'fast fashion'. https://www.unenvironment.org/news-and-stories/press-release/un-alliance-sustainable-fashion-addresses-damage-fast-fashion. Accessed 14 Apr 2020
5. Nosotros HX (2020) How do we incorporate sustainability to the Latin American Textile Industry. https://en.hechoxnosotros.org/post/how-do-we-incorporate-sustainability-to-the-latin-american-textile-industry. Accessed 14 Apr 2020
6. Sostenibilidad Para Todos (2019) ¿Que Es La Moda Sostenible? https://www.sostenibilidad.com/desarrollo-sostenible/que-es-la-moda-sostenible. Accessed 27 Ap 2020
7. Brismar A (2019) Seven forms of sustainable fashion. available via green strategy. https://www.greenstrategy.se/sustainable-fashion/seven-forms-of-sustainable-fashion. Accessed 31 May 2020
8. Niinimäki K (ed) (2018) Sustainable fashion in a circular economy. https://shop.aalto.fi/media/filer_public/53/dc/53dc45bd-9e9e-4d83-916d-1d1ff6bf88d2/sustainable_fashion_in_a_circular_economyfinal.pdf. Accessed 29 Apr 2020
9. European Sustainable Business Federation (2019) Bold policies needed to mainstream sustainable fashion. https://ecopreneur.eu/2019/03/28/press-release-bold-policies-needed-to-mainstream-sustainable-fashion/. Accessed 14 Apr 2020

10. Gardetti MA, Muthu SS (eds) (2020) The UN sustainable development goals for the textile and fashion industry. Springer, Singapore
11. UNToday (2020) Working towards sustainable fashion. UNToday #801

An Alternative Circular Business Model: Pineapple Waste for the Production of Textile Fiber for Rope Confection in Costa Rica

Esteban Valverde, Luis Torres, Rodrigo Chamorro, Paola Gamboa, Carolina Vásquez, Diego Camacho, Anthony Hallog, and Roberto Quirós

Abstract Costa Rica is a world-leading exporter of fresh pineapple; thus, it has the largest amount of pineapple waste. In general, pineapple cultivation produces a negative environmental impact, damaging protected areas of riverbeds, contributing to deforestation, increasing soil erosion, rising sediments in rivers, and affecting the landscape. In this work, a novel assessment of a sustainable business model that uses the pineapple leaf stubble waste for the creation of agricultural rope is proposed in order to address environmental and social issues in the rural community of the country. The decision to use this assessment was based on the methodology proposed by Veloz and Parada (Veloz C, Parada Ó (2015) Procedures for the selection of ideas and sources of financing for enterprises. UNEMI Science Magazine), which evaluates different business ideas grounded on their comparison against different criteria to select a final entrepreneurial idea. Furthermore, in order to quantify different indicators of the project such as commercial, financial, and environmental viability, this work defines aspects related to the production process, requirements of raw material, equipment, and spatial elements. In addition to the replacement of polypropylene rope, the proposed approach integrates the workers' association and the owners in the rope production process to create a greater social impact by providing job opportunities and economic benefits for the community. Finally, this circular business model has a product-level circularity indicator of 91% and an acceptance rate of 90% among the pineapple producers and the worker's association as well as showing profitable results from an economic perspective.

E. Valverde · L. Torres · R. Chamorro · P. Gamboa · C. Vásquez · R. Quirós (✉)
School of Industrial Engineering, University of Costa Rica, Escazú, Costa Rica

D. Camacho
School of Forest Engineering, Technological Institute of Costa Rica, Cartago, Costa Rica

A. Hallog
University of Queensland, Brisbane, Australia

R. Quirós
Sostenipra Research Group, Institute of Environmental Science and Technology (ICTA), Universitat Autònoma de Barcelona, Edifici C Bellaterra, 08193 Barcelona, Spain

M. Á. Gardetti and R. P. Larios-Francia (eds.), *Sustainable Fashion and Textiles in Latin America*, Textile Science and Clothing Technology,
https://doi.org/10.1007/978-981-16-1850-5_10

Keywords Pineapple · Textile · Business model · Circularity

1 Introduction

Agriculture has always been one of the main human activities. As such, it has put great pressure on the environment, compromising the quantity and quality of resources and future food production. The notable increase of agricultural production throughout history occurred mainly due to the expansion of the cultivated area up to the middle of the twentieth century and the rise of the intensive use of external agricultural inputs during the middle of the last century. Therefore, this resulted in soil degradation, loss of habitats, and contamination because of the high use of agricultural inputs such as fertilizers, pesticides, and fossil energy stand out [1]. In Costa Rica, the National Chamber of Pineapple Producers and Exporters [2] indicates that 45 000 hectares of pineapple have been planted, and, according to the University of Costa Rica [3], 250 tons of waste are produced for each hectare. This waste includes 110.25 million tons of stubble a year. The management of this creates a negative environmental and social impact because of the excessive use of agricultural chemicals. Another problem coming from the poor waste management is the appearance of the "Stomoxys calcitrans" fly that affects the livestock and, consequently, the economic stability of the region.

The environment has been affected to such an extent that the production and consumption of eco-friendly products have increased in order to counter the negative effects of agriculture in the environment [4]. "Piñatex" is one of the main references for the approach behind this article, since it corresponds to an innovative natural textile created from the fiber of the pineapple leaf. Moreover, from the initial sampling to the development of a viable supply chain, Piñatex journey has been inspired by the principles of a circular economy [5]. Because of all the damage agriculture produces in the environment, it is important to take measures to stop actions that directly impact the environment in a negative way such as mismanagement of residues from agricultural activities. The international regulatory framework on sustainable development has been transforming the role of agriculture and, more specifically, the policies and strategies of the circular economy and bioeconomy. Additionally, the industry has developed new and better techniques for agricultural waste recovery based on industrial innovation and high technology, which contributed to guaranteeing resource efficiency, sustainable production and consumption, and the reduction of the negative environmental impact [6]. To partially mitigate the negative impact that the incorrect management of pineapple stubble produces, different authors have proposed various alternatives to make this waste valuable like using them to create textiles, ropes, paper, energy, compost, silage, biorefinery, and chemical products, which will be described later. The objective of this work is to evaluate these alternatives and select one that allows the implementation of a circular business model that takes advantage of this waste and partly reduces the environmental problems associated with its inadequate handling.

In order to determine the characteristics of the product, calculate the requirements of raw material, and determine the necessary equipment and space needed for its production, a sampling of stubble fiber extraction was carried out in collaboration with the School of Forest Engineering of the Technological Institute of Costa Rica. With this analysis of the samples, this article examines whether or not this proposed business model presents the benefits respecting the commercial, financial, and sustainable development aspects. Following, the business model proposal for the commercialization of the final product was designed with the information obtained from the analysis previously mentioned. To carry out this, this work considered the aspects of marketing, human resources, and internal and external factors about the business as well as how well the model would integrate in the sustainable development of the focus community.

The importance of innovative projects that link waste recovery processes and radical economic models lies in the fact that they are tools that open a sustainable development, which greatly improves the environmental, social, and economic situation of a community [7]. One example of such an innovative project is the circular economy. Based on multiple criteria analyses, the best alternative of pineapple stubble valorization with the greatest commercial advantage and sustainable impact in Costa Rica is the rope for agriculture. Also, the rope presents a circularity index of 91% in relation to the flow of materials in its production process and presents a circularity index of 91%, highlighting the benefits of this alternative.

2 Methodology

The first important element of this research is a multi-criteria decision-making problem: the selection of one of the alternatives for the valorization of pineapple stubble. Therefore, the implemented tool for the selection of the substitute of the pineapple stubble is the multi-criteria decision matrix, since, according to Veloz and Parada [8], constitutes an applicable technique for making rational decisions between different alternatives. This matrix is characterized by the objectivity of the selection process because of its systematic and repeatable methodology that ends up in quantitative results. In addition, to do this, the matrix first weights the degree of compliance that the options have in terms of each of the different criteria that could affect the ending result of the study against each other. Next, this degree of compliance of the different criteria is integrated into a single overall rating (score) for each option. The comparison of the global ratings represents a rational criterion to select the most appropriate idea, which is the one with the highest global score. Finally, after the selection of the alternative, the next step is the preliminary evaluation of the feasibility of the project concerning operational, economic, social, and environmental factors. Sapag and Sapag [9] indicate that a project to create a new business, launch a new product, take advantage of natural resources, among others must be evaluated in terms of convenience to ensure that it will solve a human need in an efficient, safe, and profitable manner. In other words, the intention of a project is to provide the best solution for the economic problem of the community by ensuring the necessary conditions and information to rationally allocate scarce resources for the most efficient and viable alternative solution to human needs. For this reason, the article carries out a study of the technical, economic, and sustainable impacts of the use of pineapple waste to produce textile fiber for rope confection in Costa Rica.

2.1 Waste Treatment Alternatives

The opportunity to investigate alternatives for the use of waste such as the pineapple stubble comes from the increase of pineapple production and, therefore, the increase of waste and environmental and economic negative impact. Following the methodology proposed in this study, the authors conducted an extensive research to find different ways to add value to pineapple waste. Moreover, the research includes bibliographical research and interviews with experts in the field that worked in developing alternatives in the country. There are different examples of uses for pineapple waste inside and outside of Costa Rica, some more developed than others. Specific research was done on the defined group of alternatives in order to determine the level of growth and application in the country or other areas of the world. The data gathered will be described in the theoretical framework; furthermore, Table 1 shows a compilation of the bibliographic references.

2.2 Selection of the Alternative

A focus group integrated by students and professors from the Industrial Engineering major from the University of Costa Rica and the Technological Institute of Costa Rica determined four fundamental criteria to use as benchmarks in the comparison between the alternatives. This selection process looks to choose the alternatives that best fulfill the objectives and interests of the project and the team working on it [20]. The defined criteria consist on:

- Information accessibility: how accessible the information of each alternative represents a key element to consider in order to gather data through research, tests, and people involved in the use of the resource.
- Technology simplicity: each alternative must have modern but accessible technology for its viability of the project in the country.
- Updated literature: the studies, tests, and research available about each alternative must be up-to-date and have valid and results to determine their viability for the proposal of this article. In addition, information regarding the degree of success of each alternative is another crucial element to determine the feasibility of the option.

Table 1 Bibliographical review of alternatives

Alternative	Bibliographic references
Textile	• Camacho [10]. Uses of Pineapple Stubble. Dissertation, Technological Institute of Costa Rica. • Serrano [11]. The Pineapple Fiber-based Leather that will Revolutionize the Textile World. In: VICE Media Group. Available via DIALOG https://www.vice.com/es_co/article/9b4jm5/el-cuero-a-base-de-fibras-de-pia-que-revolucionar-el-mundo-textil. • Potts [12]. Global Textile Trends Round-Up: Fabrics from. Euromonitor International.
Paper	• Araya [13]. Use of Pineapple Stubble (Ananas Comusus) to Obtain Pulp for Papermaking. University of Costa Rica, San José.
Rope	• Camacho [10]. Uses of Pineapple Stubble. Dissertation, Technological Institute of Costa Rica.
Compost	• Acuña [14]. Pineapple Waste: A Headache for Growers. In: With an eye on the pineapple. University of Costa Rica. Available via https://www.ucr.ac.cr/noticias/2018/06/21/desechos-de-la-pina-un-dolor-de-cabeza-para-productores.html. • Amador [15]. Pineapple activity in Costa Rica. Energy potential of pineapple stubble. Dissertation, State Distance University.
Chemicals	• Bermúdez [16]. Use of pineapple stubble. Dissertation, University of Costa Rica. • Mata [17]. Options for using stubble in Costa Rica. Dissertation, University of Costa Rica.
Energy	• Amador [15]. Pineapple activity in Costa Rica. Energy potential of pineapple stubble. Dissertation, State Distance University.
Biorefinery	• Uribe [18]. Pineapple activity in Costa Rica. Valorization of pineapple stubble under a biorefinery approach. Dissertation, University of Costa Rica.
Silage	• Peña [19] Pineapple stubble silage: a viable option for agricultural producers. In: With an eye on the pineapple. University of Costa Rica. Available via https://www.ucr.ac.cr/noticias/2018/06/21/ensilaje-de-rastrojo-de-pina-una-opcion-viable-para-productores-agropecuarios.html of subordinate document. Accessed on September 15, 2018.

- Raw material: each alternative uses the waste to develop different products, which reduces its environmental impact, but all of them take advantage of different percentages of the waste. Therefore, it is important to determine the amount of waste each option uses.

The alternatives are evaluated in terms of these criteria and the research done previously that includes bibliographic review and interviews with experts in the field. The methodology behind the tool used to evaluate the alternatives is based on two different rubrics. First, the type of evaluation scale used is numerical, ranging from 1–5. In this case, 1 represents the most negative punctuation while 5 represents the most positive one. Second, considering how much each alternative meets the project criteria, a relative percentage is assigned to the comparative criteria used in the selection process. The team members assigned these relative percentages using their own criteria. Moreover, the percentages assigned to each criterion in the global score [20] are:

- Information Accessibility: the information regarding the alternatives represents 35% of the global score of this work because of its great importance as a required element in the justification and development of an entrepreneurial project.
- Technological simplicity: the use of complex technology in the alternative may represent a challenge for the development of an entrepreneurial project if the workers do not have extensive knowledge regarding the handling of the tools that the project demands in the process of production. Thus, this research looks for a certain degree of simplicity that fits with the business proposal model in the technology each alternative needs. Because of its major role in the selection of the alternative, this criterion represents 35% of the global score.
- Updated literature: as mentioned previously, information is the key while developing an entrepreneurial project. Its relevance lies in the fact that research about projects carried out in the country works as an indicator of the technical feasibility of each alternative; however, the criterion percentage in of the global score is 20% because this is not a limiting factor while determining the most suitable alternative since an innovative project may not require major academic and industrial research and findings.
- Raw material: as long as a proper waste management plan is established for the waste of the project, it is possible to add a value to a product that meets the objective of this work made mainly out of stubble, without using the entire leaf, is feasible. The overall percentage of this criterion in the global score is 10%.

Based on the results of the global score, two final options were selected from the initial set of alternatives. Fiber extraction tests were developed with these two options to compare the technical aspects of each sample in the production of agricultural rope and textile. With this new analysis of the fibers and the technicality of their application, the researchers selected the one used for the actual design of the business model proposed in this work.

2.3 Preliminary Analysis: Fiber Extraction

In order to detail the current conditions of the extraction process, a 54 kg sample of stubble leaves processed with a decorticator or shredder machine developed by the Technological Institute of Costa Rica were used as samples in the study. After its extraction, the fiber is manually cleaned by removing the remaining leaves. To remove them, the stubbles must go through a drying process that removes the moisture and dries the leaves, so workers can easily clean it by using combs [21]. In addition, Table 2 shows the details of the sample used.

The tests measure the weight of the leaf sample, the weight of the extracted fiber, the weight of the processed fiber, and the volume and length of the leaves. Moreover, the findings of these tests were part of the comparative technical study between both alternatives used to determine the final focus of the proposed project, and later define the business model. In order to continue, the research studies some aspects

Table 2 Sample details

Planting location	Los Chiles, Alajuela
Pineapple variety	MD-2
Age at harvest	124 weeks
Leaves weight	54 kg
Harvest volume	0.48 m^3
Harvest time	2 h
People who carried out the harvest	1

from the two final activities such as the required raw materials, specialized equipment requirements, market of the product, and its effect on sustainable development.

2.4 *Business Model*

According to Osterwalder and Pigneur [22], a business model expresses the logic by which a company earns money. To do this, it generates and offers a value to one or more customer segments. Besides a business model involves the relationship between a set of elements to create, manage and control the offered value. The model business in this article attempts to solve the pineapple stubble problem in Costa Rica by joining forces with pineapple producers. This kind of business is an opportunity for the producers to benefit from a sustainable development model that uses pineapple waste, leading to economic, social, and environmental improvement [23]. A comprehensive approach based mostly on the design thinking methodology was used in the design of this business model [24]. One of the benefits of using this approach is the constant communication and feedback from the producers, which leads to a realistic design that fits with the strategy of the business. In addition to the design thinking methodology, other traditional techniques were considered to develop a valuable business model proposal such as the business model canvas [22]. The production process, an element of the business model, was designed considering different factors such as the results of the analysis and characteristics of the fiber samples, the production rates, and the space the machinery needs to process the leaves. The design production process accounts for the daily needs of the agricultural rope the pineapple producers need. Furthermore, the machinery and economic results of the project reflect the growth of the production.

2.5 *Customer Evaluation*

The research uses a client acceptance evaluation in order to find out if the proposed product based on a sustainable business model is of interest to real pineapple producers. This evaluated the opinion of producers' and the workers' association that represents a population that could potentially implement the proposed business model project. While the pineapple producer is the landowner of the pineapple plantation, the supplier of pineapple stubble, and the buyer of the rope, the workers' association is the one in charge of the funding and operation of the production process. In addition, the evaluation of the two populations was a questionnaire with qualitative responses that were translated to quantitative by using a Likert scale. This scale states that attitudes can be measured because they are linear. In this survey, a total disagreement by the population represented a value of 1, and a total agreement represented a value of 5. Overall, the evaluation had a value of 70% in terms of its importance when accepting the proposed business model.

2.6 Circularity Indicator

For this research, the authors used the Material Circulatory Indicator by the Ellen McArthur Foundation [25], which is a product-level circularity index that measures the circularity of the textile rope. This index generates a comprehensive analysis of the product's life cycle because it clearly includes the flow of the material from cradle to cradle in the measurements. Linder et al. [26] argue that this approach might include protected data from companies, making the results untraceable. However, this is not the case with this research because the data regarding the flow of the material are more crucial to determine the circularity of the product than other data. The calculations included real results from the field extraction sample.

2.7 Economic Assessment

The article details the cash flow estimation of the project through a defined timeline in order to determine the financial viability of the project. This estimation included predictions and projection of income and expenses related to the proposed business model where the stubble is transformed into agricultural rope. The cash flow was measured by using financial indicators calculated to measure the possible success of the project. These indicators were the Net Present Value (NPV), the Internal Rate of Return (IRR), the Recovery Period, and the Desirability Index (ID). The NPV measures the resulting surplus after obtaining the desired profitability and recovering the entire investment. Furthermore, the IRR measures the profitability as a percentage; this indicator is a measure of the maximum eligible rate of the project [27] and must be greater than the capital cost of the project for the model to be considered viable. In addition, the Recovery Period is an indicator that calculates the amount of time that would take for the project to recover the initial investment [28]. Finally, a sensitivity analysis is performed as well to measure how the project deals with key variables of the business model.

3 Alternatives Context

The opportunity and need to do research with the objective of exploring different alternatives for the use of pineapple waste rise because of the increase of pineapple production and Costa Rica, which increases the generation of waste and, consequently, the environmental and economic impact of the pineapple plantations. By using a bibliographic review and interviews with experts on the subject, this article identifies and describes different uses of the pineapple stubble.

3.1 Silage

According to University of Costa Rica researchers, pineapple stubble has a nutritional value suitable for feeding cattle, both dairy and fattening. Silage is a type of storage of plant-based material that can be used to feed ruminants, taking advantage of excess non-valued material. This corresponds to an excellent feeding alternative when the forage is scarce [19]. The silage proposal is attractive since the degree of use of the stubble is 100%; however, it requires complements such as dehydrated citrus pulps, molasses, and pineapple wreath [29]. Currently, silage is made using only the pineapple crowns [10], where the thickness of the foliage makes the silaging of the crown easier than the process of silaging the entire sheet. Also, most ranchers keep cattle grazing with additional nutrients that come from using stubble, which represents an extra effort since the production of silage demands the use of other products as its component [30]. Due to farmers' interest in simplicity and lower costs, this alternative is less appealing for them.

3.2 Energy

The biogas program of the Costa Rican Electricity Institute (ICE) is currently developing tests for the use of the liquid fraction of stubble as an energy source. The institute estimates that 5517186 tons of stubble fresh per year would theoretically produce 147 027 661.28 m^3 of methane, which can approximately generate that 31.13 MW. These data are promising when taking into account that the currently installed capacity of biomass in the country is 2.72 MW [15]. Despite the existence of a consolidated project for the development of biogas plants by ICE, the alternative is not suitable for the objectives of the project. Due to this project being an incipient research, the technology developed is not optimal and requires expertise and knowledge in the chemical field. In addition, this alternative has a moderate use of stubble [10] since the extractable liquid of the leaf is modest, leaving the rest of the stubble unused. In the same way, Piñar [30] corroborates that the energy extraction yields for stubble are low in number. In general, the low performance obtained in terms of the criteria previously established and the little research available on this topic proved it to be an impractical alternative for the team to develop a business model based on it.

3.3 Biorefinery

The use of the solid fraction of stubble for biorefinery is the sustainable processing of biomass to convert it into a variety of biocomposite products (food, chemicals, and raw materials) and bioenergy (biofuels, power, and heat) [31]. The potential use of waste in biofuels is currently under study at the University of Costa Rica's Center for Agronomic Research (CIA). Results of their research have been achieved from the use of the solid fraction of the stubble that is put under the hydrolyzate phase, followed by a lipid extraction phase to turn the stubble from biomass to biodiesel synthesis [18]. There is moderate access to information about this alternative in accordance with the research made at the UCR's CIA. Despite being an acceptable use of the raw material, it has disadvantages such as the fact that the current investigations carried out to obtain yields are much lower than those of other raw material options. According to Uribe [18], it is necessary to widely improve the process to obtain better results, which means that it is not a viable option to develop at a national level.

3.4 Chemicals

Among the alternatives for chemical products obtained from pineapple stubble are the extraction of alcohol, the creation of enzymes for the pre-digestion of food, and the development of biochar. The proteolytic enzymes of pineapple were used to predigest food in 1984; subsequently, work was carried out on the saccharification of pineapple stubble for the extraction of ethanol [17]. Moreover, the effort to harness the creation of ethanol continues at the National Center for High Technology (CENAT). This alternative is attractive to evaluate even if there is not much information about it in the market since the option presents a profitable opportunity related to the potential profit margin obtained with its final product [16]. The idea of chemical products produced from pineapple stubble has been developed mainly in predigestion and generation of ethanol products. However, in the case of this by-product, not much information is known about its commercial viability, compared to conventional paper. This limits the analysis of this alternative.

3.5 Compost

Another possible use for this waste is the production of compost through the aerobic digestion of microorganisms, which can be used as an alternative to mitigate the loss of the fertile soil layer that generates degradation and erosion in the soil due to the pineapple activity [14]. This alternative is being evaluated in terms of the use of the solid fraction of the stubble that would not be used in biogas generation programs such as the one currently being developed by the Costa Rican Electricity Institute [15]. This option allows the waste to be degraded for later use. Furthermore, for this alternative, market feasibility studies and marketing aspects are unknown. Composting from stubble is one of the treatment alternatives that is currently practiced by some pineapple farms. One of the great disadvantages that this alternative presents is the lack of studies supporting its viability as a compost that allows the improvement of the quality of the soil on which the stubble is reintegrated. Another fundamental element of the alternative is the little added value that represents a project that requires not only research but also an added value proposal for the use of the waste.

3.6 Textile

A textile fiber must meet three main characteristics: flexibility, fineness, and length, which determine the application and end-use for which they were naturally and chemically developed. However, it is important to create awareness and establish parameters that allow the measurement of the short- and long-term impact of textile fibers. It is precise because of this that the development of technologies for the creation of new fibers comes to life, and, as a result, Eco-fashion: a fashion trend designed with ethics and ecological awareness is born [32] Textiles with characteristics similar to animal leather can be made with the fiber extracted from the pineapple stubble that corresponds, approximately, to 2% of the pineapple waste [10]. Using this waste is more environmentally friendly than animal leather production since the latter generates a greater environmental impact. For example, for every metric ton of rawhides processed, 800 kg of waste is produced. Also, the production of 200 kg of leather needs from 45,000 to 50,000 l of water in addition to the amount of toxic chemical residues discarded in the process [33].

To create this textile, the cellulose fibers of the stubble must go through chemical, physical, and non-woven processes. The fibers are bonded together like silk for them to have an extremely flexible and resistant fiber, similar to that of conventional leather [11]. Furthermore, it is important to mention that pineapple has between 30 and 40 leaves, which later become stubble [10]. To produce 1 m^2 of this textile, about 480 pineapple leaves are needed: approximately 15 pineapples [11]. Regarding the market price, it is estimated that pineapple-based textiles can cost 18 €/m^2, while the traditional leather's lowest price is around 30 €/m^2 [12]. In the case of rope and textiles, information about the production methods developed by a professor at the Veritas University, Juan Guillermo Chica, is accessible. He has developed and tested different methods for producing textiles and rope from natural fibers. As Chica [34] indicates, the production methods of both products need very low technological complexity and low-cost machinery, especially in the case of agricultural rope, also, the products developed are good quality. This alternative can produce textiles as well as rope of different thickness, showing the technical feasibility and the potential for automation of both alternatives.

3.7 Agricultural Rope

In the agricultural sector, the use of ropes is required to provide support and protection to the crop. Commonly, the synthetic plastic rope is used as a tying and protection mechanism for crops, which must be discarded or recycled with time. There are natural type ropes such as hemp, cabochon, cotton, among others as well as synthetic type ones like nylon, polyester, polypropylene, polyethylene, among others [35] For the most part, synthetic ropes are used because of the cost and product access, specifically polypropylene. The natural fiber rope is biodegradable, which makes the alternative attractive for the business model project. This is why pineapple waste is used in the making of natural fiber rope. Finally, technical tests have been carried out for the development of this type of rope; however, the existence of market studies related to this alternative is unknown [10].

4 Selection

As explained in the methodology, different alternatives were compared in the case of Costa Rica, while considering the factors of information accessibility, technological simplicity, updated literature, and raw material by using a multi-criteria comparison matrix. These critical factors are the key to the development of a feasible project, taking into account the context in which the project should be developed. The comparison is shown in Table 3.

The results provided by the comparison highlight the significance of the two alternatives; textiles and agricultural rope obtained a threshold value of 3. The textile and the agricultural rope have a low utilization value of raw material because they only require the fiber in the pineapple stubble that represents approximately 20.3% of the total amount; these results are similar to Camacho and Mora [21]. These two alternatives have a common starting point since both require the extraction, drying, cleaning, and combing of fiber. Subsequently, as Chica [34] explains, the process focuses on each one of the stages. This demonstrates that workers on the pineapple fields can replicate the procedure with the correct machinery.

The advanced state of the art in these two alternatives is due to the research that experts from the Technological Institute of Costa Rica have done since they designed an agricultural rope along with pineapple farmers, mainly in the northern region of the country [21]. In the case of the textile, a researcher from Veritas University previously worked on the development of plain weave textile and leather-like material from natural fibers, including pineapple stubble fiber [34]. Nonetheless, neither one of the

Table 3 Multi-criteria comparison matrix

Alternative	Information accessibility (35%)	Technological simplicity (20%)	Updated literature (35%)	Raw material (10%)	Global score
Textile	4	4	4	1	3.70
Agricultural rope	3	4	3	1	3.00
Silage	3	3	2	5	2.85
Energy	3	3	2	3	2.65
Biorefinery	3	2	1	4	2.20
Chemicals	3	1	2	2	2.15
Paper	1	2	2	1	1.55
Compost	1	4	1	4	1.90

studies mentioned above developed a proposal considering an industrial approach nor a complete business model; this denotes a niche to design a business model that addresses the correct and viable waste management. The previous research groups supplied valuable information for the development of the project, increasing access to information.

In the case of the other alternatives, many were in the early research stage, diminishing not only the advancement of state of art but also the information access. Although at this stage, the silage and compost are not the main focus of this research, they should represent complementary alternatives for related projects because both have an elevated utilization of the pineapple stubble and require a simple manufacturing process without specialized personnel or machinery. Even though the Costa Rican Electricity Institute (ICE) made some experimental progress using the liquid fraction of the pineapple leaf stubble [15], energy production requires further study.

The final selection between the two most promising alternatives requires careful consideration of the context for the production of any of the two products. The plain acquired from the pineapple stubble resembles the appearance and texture of the linen textile; as pointed out by Chica [34] this leads to the research of the linen textile market in the country. According to the European Commission [36], the maximum annual consumption of linen in Costa Rica from 2011 to 2018 was €31 500, and Costa Rica mainly imports this material from countries of the European Union. This importation value makes it infeasible to produce a sustainable linen-like textile, so exporting to the European Union is part of the business model as this market focuses mainly on wearable linen textile (61%) and has an estimated linen textile market of around €128 187 192 [36].

On the other hand, the agricultural rope is extensively used for the production of national pineapple; in fact, the national productive pineapple terrain is approximately 44 500 ha [2]. In a pineapple farm in the northern region of the country, the usage of the agricultural rope is 16620.4 m/ha annually [23]. Under the supposition of a similar consumption of the rope in the country, the estimated market would be €1

103 104. The value of the final product is smaller for the rope; however, the process of the linen textile requires additional specialized machinery, which represents a more substantial initial investment.

As one of the main objectives of the project is to create a product that partially mitigates a national problem with a sustainable proposal, the usage of the pineapple leaf stubble as the agricultural rope is more attractive because it substitutes the polypropylene rope with a biodegradable alternative. The agricultural rope has a more eco-friendly impact since it replaces the polluting synthetic fiber that directly affects the production process. The rope has a lower initial investment because it requires less specialized machinery, which has a positive impact on energy consumption in the manufacturing process. This lower investment is more propitious for an innovative project that could expand later due to demand and profit. Therefore, manufacturing a type of agricultural rope that later can expand to generate some linen-like textile is the selected alternative to design the business model of this project.

5 Business Model

One of the main objectives of this project is the design of a business model that provides a real and profitable opportunity to benefit both pineapple farming and workers integrally. A proposal that involves the production department and the workers' association was created, taking into account the real limitations and opportunities found in the waste valorization process. Traditional approaches were employed; for instance, the CANVAS model that Osterwalder and Pigneur [22] developed. Nonetheless, the methodology was mostly an iterative, agile process based on the Lean Startup technique as explained by Ries [37], aiming toward an agile design in high uncertainty environments, through the design and feedback directly from the stakeholders. The proposal aims at supplying the annual and local demand for polypropylene-based agricultural rope in the sector of pineapple farming and intends to substitute this product with a biodegradable alternative. The model grants the establishment of a replicable and scalable plan because the amount of pineapple leaf stubble that is required to fulfill the local need for the production of agricultural rope is much smaller than the annual waste generation since it represents around 10.7% of the refuse (further detail in Sect. 6); the remainder of stubble can be manufactured progressively, expanding the market by selling this biodegradable rope to other agricultural sectors. In the case of the specific farm for this project, the monthly consumption of the agricultural rope is shown in Table 4.

At first, the proposed model consisted of transporting the leaf stubble to a processing location; however, according to the University of Costa Rica [3], the available amount of raw material produced per week is 10 ha or 2500 tons as the conversion factor is 250 ton/ha, which represents a logistics problem regarding the transportation of these materials. The pineapple stubble represents 70% of the total waste, and the stem composes the other 30% as calculated by Solano [23], leading to a result of 1750 tons available per week. Table 4 indicates that on average, the weekly

Table 4 Monthly agricultural rope consumption

Consumption (units)[a]	350
Equivalent rope (m)	740 250
Cost ($)[b]	3 384.50
Planted hectares required	73.1

[a]Units of 2105 m of polypropylene agricultural rope
[b]With 1 USD = 585 CRC as of 30 October 2019

demand is 81 (4.37 weeks/month), which translates to 131 183.5 kg of pineapple leaves stubble that would have to be transported (374.81 kg per rope unit, explained in Sect. 6); this amount would require a colossal investment in fuel to be delivered to a pickup station and later displacement to the processing plant, making the external processing plant an infeasible option for such a low priced product in a pioneering and high uncertainty new business in the country. This constraint limits the business model since it has to be carried out within the farm. The project is difficult to fully document in each stage due to different changes in this agile environment and to the feedback from the stakeholders. Figure 1 explains the final business model in a nutshell.

The designed business model provides an internal alternative that reduces the main risk of the process because the farm agreed on buying the biodegradable rope created; the worker's association through an in-situ factory will consume the farm's agricultural rope made out of the farm's waste—the pineapple leaves stubble. A key aspect of the process is that the farm will have an internal provider of the product being created out of the waste of their main activity (pineapple production), creating a circular product. The substitute product created will have the same presentation as the substituted polypropylene agricultural rope (length: 2105 m, diameter: 00.5 cm).

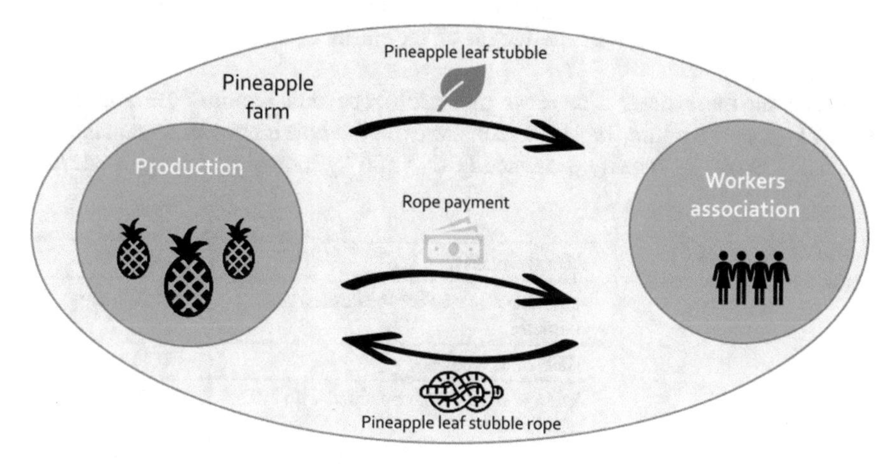

Fig. 1 Business model overview

A biodegradable alternative substitutes the polypropylene rope, creating an eco-friendly and economic product since the polypropylene rope requires recollection and cleaning processes before disposal ($4/unit). This benefits the members of the workers' association because the profits will be distributed amongst them, enhancing the living conditions in the community. This aspect highlights the bidirectional benefits of this business proposal since its advantages can be perceived in the social, economic, and environmental sectors of a given society.

The perception of pineapple farming among the Costa Rican population has been in constant decay due to different factors, ranging from inhumane working conditions to the destruction of the land caused by the excessive use of polluting chemicals; the effort of the farm to create a circular product with special care for environmental impact would greatly improve this corporate image by saving costs on the recollection and cleaning processes in manufacturing the rope.

6 Process Design

Table 5 shows the amount of raw material, leaves stubble, required to produce each unit of the agricultural rope.

The sample needed to obtain the results in Table 5 was 30 samples. The estimations are based on the average length as this measurement will not be highly dependent on the production process machinery but on the intrinsic fiber characteristics of the pineapple stubble. The decortication, combing, and drying processes affect the resulting fiber mass because this element can reduce residues of the plant and moisture more or less effectively depending on the machinery. Therefore, making mass conversions a less reliable measure amongst a variety of processes can only be fully analyzed if experimentation with each machine is performed. This discovery leads to the consideration of the length of the farms as a factor to determine raw material requirements.

Based on these data, it is clear that each rope unit requires the amount of 3750.81 kg. Furthermore, Table 5 demonstrates that stubble represents a scant amount of the 1 750 000 kg weekly produced (7 647 500 kg monthly); the required mass

Table 5 Raw material required

Average mass of 1 leaf	0.07 kg
Amount of leaves for the desired rope diameter	1.50 leaves
Mass of 1.5 leaves	0.11 kg
Average rope length obtained with 1.5 leave's fibre	0.59 m
Groups of 1,5 leaves for 2115 m rope	3569.60 groups
Leaves weight for 2115 m rope	374.81 kg
Leaves required for 2115 m rope	5354.39 leaves

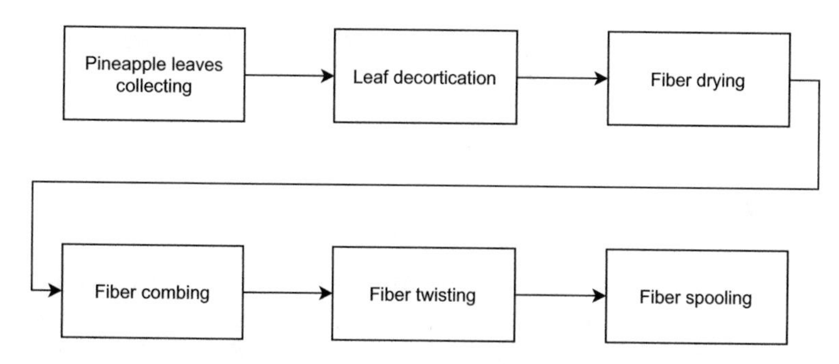

Fig. 2 Production process activities

corresponds to 350 units per month, demanding 131 1830.5 kg of leaf stubble and leading to 10.7% of utilization of the monthly availability. This level of production is sparse, but it depends on the consumption of the farm, which is relatively lower than the availability. Since the project aims for scalability, the expansion of the operations for posterior external sales enhances the span of this project.

The next step is the identification of the activities of the process, which are depicted in Fig. 2 [34].

The previous flowchart was used to select the different machines needed for the production stage. Each of the activities will be described in further detail to specify the design choices that led to the final process. The flowchart should help to determine the economic viability of the proposed business model to create a circular product. This project is the first documented example found of a design project with an industrialized approach.

6.1 Collection of Pineapple Leaves

A fundamental element of this proposal is that the leaf stubble is treated as waste and increases the investment in the different pineapple farms, which varies depending on the different techniques applied, ranging from burning (irresponsible farms) to composting (responsible farms), as explained by Solano [23]. In negotiation with the operations engineer of the farm and the workers' association, there is an agreement that if they assume the cost of transportation and agree on selling the product only to the pineapple farm, the members of the association will receive the leaf stubble for free. The last agreement relies on the thought that the pineapple farm would charge the employees that belong to the association if they profit outside the farm. However, further negotiation will be required to consider the appropriate distribution of the revenue and the price for the stubble.

The collection process will be carried out by specialized personnel that will be in charge of cutting and moving the leaf stubble to the vehicle (lent by the farm and with fuel and maintenance covered by the association) in three rounds because the space in the trucks is limited to 60 m^3.

6.2 Decortication

The decortication is the process of extracting the fiber from the leaf, in this case, this is performed by the selected Zhanjiang Weida Machinery Co pineapple fiber decorticator machine [38], whose main characteristics are described in Table 7. One of the relevant characteristics of this machine is that due to its production capacity, it is going to operate in three bursts, one for each raw material arrived.

6.3 Fiber Drying

Fiber drying is a pivotal activity to remove the moisture content from the fibers, and this process is performed in a drying area, which was designed to minimize the construction area, as the cost of construction is a limiting factor to maintain a low initial investment. This drying area was selected instead of an industrial one to reduce the costs because industrial dryers found had prices close to $60 000. This area is a multilevel, open, drying space with corridors to hang and collect the fibers from a cable with a roof at the top to maintain warmth and protect the place from the rain. The design consists of a basic unit of two parallel cables hanging from two poles for each extremity and the dual cable line separated by 30 cm, with a cable length to be determined by the optimization model. The observation showed that 54 kg of leaf stubble, processed as fiber, can be hanged in 15 m of cable. This modular element is the base for the design, alongside 10.2 m halls for the hanging and collection of the fibers.

This design is multilevel as it has four levels and fibers with a vertical length (half of the length and hanged from center) of around 30 cm (see Table 5), mainly to reduce the production area and traversal distance for hanging and collecting. The optimization was performed with the modular unit and the length of the elements, including halls. Due to the nonlinear nature of an area optimization problem, it was carried out in Microsoft Excel GRG Nonlinear Solver, as it is a nonlinearly constrained optimization problem. The following equations present the described model.

$$min A = L * W$$

$$L = y * z + 2.4\text{m}$$

$$W = x * 0.3\text{m} + (x + 1) * 1.2\text{m}$$

Table 6 Optimization results

Variable	Value
X	8 modules
Y	22 modules
Z	10.22 m
L	290.4 m
W	130.2 m
A	3860.29 m^2

$$\textbf{\textit{Restricted to}}: \quad x * y * z * 2 = 429.5\,\text{m}$$

where:

- A: Area of the drying space.
- L: Length of the drying space.
- W: Width of the drying space.
- x: number of modules in width direction.
- y: number of modules in length direction.
- z: length of the module cable.

These equations demonstrate that 54 kg of pineapple stubble (10.08 kg of fiber) could be hanged in 15 m of cable, the same proportion aids to determine the daily production requirements, where approximately 61840.37 kg of leaf stubble would be processed, hanging 17170.88 m. As the design proposes using four levels, the amount of cable required per level would be 4290.5 m, as expressed in the restriction for the model. The construction cost is assigned by the area constructed, so the optimization is carried out for planar construction area, where each level will replicate the arrangement of the optimal area found for the base; Table 6 summarizes these results.

Table 7 Main characteristics of the machines

Machine	Capacity	Power (W)	Required operation time (h/day)	Price ($)	Quantity	Total cost ($)
Decorticator	1 500 kg/h	7 500	3	11 000	1	11 000
Comber	300–700 kg/h	7 500	1	13 500	1	13 500
Twister	700 m/h	120	6	150	8	1 200
Spooler	180 000 m/h	1 500	0.2	1 350	1	1 350
Total						50[a]

[a] Value without transportation costs, included when performing the economical engineering of the project

6.4 Fiber Combination

Combing is one of the essential stages of the project as it removes the extra residues after the drying process, but this step mainly attempts to separate fibers that come intertwined; the process joins the end and the start of the different fibers. Moreover, if the fibers are carefully loaded in the machine, the outcome would be a uniform and continuous strand of fiber that can later be twisted into the correct shape to become the desired rope. The selected comber is the fiber comber from Zhanjiang Weida Machinery Co [38], as shown in Table 7. The output of the comber machine will be separated into 16 barrels that will later feed the twisting machines; these containers are production buffers because the capacity of the combing machines is greater than the capacity of the twisting machines. The twisting machines require two barrels because these machines demand two sources of fiber to twist them against each other.

6.5 Fiber Twisting and Spooling

No fiber twisting machine specifically designed to obtain pineapple fiber was found; hence, the machine used to process rice–wheat and hay stalk was modified to ensure its correct functioning, as Table 7 shows [39]. In the case of the spooling machine, the selected one is the spooling machine from Pang [40], whose characteristics are shown in Table 7. In the case of the spooling machine, the selected one is from Pang [40]; Table 7 highlights its characteristics.

6.6 Machinery

The machinery represents the most affordable option that can sustain the daily production; the results and production programming were performed with the data from the vendors.

6.7 Facility Layout

The facility will operate with four plant workers and one administrator to use the minimum workforce required, as the design is highly constrained by an increase in costs, which affects the product price. The design layout takes into account that a natural drying station would be a time-intensive activity that would separate the incoming flow of raw material from the outgoing finished product, which would be fed from the dried fiber on the last production day. This creates a U material flow;

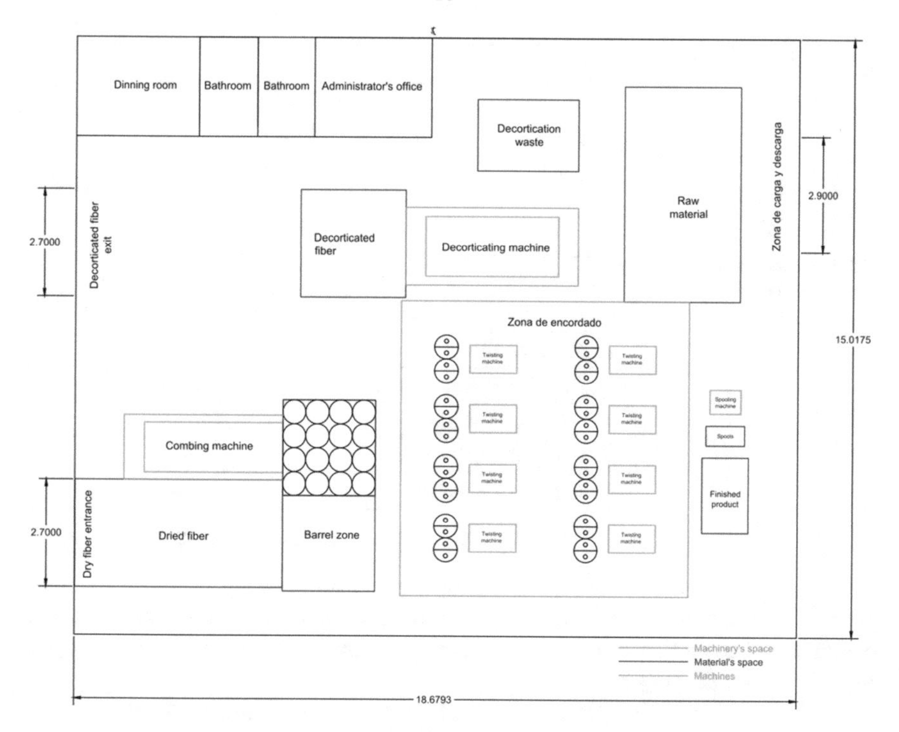

Fig. 3 Facility layout

as shown in Fig. 3, the layout respects at least the 10.2 m halls recommended in the country, and the extra space required for machinery manipulation. The facility is followed by the drying area on the left of the figure, but it is not included in this view of the layout.

7 Results

The analysis of the economic viability of the process designed was performed in much greater detail than presented, but the main aspects to be considered such as the initial investment, economic indicators, and conditions will be explained. The initial investment considered different construction prices depending on the area (production plant or drying space), which led to the final initial investment presented in Table 8.

Table 9 reviews the calculated costs of operation, salaries, electricity, and depreciation of the economic indicators of the project. The price for the final product is $17 whereas the polypropylene alternative costs $10; the price was selected so that the economic indicators resulted in positive because of the support of the company. The main reason why the company endorses the higher cost is because of the alternative

Table 8 Initial investment

Category	Cost ($)
Machinery	55 5570.74
Facility	194 0780.05
Total	249 50.78

Table 9 Economical evaluation indicators

Category	Value
Project's NPV without financing	$51 0860.83
Project's NPV with financing	$66 6660.09
DI without financing	10.95
DI with financing	10.26
IRR without financing	15%
IRR with financing	10%
Return period	9 years

price. The difference seems trivial to the company accounting for the new environmental benefits, and $4 savings per unit in recollection costs offered by the proposed rope as Solano [23] explains.

As described in Table 9, the project results are profitable with or without financing, but as the indicators show, financing 80% of the project is the leading scenario. The return period results influential, but considering the context in which this project develops, it gives a stable complementary source of income for the workers in the pineapple farm, leading to a more favorable socioeconomic situation. The design of the project was made alongside the operations engineer, the worker's association representatives, and the pineapple farm directives. The suppositions regarding raw material, transportation considerations, localization, and the price of the rope ($17) were reviewed and endorsed by both parts, giving a true implementation capability for the design.

In the case of the circularity of the product obtained, the selected indicator is the circularity indicator proposed by the McArthur Foundation [25], which is based on the flow of materials; other authors have also proposed different measures, as studied by Linder, Sarasin and van Loon [26]. These authors' arguments focus on the idea that several details of the production process are hidden in the numbers presented for the circularity of the index chosen due to confidentiality clauses, making these indicators unreliable under this condition. In the case of this investigation, no detail is private, and the aim is to provide as much information as possible and to give a more illustrative and direct approach to the circularity of the product, revealing key elements of the effective recycling of some steps concerning the process.

The mass of the finished product corresponds to the percentage of fiber contained in the raw material per unit (2%), obtaining approximate of 70.5 kg. The calculation of the material circularity index starts by calculating the mass of the virgin material required for the product, using Eq. (1).

$$V = M(1 - F_R - F_U) \tag{1}$$

where:

- V = Mass of virgin material used.
- M = Mass of the finished product.
- F_R = Fraction of raw materials from recycled products.
- F_U= Fraction of raw materials from reused products.

In the case of this research, the mass comes from a recycled product, which is the pineapple leaf stubble, leading to a virgin material mass that equals to 0 kg. The next element in consideration is the mass of the final product that cannot be recovered by reusing or recycling; consequently, this mass is called irrecoverable and is calculated with Eq. (2).

$$W_0 = M(1 - C_R - C_U) \tag{2}$$

where:

- W_0 = Irrecoverable mass of the final product.
- C_R = Fraction of the finished product that will be recycled.
- C_U = Fraction of the finished product that will be reused.

The totality of the finished product can be recycled as it is biodegradable, thus leading to 0 kg of irrecoverable mass in the finished product. The total mass of the waste generated in the recycling process is calculated with Eq. (3).

$$W_c = M(1 - E_c)C_R \tag{3}$$

where:

- W_c = Waste's mass due to the recycling of the finished product.
- E_c = Recycling process efficiency.

The recycling process is a natural aerobic decomposition that will be efficient for the recycling process, generating zero waste mass due to the recycling of the finished product. Another element in consideration is the waste mass-generated due to the process required to obtain the raw material, as calculated with Eq. (4).

$$W_F = M\frac{(1 - E_F)F_R}{E_F} \tag{4}$$

where:

- W_F = Waste mass-generated due to the process required to obtain the raw material.
- E_F= Efficiency of the recycling process to obtain the raw material.

In the case of W_F, the fact that the pineapple farm will take care of the stem does not guarantee the proper treatment of it; this corresponds to 30% of the pineapple stubble mass. This circumstance will be considered in a pessimistic case where the totality of the stems is mistreated, yielding a recycling process efficiency of $E_F =$ 70%, which gives a waste mass of $W_F = 30.21$ kg. In the previous data, the total waste W is calculated with Eq. (5).

$$W = W_0 + \frac{W_C + W_F}{2} \tag{5}$$

Using the previous calculations, $W = 10.605$ kg. The linear flow indicator (LFI) is determined with Eq. (6).

$$LFI = \frac{V + W}{2M + \frac{W_F - W_C}{2}} \tag{6}$$

The LFI, for this product's data, corresponds to LFI = 00.10. The utilization factor is calculated with Eq. (7) and is composed of two elements: product usage and lifespan.

$$X = \left(\frac{L}{L_{ave}}\right)\left(\frac{U}{U_{ave}}\right) \tag{7}$$

where:

- XProduct utility.
- LProduct lifespan.
- L_{ave}Similar products average lifespan.
- UProduct utilization in functional units.
- U_{ave}Similar products average utilization in functional units.

As the methodology states, one of the two elements should be selected to carry out the calculation. The functional unit selected is the amount of crop tying it supports. The lifecycle of the tying ropes used in this industry corresponds to one tie per usage. With this type of usage, the utilization of the product will be similar products, which means the utilization of X1. The utilization function is selected to balance the effect of a change in LFI with a corresponding change in utilization X, as defined in Eq. (8).

$$F(X) = \frac{0.9}{X} \tag{8}$$

The $F(X) = 0.9$ because the utilization X = 1. The MCI preliminarily is calculated with Eq. (9).

$$MCI_* = 1 - LFI * F(X) \tag{9}$$

This calculation is preliminary mainly because it depends on values that can turn negative; the definition of the MCI is given in Eq. (10).

$$MCI = max(0, MCI_*) \tag{10}$$

For this product, $= MCI_* 00.91$, meaning that the $= MCI 91\%$. This value in the Material Circularity Indicator means that the designed product, because of the material flow, is highly circular. According to this indicator, if the pineapple farm assures a proper treatment, the circular product will be obtained.

8 Conclusions

A replicable and scalable business model that provides social, economic, and environmental benefits allows the creation of a highly circular product (MCI91%) from the waste of the pineapple farming industry. This study provided the first methodological approach to a longstanding problem that Costa Rica faces due to the incorrect waste management of the pineapple crops, creating an opportunity for this sector and its workforce. As the research was conducted side by side with the farm, it not only constitutes theoretical investigation but also a practical study of the viability of implementing the presented proposal that was accepted by the executives of the farm and the workers' association. This study was conducted as an effort to solve the environmental and social problems in Costa Rica regarding pineapple farming.

Studies in different countries can broaden the perspective of the selection process because the context can change the alternatives and the final product. The appearance of new technologies and processes will considerably assist new designs, but further investigation is needed. Few documented research have been done on the valorization of pineapple leaf stubble in Latin America; this limited the amount of information available to conduct this study.

Practical implementations and study cases in the manufacturing and distribution processes of the proposed model would improve the body of knowledge in the field. Considering that fiber in pineapple stubble represents only 2% of the leaves, further investigations on the complementary alternatives that lead to more extensive usage of this stubble and particularly relevant research on machinery, process, and product quality would enrich this study. Alternatives such as silage and compost constitute suitable complementary alternatives for this proposal, but further examination is needed to determine the properties and conditions of these options.

References

1. Andrade F (2016) The challenges of agriculture. International Plant Nutrition Institute, Balcarce
2. National Chamber of Pineapple Producers and Exporters (CANAPEP) (2016). CANAPEP Statistics. https://canapep.com/estadisticas/. Accessed 4 Nov 2018
3. University of Costa Rica (2018) With an eye on the pineapple. https://www.ucr.ac.cr/noticias/2018/06/21/ucr-investiga-y-aporta-soluciones-a-polemico-cultivo-en-costa-rica.html. Accessed 15 Jan 2019
4. Arroyave C (2015) Trends in organic production and consumption in Antioquia. University of Medellin, Medellín
5. Ananas Anam (2018). https://www.ananas-anam.com/about-us/. Accessed 2 Sep 2018
6. Corlett R (2020) Agricultural waste: review of the evolution, approaches and perspectives on alternative uses. In: Global ecology and conservation. Elsevier. https://www.sciencedirect.com/science/article/pii/S2351989419307516. Accessed 10 Apr 2020
7. García C (2016) Circular economy and its role in design and sustainable innovation. UNIMAR Publishing Books, Pasto
8. Veloz C, Parada Ó (2015) Procedures for the selection of ideas and sources of financing for enterprises. UNEMI Science Magazine
9. Sapag N, Sapag R (2008) Preparation and evaluation of projects. McGraw Hill, Bogotá
10. Camacho D (2018) Uses of pineapple stubble. Dissertation, Technological Institute of Costa Rica
11. Serrano M (2015) The pineapple fiber-based leather that will revolutionize the textile world. In: VICE Media Group. https://www.vice.com/es_co/article/9b4jm5/el-cuero-a-base-de-fibras-de-pia-que-revolucionar-el-mundo-textil
12. Potts E (2015) Global textile trends round-up: fabrics from. Euromonitor International
13. Araya R (1998) Use of pineapple stubble (Ananas comusus) to obtain pulp for papermaking. University of Costa Rica, San José
14. Acuña O (2018) Pineapple waste: a headache for growers. In: With an eye on the pineapple. University of Costa Rica. https://www.ucr.ac.cr/noticias/2018/06/21/desechos-de-la-pina-un-dolor-de-cabeza-para-productores.html
15. Amador R (2018). Pineapple activity in Costa Rica. Energy potential of pineapple stubble. Dissertation, State Distance University
16. Bermúdez E (2018) Use of pineapple stubble. Dissertation, University of Costa Rica
17. Mata J (2018) Options for using stubble in Costa Rica. Dissertation, University of Costa Rica
18. Uribe L (2018) Pineapple activity in Costa Rica. Valorization of pineapple stubble under a bio-refinery approach. Dissertation, University of Costa Rica
19. Peña M (2018) Pineapple stubble silage: a viable option for agricultural producers. In: With an eye on the pineapple. University of Costa Rica. https://www.ucr.ac.cr/noticias/2018/06/21/ensilaje-de-rastrojo-de-pina-una-opcion-viable-para-productores-agropecuarios.html. Accessed 15 Sep 2018
20. Gamboa P, Vásquez C, Camacho C (2019) Criteria and weights for evaluating alternatives. Dissertation, University of Costa Rica, Technological Institute of Costa Rica
21. Camacho D, Mora R (2012) Production and industrialization of natural fibers obtained from the pineapple bush (Ananas comunus) grown in Costa Rica. Dissertation, Technological Institute of Costa Rica
22. Osterwalder A, Pigneur Y (2011) Generation of business models. Deusto editions, Barcelona
23. Solano L (2019). Business opportunity from pineapple stubble. Dissertation, Technological Institute of Costa Rica
24. Plattner H, Meinel C, Weinberg U (2019) Design-thinking. Finanzbuch Verlag, Múnich
25. Ellen MacArthur Foundation (2015) Circularity indicators. Ellen MacArthur Foundation. https://www.ellenmacarthurfoundation.org/resources/apply/circularity-indicators. Accessed July 2019
26. Linder M, vann Loop P (2017) A metric for quantifying product-level circularity. J Ind Ecol 545–558

27. Sapag N (2011). Investment projects: formulation and evaluation. Pearson Education, Santiago
28. Vaquiro J (2010) Investment payback period-PRI. Fut SMEs Mag 45–92
29. López M, WingChing R, Rojas A (2018). Bromatological composition of pineapple crown silage with citrus pulp, hay, and urea. In: Mesoamerican agronomy. University of Costa Rica. https://www.redalyc.org/jatsRepo/437/43743010004/index.htm. Accessed 15 Feb 2019
30. Piñar J (2019) Energy recovery of pineapple stubble. Dissertation, United Nations Program, UNDP
31. Hernández R, Prado L (2018). Biorefinery impact and opportunities of agricultural waste. UNED Res J https://revistas.uned.ac.cr/index.php/cuadernos/article/view/2059. Accessed 5 May 2019
32. Villegas C, González B (2013) Sustainable natural textile fibers and new habits of consumption. Leg Mag Archit Des. Autonomous University of the State of Mexico
33. Famielec S, Wieczorek-Ciurowa K (2011) Waste from leather industry. Threats to the environment. Czasopismo Techniczne 43–48
34. Chica J (2019) Production process for natural textiles. Dissertation, Veritas University
35. Álvarez B (2018) Orientation in the field and mountaineering techniques. In: Ebooks EUNED. https://ebooks.uned.ac.cr/product/orientacin-en-el-campo-y-tcnicas-de-montaism. Accessed 13 June 2019
36. European Commission (2019). Trade market access database. https://madb.europa.eu/madb/statistical_form.htm. Accessed 12 Sep 2020
37. Ries E (2011) The lean startup: How today's entrepreneurs use continuous innovation to create radically successful businesses. Currency
38. Zhanjiang Weida Machinery Industrial Co. (2019) Wei Jin Pineapple fiber combing machine. https://www.weidajixie.net/20160515/817.html. Accessed 1 Jul 2019
39. Alibaba Group (2019). Alibaba.com rice wheat hay stalk straw rope making machine for sale. https://www.alibaba.com/product-detail/Rice-Wheat-Hay-Stalk-Straw-Rope_60024956051.html?spm=a2700.galleryofferlist.0.0.23464337MkNOii. Accessed Oct 2019
40. Pang SQ (2019) Yarn machine winding machine traverse winder rope cone coil winding machine. https://www.alibaba.com/product-detail/Yarn-machine-winding-machine-traverse-winder_60820315334.html?spm=a2700.7724838.2017115.40.2aa55936ucnvv1&s=p. Accessed 1 July 2019

Fashion, Design and Sustainability. New Horizons in the Ways of Conceiving Production Processes

María Eugenia Correa

Abstract At the present, the market of goods that surrounds us reveals an urgent reality: the rethinking of the current production model. Hundreds of products fill shops waiting for their commercialization, inviting consumers to try and sense the shopping experience. But the reality shows that the products that overflow the stores exceed their own ability to acquire them, leaving in many cases the reality of discarding as a way of releasing products that have not been sold and thus being able to continue with the wheel of production and consumption. A truly alarming situation if we consider the level of pollution that this generates, and not only this, but also the manufacturing conditions that prevail behind. Specifically in the case of fashion, there is evidence of the need to review not only the current modes of consumption, but the conditions of production that operate behind the garments and accessories offered. In this sense, continuing with the parameters of a 'fast fashion' that season after season—or precisely, after micro-seasons—produces garments that exceed the consumption capacity, raises questions that lead us to reflect on this evident reality: How is it possible to sustain an adequate integration between production and consumption? How can it be produced taking care of the planet's resources, without affecting or compromising future generations? Is it possible to think of a sustainable design that respects the environment as well as the productive chain that shapes these goods? We understand that a design conceived from the logic of sustainability promotes a new and disruptive paradigm with respect to the one already installed, where prevail massive productions that do not usually contemplate the care of the natural and social environment. In relation to this, in the last decade, in Argentina, designers and companies have begun to conceive their services and productions from a greater social awareness, enabling new practices, such as the use of raw materials not harmful to the environment, the reuse of materials or recycling thereof. This conception highlights a clear new trend in product development. In this emergency context of these types of productions that claim senses of respect and care for the environment, as well as the human resources that enable their developments, projects are built on

---

M. E. Correa (✉)
Gino Germani Research Institute, University of Buenos Aires / CONICET (National Council for Scientific and Technical Research), Buenos Aires, Argentina
e-mail: eugeniacorrea@sociales.uba.ar

M. Á. Gardetti and R. P. Larios-Francia (eds.), *Sustainable Fashion and Textiles in Latin America*, Textile Science and Clothing Technology,
https://doi.org/10.1007/978-981-16-1850-5_11

designs that articulate sustainable practices, seeking to convey social awareness and values such as responsible consumption, decent production and fair trade. In this sense, this work aims to present cases of sustainable design enterprises of clothing and accessories developed in the city of Buenos Aires, and to analyse, according to these, productive modalities that promote sustainability, as well as the care of natural and human resources, configuring new models of responsible production and consumption.

Keywords Fashion design/designers · Sustainability · New productive paradigm · Ethic fashion · Responsible consumption

1 Introduction

At the present, the market of goods that surrounds us reveals an urgent reality: the rethinking of the current production model. Hundreds of products fill shops waiting for their commercialization, inviting consumers to try and sense the shopping experience. But the reality shows that the products that overflow the stores exceed their own ability to acquire them, leaving in many cases the reality of discarding as a way of releasing products that have not been sold and thus being able to continue with the wheel of production and consumption.

This is a complex situation if we consider the level of pollution that this reality generates, and not only in this ecological terms, but also the manufacturing conditions that prevail behind.

Specifically, in the case of fashion, there is evidence of the need to review not only the current modes of consumption but the conditions of production that operate behind the garments and accessories offered.

In this sense, continuing with the parameters of a "fast fashion" that season after season—or precisely, after micro-seasons—produces garments that exceed the consumption capacity, raises questions that lead us to reflect on this evident reality: How is it possible to sustain an adequate integration between production and consumption? How can it be produced taking care of the planet's resources, without affecting or compromising future generations? Is it possible to think of a sustainable design that respects the environment as well as the productive chain that shapes these goods?

We understand that a design conceived from the logic of sustainability promotes a new and disruptive paradigm with respect to the one already installed, where prevail massive productions that do not usually contemplate the care of the natural and social environments.

In relation to this, in the last decade, in Argentina, designers and companies have begun to conceive their services and productions from greater social awareness, enabling new practices, such as the use of raw materials not harmful to the environment, the reuse of materials or recycling thereof. This conception highlights a clear new trend in product development.

In this emergency context of these types of productions that claim senses of respect and care for the environment, as well as the human resources that enable their developments, projects are built on designs that articulate sustainable practices, seeking to convey social awareness and values such as responsible consumption, decent production and fair trade.

In this sense, this work aims to present cases of sustainable design enterprises of clothing and accessories developed in the city of Buenos Aires and to analyze productive modalities that promote sustainability, as well as the care of natural and human resources, configuring new models of responsible production and consumption.

The information was produced from a qualitative approach. On one hand, it was applied the technique of in-depth interviews to designers/owners/creators of the brands. On the other hand, some data were obtained from designers' social nets accounts—as Facebook or Instagram—and from their sites in Internet, as part of a virtual ethnography method. Also, some relevant information was generated from certain public interviews or talks given by designers at different events.

2 Rethinking Fashion Design in Terms of Sustainability

When talking about fashion, we inevitably think about the model that currently prevails in this industry, which aims at the unlimited production of garments and accessories that are offered in a market already saturated with products, for consumers who, tempted by marketing and advertising strategies, they buy without even stopping sometimes to think about whether they really need what they are buying.

This is, the current model behind the way we dress—and the way we consume and renew our cabinets with some frequency—responds to a "fast fashion" scheme where the times when garments and accessories are produced respond to productive canons that do not usually contemplate the ethical norms involved, both the decent working conditions or the materials used to make clothes. What does this imply? First of all, not considering respect as a rule when recognizing workers, based on wages and decent production conditions, allowing real and visible traceability, that contemplates from beginning to end the links of the productive chain and its ways of operating. Second, the materials with which it is made should be considered from an ecological perspective, of environmental care, avoiding harmful components or polluting residues of water, air and soil.

We talk about issues that are not covered—or if they do it, almost nil—by the prevailing model of production in the textile industry, which one, instead of privileging or considering, at least, this reality advocates producing not only for seasons but for micro-seasons, to guarantee a greater renewal of stock in less time. This situation promotes, in return, greater accumulation and therefore, greater discard, generating a large number of products that leave the market without being sold and thus its final destination, in the case of advanced economy countries, is often incineration, the landfill, or in the best of cases their transfer to emerging countries, where, sometimes, they are sold as sales.[1]

That is, in general terms, we are immersed in an unprecedented reality: tons of products become waste daily in people's everyday lives. This reality is clearly visible in the field of fashion, in which the seasonal productions of garments and accessories exceed the demands of the consumers themselves, generating surplus products that, after final settlements, find their destination practically as waste. As expressed by Elena Salcedo:

> Fashion is a system that attracts consumers to buy something new every season, sometimes with new and interesting offers, but most of the time with designs that are just a slight twist on the best-selling designs from the previous season. This system generates a need for constant change, continuous consumption and an increasing accumulation of clothing waste. [10]: 41

"Where there is design, there is waste" expressed Bauman (2015, in Cambariere [2]: 24) observing a production system that privileged—in contrast to the guiding principle of "less is more"—an incessant incorporation of new goods, hardly modified in their designs, which would allow them to be continually replaced.

A whole panorama of excess products in a market that is already saturated by a constantly expanding visual and material culture presents itself to us, leaving us the option, the question or the imperative need to build a better model. In line with this, Saulquin raises:

> At the same time that it is now essential to promote all those actions that seek to strengthen ties of collective action aimed at the common good, increasing attention to environmental problems begins, unthinkable a decade ago. Thus a new vision is installed that departs from the consumption of fashion and that made waste a constant. [11]: 44

This opportunity that we have to build as consumers, professionals in the area, producers, entrepreneurs, designers, in general, as citizens, attentive to environmental problems, a new consumption model—since it is ultimately what feeds and allows the production wheel to circulate—it is certainly significant, in a context as particular as the one we are currently experiencing.

I write these lines and we are—the whole world is—submerged in a pandemic caused by a highly contagious virus that has humanity on edge. One of the most important issues that we have seen in recent times, since we are facing this uncertain situation, is the real damage we are doing to the planet, to our land, with this unstoppable model of production and consumption in excessive and abusive terms. Damage that will take many, many years to be repaired.

---

[1] Such is the case of the HyM brand, which, in June 2019, began selling its products in Argentina, at The Luxury Outlet shopping centre in Luján, Province of Buenos Aires. The Luxury Outlet is a commercial space where clothing products, footwear and accessories from top brands that belong to previous seasons or present some faults, notorious or imperceptible, are sold. In the case of the HyM products that are sold in Argentina, they respond to "surplus stock that could not be placed in other markets", as mentioned in the note in the newspaper La Nación. "The HyM Brand arrives in Argentina in June". June 25, 2019. https://www.lanacion.com.ar/economia/la-marca-hm-llega-argentina-junio-nid2261272.

The need to care for the planet, the soil, the air, the water, our oceans, is urgent, it is real and vital. We need more than ever to adopt a new model of production in the textile industry, and mainly, of consumption. We need a change within the production system, according to the premise of sustainable design, to align ourselves with the sustainable development goals promoted by the United Nations. Change is urgent, and it is already beginning to manifest.

3 Sustainable Logic: Perspectives Beyond the Market

In the last decade, the word sustainable, as well as sustainability, has become more famous, referring to an emerging production model that claims the care of natural resources, so that they are available to future generations. As Lipovetsky and Serroy express:

> There are also ecological values, which, in the name of the Earth threatened by techno-commercial madness, call to stop the irresponsible consumer party. Faced with the dangers and catastrophes that lie ahead, an ethic of the future grows that demands the will not to compromise the living conditions of future generations. The primacy of consumerism in the present is thus stigmatized in the name of an ethic of long-term responsibility. [8]: 333

Now, in relation to the design of clothing and its conception from a perspective oriented to the care of the natural and social environment, it is worth noting, as we mentioned earlier, towards the beginning of the new millennium, the issue of environmental impact begins to gain more public relevance in Argentina, giving rise to the start of initiatives focused on the sustainable design of garments and accessories. At the same time, several companies, as a consequence of this greater relevance acquired by the sustainable theme, begin to incorporate the care of the environment as one more component that is integrated into their production modality and their added value offered to the market, specifically to the users of their products.

In this sense, we can express, in terms of Lipovetsky and Serroy, that:

> The time is witness to the appearance of a "durable design", in charge of creating a new industrial world. (…) natural materials, eco-objects, durable and recyclable products: it is the moment of biodesign, of sustainable design that no longer only raises the question of the conception of objects in terms of aesthetics and functionality, but also in terms of impact on the environment. [8]: 217

This is part of the growing concern at the global level about sustainable development, based on the specific problems that the growth of the world population implies and the unlimited use of resources that are limited. Given this, it is important to think that "the preservation of the environment and fundamental human values, [as well as] the notion of sustainable development, become increasingly visible concerns" [1]: 202. In this sense, "sustainable development represents growth capable of satisfying" the present needs without compromising the own needs of future generations (Definition of the United Nations World Commission on Environment and Development, 1987—"Brundtland Report"—in Bony [1]: 202. This report mentions that "development involves a progressive transformation of the economy and society" and highlights "social considerations on access to resources and the distribution of costs and benefits of development" [6]: 96.

It should be noted that the Brundtland Report represents the starting point for discussions on sustainable development worldwide, thus shaping an important political change around it (Mebratu 1998, in Gardetti et al. [7]).

In this scenario, sustainability has begun to be thought of as an element that considers the economic, social and environmental dimensions, configuring a new perspective regarding the development of products, not only with the emphasis placed on them but also on people, in the final users, and in their environment.

Among the objectives of the practice of sustainable design is included the idea of conceiving products which improve the quality of people's life, attending at the needing of the care of the natural and human resources required for their production. At the same it's important to conceive productive developments with a low level of harmfulness and contamination for the planet.

That is, "in the face of a diverse market, in which traditional practices survive, legitimized by an industrial model of wealth and accumulation—as well as exploitation—new models of development emerge, more conscious, and why not? More free" [3]. We refer to those cases in which an alternative to the prevailing way of producing is proposed, from the paradigm of sustainability, promoting dynamics of care of the environment, of the workers, betting on ethical creativity and innovation from the search for materials and technologies to produce.

In relation to this, there is an appeal to a productive dynamic that positively impacts the environment. Namely, techniques such as the recycling of discarded textile material, the reuse to give continuity to the life cycle of the product, garment or accessory, the preparation with sustainable and ecologically friendly raw materials, the planning of product cycles, contemplating their final destination and informing users of the possibility of reuse or recycling.

In the specific case of clothing production, as Saulquin puts it: "The really novel thing is having to think of each garment as a totality that must attend to the entire life cycle, from obtaining the material to its degradation" [11]: 90. This question is certainly new, as the author puts it, given that the post-use/consumption stage of the created product was not contemplated in the conception and production, until now, which also begins to be raised in the production planning stage, what to do with the product once its life cycle ends.

Thus, this new horizon focused on sustainable design proposes a certainly significant change in relation to the production of new practices that are installed in the productive scene. It is not only the practice of the producer that seeks to signify itself based on these new senses and principles to be contemplated but also that of the user, who in his daily use has a great impact on global warming and environmental pollution.

It is that, precisely, it is about thinking on this problem as a consequence of human action, and it is from the hand of these behaviours how it can be reversed, or, at least, mitigated. As Riechmann puts it: "The ecological crisis is not an ecological problem, it is a human problem. It's about anthropogenic climate warming, overconsumption of resources by human societies, mass extinction of species due to human behaviour" (Riechmann 2008, in Gardetti et al. [7]: 36). This is why both actors, both producers/industrialists and the consumers themselves, as members of a complex system that requires integrating a more contemplative and supportive dynamic with these processes, are appealed to review and update their practices, according to a more committed social and environmental vision.

4 Sustainable Fashion: Towards a New Paradigm of Awareness

As we have been pointing out, the need to think about new modes of production and consumption within the framework of current economies becomes a priority. In addition to the over supply situation, environmental pollution, toxicity as a result of current production methods, which pollute rivers, clean waters, air and soil, are added in order to discard their waste [5]. The issue of waste treatment has become a problem for manufacturing countries, and many of these "have found" the solution by moving their production plants—and its consequences, of course—to emerging countries. The reality of what happened at Rana Plaza in 2013 accounts for this.

Today little has changed for the textile sector since that tragedy, in terms of production carried out by large companies, but what has begun to gain greater visibility is the awareness that has begun to be created around the current production model in the textile industry and that begins to manifest between independent producers and consumers. According to Saulquin: "it can be affirmed that the massive and unscrupulous consumerism of fashions, which had elapsed since the consolidation of industrial society, is slowly evolving towards conscious consumption that begins to think about the common good" [11]: 45.

In this sense, various ventures have faced in recent year productions based on the philosophy of sustainable design as a way of assuming a new development model more committed to the social and natural environments. As Saulquin expresses: "As the current century progresses and complying with its guiding ideology, the importance of sustainable design is increasingly established among independent creators, with productions based on ethics and social responsibility" [11]: 90. Thus, an incipient—but gradually increasing—number of designers have assumed the commitment to orient their projects towards a more conscious look at the resources used, their value, and the effects generated by a type of non-responsible production.

That is, we are talking about new patterns and modes of production, attributable to the new times, to the new demands of a market, and precisely of a constantly changing consumer public. As Lipovetsky and Serroy reflect:

> The intensification of competition and new consumption expectations have translated into the advent of a post-Fordian economy characterized by the imperative of innovation and hyper-diversification of products. (…) But it is an expanding logic that must be increasingly integrated into the ethical dimension of respect for the environment, and this parameter is

new. After the age of carefree creativity, that of eco-responsible creativity prevails or will prevail. [8]: 189–190.

It is, in this context within which a new consciousness begins to be assumed, where a new disruptive design begins to emerge that proposes a discourse according to the current time of greater awareness of environmental problems. In line with this, we can argue that "Eco-materials, ethical production and eco-efficiency have been popular themes in the textile industry in recent years. Moreover, at the beginning of the twenty-first century, several designers have made use of the concept of reuse and redesign in designing trendy products. However, a new sustainable mindset is still waiting to emerge at large, as we continue to design and manufacture textiles and clothing mainly in traditional ways. As Fletcher (2008: 121) describes the current situation, "it uses yesterday's thinking to cope with the conditions of tomorrow"." [9]: 1878.

In line with this, we can propose, in Ezio Manzini's terms: "Yes, it is true that the gravity of the environmental problem is already too evident, as well as the limited nature of the biosphere, (…) it should lead us to a profound change of our culture of the project, of producing and of consuming." To which he adds: "The uncontrolled and uncontrollable production of forms without reason and the increase in semiotic contamination could be contrasted (…) with new territories to investigate, new horizons of common sense to adopt and new praxis. That is why today we are obliged to rethink the parameters of the so-called project culture" (Manzini 1992, in Cambariere [2]: 12–127).

Thus, in this context of the need for new actions, several ventures came to light with the firm idea of contributing to practices and messages in favour of caring for the planet, the environment and resources. Among these are designers who have started their projects aimed at the development of clothing seeking to strengthen this "sustainable mindset", still incipient, or at least, seen with more commitment in certain small companies. This, because, even the great machinery or the great industry has not incorporated major changes in their traditional productive modalities, as we mentioned in the previous lines.

Given this, it should be noted that, although designers who opt for this modality still correspond to a minority, a growth of this type of projects begins to emerge, born with the aim of developing productions cared for in ecological terms and working conditions.

Thus, we observe among these, the cases of sustainable design clothing and accessories produced by the firms Cubreme, Animana and Maydi, developed in Argentina.

In the first case, we refer to Cubreme brand, by the designer Alejandra Gotelli, who has been working with lines of textile garments and accessories in natural and organic fibres since 2005 and following the philosophy of fair trade. Coats, sweaters, blankets, accessories and textile pieces for the house are produced under the brand's sustainable production modality.

After some years living in Brazil, Alejandra returned and started the beginnings of Cubreme in Argentina, by the time she became to see a reality "that did not close" in the textile industry. This is how she began to dive into the world of fibres and see how she could develop a very low-impact national product: "I was clear that I wanted to present a different proposal than the conventional one" (Personal communication with Alejandra Gotelli in October 2017).

Her work refers to an extensive career based on the study and research of textile materials, camelid and goat fibres, wool, with small producers, as well as an intense study on the ways of making, in relation to the loom and its various variants, in order to build the production chain. Thus, fibres began to be part of her know-how, and at the same time, she became interested in the consequences of traditional cotton cultivation, the amount of agrochemicals and the use of pesticides, with their damage to the soil and the environment.

"The disaster that we are carrying out from the industrial revolution to this part with the form of production that we have, everything has been to worsen a productive circuit that in the textile sector is truly more what hurts than what it generates. All this came to my soul." Alejandra says. This is how Cubreme emerged, appealing to the nobility of natural fibres, the purity and freshness of its genuine fabric, the respect of the workers and their working conditions, taking care of nature and its times.

As Alejandra expresses: "In Cubreme, the last 'e' is one and inverted and that because it is a return, this is not something that is new, I think we must return, that we are wrong and we must be aware of that we have to rethink things differently, so 'e' is a return, a return to social equity, environmental balance and business ethics."

The materials applied in the confection—both of the clothing and interior design—are natural, from Argentine llama and sheep, Guanaco from Malargüe, alpaca from Peru, 100% organic cotton, among others, and they are worked from of artisan techniques developed by labour whose value is recognized under fair trade. In other words, the development of products with non-polluting, natural materials that do not use artificial dyes in their processes are integrated into this production, and at the same time, weavers and loom workers are trained to make clothing, recovering artisanal work techniques (Figs. 1 and 2).

Her project appeals to consumer awareness of the environmental impact, offering products from a line of natural materials, made from artisanal work, careful, respected and respectful of production times, and demanding manual production techniques, like the loom. Aware of the need to offer consumers more conscious alternatives in terms of clothing, it was proposed to generate garments for this purpose:

> "What interests me most is creating that possibility of choosing something that you create. In other words, if you really believe that there may be a change, know that you have an alternative to that change that you want. Well I want to give the customer that option." (Alejandra)

At the same time she adds:

> "Do you want to wear a 100% national sheep sweater made here by people who are paid well, who are developing a trade, who like what they do, who work with love, who know the final product and not the stages not knowing where their work ends, that the label describes the history of the product behind, the origin? If you want to dress like that, if your jean doesn't weigh on your conscience, you don't know where it came from and everything that jean produced… well, Cubreme gives you that alternative." (Alejandra)

Fig. 1 Alejandra Gotelli, untitled photograph, 2016. From cubreme.com

Fig. 2 Alejandra Gotelli, untitled photograph, 2016. From cubreme.com

Here, reference is made to the consciousness that is beginning to emerge in recent times, regarding current production patterns, the damage that they produce as part of a system that does not contemplate the times of nature, or its own cycles. In this sense, as Entwistle expresses:

> From its beginnings to date, the textile industry accelerated many of its production processes. In the agricultural sector, accelerating planting, cultivation and harvesting and modifying their cycles; in the textile production area, handling the fibers with some speed, using machinery for dyeing, drying and finishing textiles. (…) Even in the area of clothing, processes such as washing, drying and ironing of textile pieces have been accelerated by the use of machinery for this purpose. (Entwistle, 2002, in Gardetti et al. [7]: 100–101)

Similarly, it is stated that "fashion, as it unfolds today, does not respect the speeds of its environment and is largely disconnected from the effects it produces in nature" (Fletcher 2008, in Gardetti et al. [7]: 101).

In counterpart to this model, the proposal of these ventures is oriented to the care of resources, of nature, to rescue and value working time, of ancestral production techniques. As stated by the brand on its site:

> "Cubreme is a decision to create beautiful, simple and enduring designs using pure and noble textiles. Valuing its origin and nature in each step of the process and finish."
>
> "We work on the traces that traditional techniques left us. The hands with their skills make possible a careful production."

This view values the working time, the vindication of local and ancestral production techniques, justly invites us to think about this more careful and respectful development model of manual work, making it a value in itself. As Cambariere puts it: "A new paradigm appears in the world of design that promotes and advocates what they call the 'essential luxury' or 'new luxury' accompanying the slow movement (cultural movement that proposes to slow down human activities to curb the tyranny of vertigo with which we live)" [2]: 136–137.

Similarly, Miguel Ángel Gardetti, director of the Center for the Study of Sustainable Luxury, in relation to this concept, states that: "It is about returning to the essence of luxury with its ancestral meaning, to thoughtful purchasing, to artisan manufacturing (by hand) and the beauty of the materials in its broadest sense" (Gardetti 2011, in Saulquin [11]: 60).

In this sense, we can state that: "Luxury begins to be linked to new values, such as the work process, craftsmanship, [the purity of] materials, non-contamination, transparency that a brand can provide regarding its productions. Those aspects are increasingly highlighted and valued by new consumers" [4]: 42.

In this same line of thought, is the case of Animaná, a brand founded by Adriana Marina in 2009, in which garments and accessories are made with natural materials from the Andes and the Argentine Patagonia. Born in the south of the country, she decided to build a sustainable project. From a joint work with weavers, artisans, llama and alpaca keepers and small companies in Patagonia, she built an alternative model for the fashion industry. The brand offers sweaters, cardigans, ponchos, shirts and more clothes of high quality. The products are sold in stores in Buenos Aires and in Paris.

It should be noted, as stated on its site, that "animaná established itself as a sustainable luxury fashion brand, which demonstrates transparency in the production and marketing of camelid fibers". The company works especially with southern camelids:

> "We use guanaco, alpaca, llama and vicuña. On the one hand, our consumers look for quality and in our garments they find that expectation together with design. On the other hand, we have buyers who are interested in the story, they want to know what is behind the product." (Adriana Marina).

At the same time, Adriana is the founder of the NGO "Hecho por Nosotros", through which she promotes the model of sustainable development and systemic change in fashion production, together with a team of academics and professionals, lecturing and reflecting on this thematic and the consequences of the current production model in the textile industry. Adriana also works with the United Nations to deepen the transformation of the paradigm of sustainable fashion.

The principles that Animaná promotes are fair trade, the use of local raw materials and natural fibres, an ecological process in the manipulation of resources and transformation of raw materials as well as improving the quality of life of local communities and respecting the principle of no child labour, all instances that articulate greater care of resources, both natural and human, when producing.

As in the case of Cubreme, Animaná is defined by its creator as "a social company that was born to provide an alternative to the fashion industry and its methods of production and consumption". In this sense, both businesswomen share the vision of being able to offer something different in terms of consumption, products that recreate a new history, which have more transparent production modalities, which make visible the people involved in the "doing" behind the ready-made garments or accessories.

In relation to this, the positioning of the Animaná brand in the luxury goods market, or specifically in this case, of sustainable luxury, allows us to think about the transformation of this concept in recent times, as mentioned above. As Saulquin maintains: "from 2000, with the important ideological change, a shift occurs from the meaning of "luxury" as accumulation to its new resignification as transforming power" [11]: 59. In this sense, the concept of luxury is now thought of as a model or pattern of consumption that—although it is associated with exclusivity and distinction—also begins to be linked to these new values that begin to emerge strongly as a product of the new consciousness regarding the current production and unlimited consumption model.

Thus, we can think, in Gardetti's terms, the relationship between this new conscious paradigm of production and consumption with the sense of greater equity and transparency in the production line, as well as greater care for the resources transmitted by the "new luxury". This is:

> If sustainability integrates a principle of justice and social equality and some consider luxury an excess, it cannot be moral while poverty exists. (…) That is why, in the future, the highest quality products or services will be those that generate the greatest benefits to those who are involved in their entire value chain. (…) This deeper and more authentic approach to luxury

> will require companies belonging to this sector of a truly excellent social and environmental performance, which implies starting with a process of internal change, encouraging business practice sustainable in all areas of the organization and its supply chain. (Gardetti, 2011, in Saulquin [11]: 60).

In this way, greater traceability around the value chain will allow production to be "whitewashed", appealing to a development model whose principles of decent work, transparency and equitable redistribution of income are reflected in what is called a production model of "clean clothes", in which no worker suffers from conditions of exploitation or slavery, and their participation in the productive dynamic is revalued. As we mentioned, fair trade is one of the fundamental pillars of these sustainable design ventures, as mentioned by the founders of Cubreme and Animaná.

Fairtrade, born in the Netherlands in the 1950s in order to raise awareness regarding the unequal nature of exchanges between North and South [2], promotes a type of exchange based on transparency, respect and equity. It fosters sustainable development that favours better conditions for small producers and workers in unfavourable situations. Thus, among the principles promoted by fair trade are:

> An income distribution that ensures fair and decent working conditions; equal income for women; encouragement to care for the environment and sustainable productive activities; the payment of a fair price to producers through, for example, minimizing the chain of intermediaries, and a commitment to improve their living conditions, since the main objective of the movement is poverty reduction. [2]: 129.

The objective of being part of a work style that respects fair trade, of contributing to improve the quality of life of local communities, has been sustained and promoted by Animaná since its beginnings. Also, another issue that stands out, as we mentioned, is the care of ties with nature, respect for animals, evidenced, for example, in the way of obtaining the fibres from the camelids, with which they make their clothes, avoiding the suffering of these.

The balance reached in all product development is based on respect and care for the environment, whether it is the artisans themselves, weavers, workers who intervene in the process of obtaining the fibres, as well as respect for the life of animals. Valuing local communities, integrating their knowledge and traditions to the garment making process, promoting fair trade, caring for the life of animals, nature, are actions and messages that seek to be transmitted by the brand and its way of working.

According to this view, Adriana expresses: "I am convinced that by making sustainable garments we can reduce not only the environmental impact, but also recover the value of ancient techniques, respect the rights of artisans while creating unique designer garments".

In line with these working models and sustainable developments, we find the case of Maydi, a company founded by designer María Delicia Abdala-Zolezzi—alias "Maydi"—based on understanding the unique value of the fibres of the camelids of Argentina and their needs to revalue them.

After training in England, where she obtained her Diploma in Fashion Design at the London College of Fashion, Maydi lived in France and Italy, where she worked for different highly recognized fashion firms, where she learned not only about the artisan work that involves the development of high-quality garments but also product.

Maydi's career continued with her return, in 2012, to Argentina, where she began her sustainable design venture based on the manufacture of high-quality artisan fabrics. The research on sustainable materials and the quality of the fibres of the northern camelids led the designer to think of a productive development model based on respect for biodiversity, nature, its cycles and working times. At the same time, she set out to claim artisan work and fair trade, pillars of her brand.

The fabrics are made by women whose knowledge and skills in different techniques—loom, two needles and crochet—allow the designs to be translated into unique pieces, made of excellent materials and compositions. "My weavers are really artists. I work with women who come from generations of weavers and dressmakers" says the designer" (Personal communication, May 10, 2019). The pieces created refer to coats, sweaters, ponchos, dresses, and other, made with the purest fibres of camelids from the north of Argentina (Jujuy) (Figs. 3 and 4).

Among the materials used to make the fabrics, the fibres of guanacos, alpacas and llamas, and merino-organic wool and mohair stand out. The products obtained are high-quality fabrics, due to the composition of their raw materials, which are exported to countries such as Italy, Japan, France, among others, responding to the high standards of demand requested. Accordingly, Saulquin maintains: "The productive potential of camelids in Argentina, given their competitive and comparative advantages, transforms these farms into an excellent sustainable resource." [11]: 51.

The need to rethink and revise production patterns within a fast-changing industry led Maydi to create, like Gotelli and Marina, timeless lines that can be offered at any time of the year. The work carried out together with the artisans, loom makers and weavers result in unique models, which include noble and very high-quality materials, in natural colours, of high durability. Betting on a development committed to nature and animals, the brand managed to position itself, like Cubreme and Animaná, in the sustainable luxury market, based on values that promote care and respect for resources.

It is important to mention that the company has obtained the WildLife Friendly certification as a result of its contribution to the care of the species and biodiversity of the areas with which it works. Because as the designer expresses: "Everything behind it is important. People who weave in their homes, even the treatment and handling of animals. It is not only fashion. It is a lifestyle. Sustainable is a lifestyle" (Maydi's talk at Lento café, from AMSOAR—Argentina Sustainable Fashion Association- on 4/4/2019).

It is not about conceiving sustainable as a project, but as a way of expression, a way of life. In this sense, the creators of these sustainable fashion design companies bet, with conviction, on a new model, based on the integration and adoption of new values, in harmony with conscious life, with decent work and greater transparency.

In these cases, we observe how a new, disruptive model with the prevailing production system becomes possible and opens a new path in the fashion industry. These companies, and their ways of carrying out their productions, allow us to think about the possibility—become a reality—of changing the current ways of doing design. From the integration of different factors: the choice of noble resources, by opting for organic, pure, durable raw materials, high-quality fibres, for their clothing. At

Fig. 3 Maydi, untitled photograph, 2020. (All rights reserved)

Fig. 4 Maydi, untitled photograph, 2020. (All rights reserved)

the same time, they seek to replace and revalue production techniques loaded with inherited knowledge, respecting working times and the modalities of co-participation, co-creation, which lead to sustaining a fair trade model, including and making visible the agents that participate in the productive dynamics.

It is a path that opens not without obstacles, full of challenges, but that bets for a real, transformative change, towards authentic and transparent processes in the textile industry and the world fashion system.

5 Final Thoughts

In recent years, towards the beginning of 2000 onwards, production developments aimed at sustainable design as a way of producing clothing began to take place in Latin America, and specifically in Argentina,—this process occurred somewhat earlier in some countries in Europe—from a new perspective, more attentive to environmental problems and their effects on societies.

Although, as I mentioned earlier, projects oriented towards caring for the environment are not the majority, the fact that several of them have started on this path, configure a panorama of action that raises a new work paradigm: conceiving products from a sustainable view, appealing to corporate social responsibility, to raising awareness of both: the production processes that are involved and their impact in environmental and social terms.

In this way, a new production model is configured on a sustainable logic whose ecological and ethical awareness principles are integrated with a largely inclusive purpose and respect for the environment. Henceforth, the projection of productions aimed at promoting these principles of greater respect and socio-environmental care, seeks to contribute to the satisfaction of their own needs from a greater social awareness of what is consumed, how it is consumed and how it is produced, registered at a time of greater questioning to the producers about the way in which their products are made.

In this sense, it should be noted that users, from the circulating information about the ways of producing and manufacturing existing products, are expected to adopt a greater commitment and awareness not only about the environment and the impact that productions generate on it but also in terms of manufacturing modalities and developments, based on the working conditions assumed by the companies. It will be necessary, then, to begin to get involved in responsible consumption issues, not as consumers, but as responsible citizens and aware of the production modalities that intervene in what we consume. As Gardetti (2017) says, "A sustainable society is not possible without sustainable individuals".

It is interesting to think about how these new spaces are opened within a field with a productive tradition anchored in practices inherited and reproduced by the actors who compose it. This productive dynamic that drives fast fashion—unethical working conditions, unbridled consumption, new productions that constantly enter a market saturated with products, among other instances—is sustained and reproduced by the actions of tradition. In Williams' terms, it acts as "an actively configuring force, since in practice tradition is the most evident expression of the dominant and hegemonic pressures and limits" [12]: 137.

But inside this hegemonic system, this scenario of sustainable fashion highlights the incipient emergence of a new consciousness, an awareness of what is produced, how it is produced, its impact, its consumption. An awakening that reflects how we live and how we can orient ourselves towards a way of life more respectful of our environment, our land, our planet, taking care of the available resources, so that they are still available for the next generations.

The cases of Cubreme, Animana and Maydi evidence how the change can take place and can be a real possibility. Alejandra, Adriana and Maydi came to their dreams a reality, and work for a more conscious, responsible and ethical consumption. The joint work with artisans, weavers, animal keepers, and other agents, refers to collaborative work, so important at the present. Cooperation is the key to reinforce our ties, to turn society into a solitary one.

To end this article, I would like to share a fragment of the book "Ethical fashion for a sustainable future", written by Elena Salcedo. It says:

> That awareness and that knowledge are what have led companies and consumers to go one step further and change the course of the situation, seeking more sustainable solutions. Other initiatives confirm that it is possible to make fashion in a different way, that a better fashion is possible and, therefore, that it is definitely possible that a new fashion exists in the 21st century: the more sustainable fashion. [10]

Definitely, the change to a more sustainable world is underway, at a slow pace. Not without obstacles, but it is happening.

References

1. Bony A (2008) Le Design. Histoire, principaux courants, grandes figures. Larousse, Paris
2. Cambariere L (2017) El alma de los objetos. Una mirada antropológica del diseño. Paidos, Buenos Aires
3. Correa ME (2019) Diseño y sustentabilidad: hacia un nuevo paradigma en el campo de la moda. In: Zambrini L, Lucena D (ed) Costura y cultura. Aproximaciones sociológicas sobre el vestir, 1st edn. EDULP, Buenos Aires, pp 213–231
4. Correa ME (2018) Creación, subjetividad e innovación en el diseño independiente actual. In: Wortman A (ed) Un mundo de sensaciones. Sensibilidades e imaginarios en producciones y consumos culturales argentinos del siglo XXI, 1st edn. CLACSO-IIGG, Buenos Aires, pp 27–55
5. Entwistle J (2014) Sustainability and fashion. In: Fletcher K, Tham M (ed) Routledge Handbook of sustainability and fashion. https://www.routledgehandbooks.com/doi/10.4324/9780203519943.ch2 Accessed 20 Jul 2018
6. Fanelli JM (Comp.) (2018) Desarrollo y ambiente en la Argentina. Siglo XXI, Buenos Aires
7. Gardetti MÁ, Luque D, Lourdes M (2018) Vestir un mundo sostenible. LID, Buenos Aires
8. Lipovetsky G, Serroy J (2015) La estetización del mundo. Anagrama, Barcelona
9. Niinimäki K, Hassi L (2011) Emerging design strategies in sustainable production and consumption of textiles and clothing. J Clean Prod 19:1876–1883. https://doi.org/10.1016/j.jclepro.2011.04.020

10. Salcedo E (2014) Moda ética para un futuro sostenible. Guatavo Gili, Barcelona
11. Saulquin S (2014) Política de las apariencias. Nueva significación del vestir en el contexto contemporáneo. Paidos, Buenos Aires
12. Williams R (2000) Marxismo y literatura. Península/Biblos, Barcelona

Incorporating Consumer Perspective into the Value Creation Process in the Fashion Industry: A Path to Circularity

Eliane Pinheiro, Rodrigo Salvador, Antonio Carlos de Francisco, Cassiano Moro Piekarski, and Anthony Halog

Abstract Sustainability in the fashion industry is a contemporary megatrend that needs new studies addressing the myriad of possibilities that exist to reduce environmental impacts. Nonetheless, it is essential to take into account the opinions/wishes of the segments targeted by the business, a practice that has been neglected by textile and fashion companies. Value offers with greater circularity in textile and fashion businesses have suffered from a lack of consideration of the consumer's perspective in their value propositions. Therefore, the objective of this research is to suggest circular strategies for the incorporation of the consumer's perspective into value creation in the fashion industry. The methods used in this chapter encompass four steps: (i) building a theoretical background related to sustainable and circular fashion, and circular business models linked to the textile and fashion industry, which aided; (ii) building a survey to investigate issues on product design, value creation, and consumer participation in fashion companies in Brazil; (iii) identifying the main deficiencies on product design, value creation, and consumer participation in fashion businesses in Brazil; and (iv) proposing circular strategies to seek overcome the deficiencies identified. Twenty companies from three states in Brazil were surveyed. The results of the

E. Pinheiro (✉) · R. Salvador · A. C. de Francisco · C. M. Piekarski
Sustainable Production Systems Laboratory (LESP), Department of Industrial Engineering (DAENP), Post-Graduate Program in Industrial Engineering (PPGEP), Universidade Tecnológica Federal Do Paraná (UTFPR), Curitiba, Brazil
e-mail: epinheiro@uem.br

A. C. de Francisco
e-mail: acfrancisco@utfpr.edu.br

C. M. Piekarski
e-mail: piekarski@utfpr.edu.br

E. Pinheiro
Department of Design and Fashion (DDM), Universidade Estadual de Maringá (UEM), Maringá, Brazil

A. Halog
School of Earth and Environmental Sciences, The University of Queensland, St. Lucia, 4072 Brisbane, Australia
e-mail: a.halog@uq.edu.au

M. Á. Gardetti and R. P. Larios-Francia (eds.), *Sustainable Fashion and Textiles in Latin America*, Textile Science and Clothing Technology,
https://doi.org/10.1007/978-981-16-1850-5_12

survey allowed identifying gaps in the knowledge of fashion companies in relation to consumers and also with regard to issues that involve the product's life cycle from the conception, use, and disposal of post-use products by the consumer. Recommendations were proposed to incorporate the consumer's perspective into creating value in the fashion industry, seeking to pave the way to more circularity-oriented business models in the textile/fashion industry.

Keyword Consumption · Value creation · Fashion industry · Circularity · Circular business model

1 Introduction

Sustainable fashion is a term that is associated with the efforts of industries to adapt to the growing use of resources and assist in the challenge of minimizing, not generating, or reusing waste. Sustainable fashion is associated with the issues of material supply, working conditions of suppliers, use of chemicals, energy and water, transport, and, recently, attention to post-consumer waste [2, 12, 52].

These issues are linked to what is proposed by the circular economy, which tells organizations to comprehensively rethink how resources are managed in order to improve financial, environmental, and social benefits, both in the short and long terms [6], and in that context, it is emphasized that a CE in the fashion industry needs to focus on solving the concerns associated with the generation of textile waste [55].

In view of the necessary changes toward sustainability and fashion, it is important to understand the consumer, knowing their needs and desires more deeply in order to offer clothing with a longer use time or stable satisfaction with the product [40], as well as to promote the purchase of products developed in companies with sustainable fashion standards, and to recycle consumer product wastes [46].

On those notes, linear production and consumption patterns in the fashion industry (e.g. fast fashion) have been the causes of great concern. Fast fashion alone causes consumers to dispose of clothes that have not yet been worn out [4], many times just for growing tired of them [10]. Production and consumption in the fashion industry have led to consumer perception of clothes as disposable items [1], and these clothes are likely to end up in landfills or be incinerated [3]. It needs to be noted that many of these clothes are in perfect condition to be reused, others could be repaired, refurbished, repurposed, or recycled. However, for that to happen, those clothes need to be put in new life cycles, as there is still great value to be recovered from those resources. Options to recover such value can be found by establishing circular business models (CBM). CBMs allow setting circular strategies considering a business perspective [48].

Nevertheless, warnings have been raised that the fashion industry needs to tackle consumer engagement by offering higher value, promoting offers that capture consumer needs and translate them into circular products or services, and consequently make consumers help manage fashion and textile systems of production and consumption. However, Lacy and Rutqvist [29] highlight that offering products or services without being sure that the capability or technology the company offers adds value to a continuous consumer engagement is a mistake.

The role of the consumer is crucial in the transition to a circular economy [47]. Consumers are key stakeholders for creating sustainable value in more sustainable business models [8], and it is observed by Lacy, Long, and Spindler [28] that for circular fashion to take off, there is still a need for consumer pull.

The consumption of fashion products is often driven by status, the perception of effectiveness and happiness has a significant effect on consumers' attitudes toward environmentally responsible clothing [54], and consumers' environmental or ethical concerns do not always translate into their buying behavior [13]. It is, therefore, highlighted the importance of campaigns promoted by fashion businesses to encourage a more sustainable behavior towards the consumption of clothing [18, 37], as well as the inclusion of actions in product development and design in order to adapt to the desires of consumers.

Many times value is created and delivered to the consumer but from the company's perspective (see [43]). Companies use different strategies to create value from both virgin materials and from waste (thus engaging in circular initiatives, and adopting more responsible conduct), but they neglect (either intentionally or not) including the consumer perspective into the value creation process, which might result in low consumer engagement, and thus, lower value capture through revenue streams. In the existing literature, there are studies specifically addressing consumer issues (e.g. [54]), but none of them addresses the role of the consumer in the creation of value in circular business models in the fashion industry. Therefore, this study aims to suggest circular strategies for the incorporation of the consumer's perspective into value creation in the fashion industry.

This chapter is structured as follows. Section 1 presented the initial considerations on sustainable and circular fashion, the role of the consumer in the value creation process of circular business models in the fashion industry, and stated the objective of this chapter. Section 2 builds on the constructs of inquiry of this chapter, which are sustainable circular fashion and circular business models in the fashion industry. Section 3 depicts the methods used in this research. Section 4, in turn, presents the main results of the survey, highlighting the issues needed to be dealt with in order to include the consumer's perspective into the value creation process. Section 5 points out circular business models (CBM) and CBM strategies that can be set into practice in order to overcome the challenges identified with the survey. Finally, Sect. 6 presents the final remarks of this chapter as well as calls for further research.

2 Constructs of Inquiry

This section presents the constructs of inquiry in this chapter, which comprise sustainable fashion and circularity and circular business models in fashion.

2.1 *Sustainable Fashion and Circularity*

Garment manufacturers face the challenge of not generating, minimizing, or reusing wastes from the products they produce in new production processes, besides assisting in giving proper disposal of textile products after the use phase. These challenges are congruent with "sustainable fashion" and "sustainability in the fashion industry", which are increasingly common terms in the fashion field. These terms are associated with the efforts of fashion brands to adapt to the growing use of resources by the industry, as well as social and environmental impacts and the use of renewable and ecological raw materials, reduction of the carbon footprint, durability, and longevity (Hvass [19, 23]).

Sustainable fashion translates into several issues: supply of materials, working conditions for suppliers, chemical products, use of energy and water, transport, and recently, attention to post-consumer waste [2, 12, 52]. These issues overlap with a circular economy, which proposes organizations to comprehensively rethink how resources are managed in order to improve financial, environmental, and social benefits, both in the short and long terms [6].

Moktadir et al., [38] affirm that the circular economy guarantees sustainable manufacturing processes and sustainable environmental practices through the prevention and reduction of inherent wastes, that is, it promotes restoration and regeneration to a project, emphasizes the value chain, preserves and increases natural capital, and optimizes resources. In addition, Ghisellini, Cialani, and Ulgiati [16] state that issues related to disassembly, disposability without negative environmental impacts, ease of distribution, as well as return, durability, reliability, and customer success, are relevant.

However, in order to promote a more sustainable fashion system, mass manufacturing organizations are asked to change their ways of acting to produce more evenly distributed value, in addition to monetary benefits [20].

In the present study, we emphasize issues related to consumers. Thus, there was an effort to understand the relationship between the scenario that includes the fashion industry and consumer behavior. First, with a global focus toward sustainability in general, a new type of consumer understanding is essential, and the main thing is that the company creates a new type of relationship with the end-user, knowing their needs and desires more deeply to build offers that comprise greater clothing use time or stable satisfaction with the product [40]. In this sense, Table 1 presents actions to arouse consumer interest in sustainability and circular economy in the clothing industry.

Table 1 Actions identified to promote consumer interest in sustainable fashion products

Actions directed at consumers	Supporting literature
Offering workshops with the objective of sensitizing consumers towards being responsible and sustainable in relation to consumption	Peronard et al. [46]
Creating awareness and allowing consumers to include climate impact as part of their decision-making	
Including practices that reduce emissions and waste of resources by making consumers aware of the impact of decisions and purchasing habits	
Including values and motivations in fashion consumption consumers are interested in	Holtström et al. [21]
Promoting changes for consumers to understand that they are resource users	
Promoting information to consumers	Franco [14]
Promoting good prices	Franco [14] Vehmas [54]
Promoting the transition to new buying habits and consumer culture	Mura et al. [39]
Communicating circular value propositions to consumers	Hvass and Pedersen [23]
Knowing consumers and their ethical concerns with clothing	Vehmas [54]
Knowing consumers' lifestyles	Diddi et al. [9] Pedersen et al. [45]

Consumers' environmental or ethical concerns do not always translate into their purchasing behavior [13] and are often influenced by improved status. Perceived effectiveness and happiness have a significant effect on consumers' attitudes toward environmentally responsible clothing [54].

In contrast, there has been a reluctant acceptance of these products by many consumers and an apparent conflict with the existing fast fashion desires in this area [37]. Consequently, designers, manufacturers, and companies are responsible not only for the environmental impacts of the fashion industry but also for encouraging more sustainable behavior regarding the consumption of clothing [18, 37]. Therefore, it is essential to include actions in product development, such as the ones shown in Table 1.

For proposing changes in the clothing industry, it is understood that it is necessary to develop the awareness of sustainability and consumer buying patterns or enter new consumer segments [17].

Given the context surrounding the search of industries for sustainable processes and consequently the consumption of clothing, Holtström, Bjellerup, and Eriksson [21] explain that it is necessary to develop sustainable solutions for long-term survival in the clothing industry, by meeting consumer preferences for fashion, as well as their growing interest in consuming less.

Table 2 Creating value for fashion products

Actions to create value for fashion products	Supporting literature
Proposing continuous innovation (e.g. several colors and smooth aesthetics)	Pal and Aneja [44]
Providing ecological products	Fonseca et al. [13]
Sharing possibilities	
Reconditioning possibilities	
Consumer involvement in responsible behavior	
Products that enable reuse	Vehmas et al. [54]
Products that enable recycling	
Promoting the idea of recycling to produce new clothes	Vehmas et al. [54] Pedersen et al. [45]
Use of sustainable and quality materials	Pedersen et al. [45] Lewandowski [33]
Upcycling	Lahti et al. [30]

In this scenario, understanding the processes of value creation in the fashion industry involves the elaboration of consumer-oriented products, in addition to recirculating consumer product waste [46]. The value creation process emphasizes the importance of creating and adding value to a product and adjusting the offer to consumers [57]. For Zink and Geyer [58], the creation of value can occur from the waste generated in the pre- and post-consumption processes. Other actions aimed at creating value are shown in Table 2.

For the garment industries, promoting changes aimed at extending the useful life of products (e.g. repair and remanufacturing) and closing material loops (e.g. recycling) is a challenge [41]. Moreover, the new culture of consumption in a sharing economy is important to reduce product disposal through reuse,however, this is a practice to be presented and studied having consumers in mind [27]. Thus, it is essential to include actions that facilitate and promote the transition process, and the creation of value is shown to be essential in the path to be taken.

Therefore, it is important to have an appropriate product design, which allows using less raw material or energy, reducing emissions and toxic materials, extending the product's life, eliminating waste before extending the life of resources, and circulating products [33].

The development of products to create value is one of the precepts of circular economy and consequently contributes to fashion becoming sustainable, with actions and practices that reduce impacts on the production and use of clothing. In the circular economy view, the efficiency of resource use increases with reuse, remanufacturing, and refurbishing of products, as well as by conventional recycling of products' raw material value [27]. However, to achieve these levels, changes in production processes and consumer behavior are essential, as shown in Tables 1 and 2.

2.2 Circular Business Models in Fashion

The textile industry produces large amounts of waste, which include wastes generated at the stages of manufacturing, pre-consumer (clothes that are somehow damaged/unfit for sale), and post-consumer (clothes that are either worn out or disposed of for some reasons) [32].

The transition to a circular economy requires changes in several characteristics of the current society, including product design, value chains, new behavior toward production and consumption (e.g.) recovering value from waste, and also business models [11, 23]. However, such change does not come without hardships [25]. This paradigm shift calls for new business models, with a circular perspective, i.e. circular business models (CBMs), considering novel products and services, and managing these throughout their entire life cycles [56].

A CBM is a path followed by an organization to propose, create, deliver, and capture value, based on a naturally regenerative system, seeking to close, slow, or narrow resource flows, and reduce or eliminate resource leakage [48]. There are many strategies that can be used to promote CBMs in the fashion industry, including (e.g.) take-back systems (TBS) [23], recycling [49], reduce [31], reuse [36], offering repair options [5], as well as long-lasting products [55], and product-service systems (PSS) [29].

Circular initiatives aim to decouple economic growth from the sole consumption of resources. Nonetheless, for a business to exist, a meaningful value offer still needs to reach consumers. The concept of value is pivotal for a business model [23]. However, it has been often reported in the literature that companies develop value offers via circular initiatives without taking the consumer perspective into consideration, besides that, companies often pointing out the need to "educate" consumers toward wanting or accepting said value offers (see [23]). Companies offer values that consumers did not ask for, thus, no real value offer exists for them [15, 51]. For all CBMs, not only in the fashion industry, the customer perspective has been reported to have been neglected (see Salvador et al. [48]). Thus, trustful communication with the consumer is vital.

The low awareness of consumers about the value of circular initiatives is one of the main barriers in CBMs [26]. It has been reported the quest for consumer collaboration towards garment collection, as a means for recycling [7], but it is necessary to connect with consumers in new ways [50]. Thus, innovation is an important driver for incorporating CE in the fashion industry [53]. It is known, for instance, that knowledge of environmental issues has influenced the intention of purchase, thus revealing a form of providing value to consumers [42], hence communication is once again signaled as a key issue.

Slow consumption is undoubtedly a challenge in the fashion industry [5]. Circular initiatives are difficult to be implemented in the fashion industry as they somehow usually go against the spread idea of fast fashion [5]. Consumers often lack finding long-lasting value in garments, which leads to shorter product longevity, as consumers grow inclined to dispose of them [24]. Therefore, James, Reitsma, and Aftab [24] state that the design of products is pivotal to tailor the value offer toward perceived value from consumers. Drawing consumers attached to items by (e.g.) offering long-lasting value, via quality products, and offering customization options [42], will make consumers less likely to dispose of them [24]. Nonetheless, it can be observed that creating value is a difficult task for designers since the value is judged and recognized on a personal level,especially when it is also argued that circular initiatives need to be incorporated into the company's core, thus not being only a "short-term campaign" [23]. Hvass and Pedersen [23] highlight the challenge of managing circular initiatives on the inside of companies since the interpretations of value propositions can vary among departments and personnel, besides the need for establishing partnerships along the supply chain for a successful system. Nonetheless, considering a circular perspective, products should be designed already considering more than one user and life cycle [29].

While some companies see establishing circular practices as disruptive (for business as usual), others see it as opportunities for business extension [55]. An example of a novel way for considering the consumer perspective has been Adidas (2017)'s "knit for you", where 3D knitting was tested, with clothes being tailored to fit the consumer and being made in a few hours. One other example of value co-creation in the fashion industry, with the participation of consumers, is illustrated by Pedersen, Earley, and Andersen [45] through the service shirt, which proposes that shirts could be overprinted in-store. Furthermore, there are examples of fashion companies worldwide deploying efforts toward other explicit circular initiatives (see H&M [22]; MUD Jeans [35]).

Accounting for the disruption companies might encounter for implementing new business practices, experimentation with new business models appears as a possibility to investigate new modes of production and consumption, rather than launching an already established approach [5]. Therefore, fashion companies could first experiment with novel circular initiatives before making a "no going back" transition.

3 Methods

This section presents the methods used to conduct this research, including the workflow and activities at each stage, as shown in Fig. 1.

Step 1—Literature Review

First, a literature review aimed to contextualize the importance of the consumer in the sustainable fashion scenario, including actions identified to promote consumer interest in sustainable fashion products, creation of targeted value for fashion products, and circular business models in the fashion industry. The knowledge gathered during the review helped to finding ways to incorporate the consumer's perspective in the value creation process in the fashion industry.

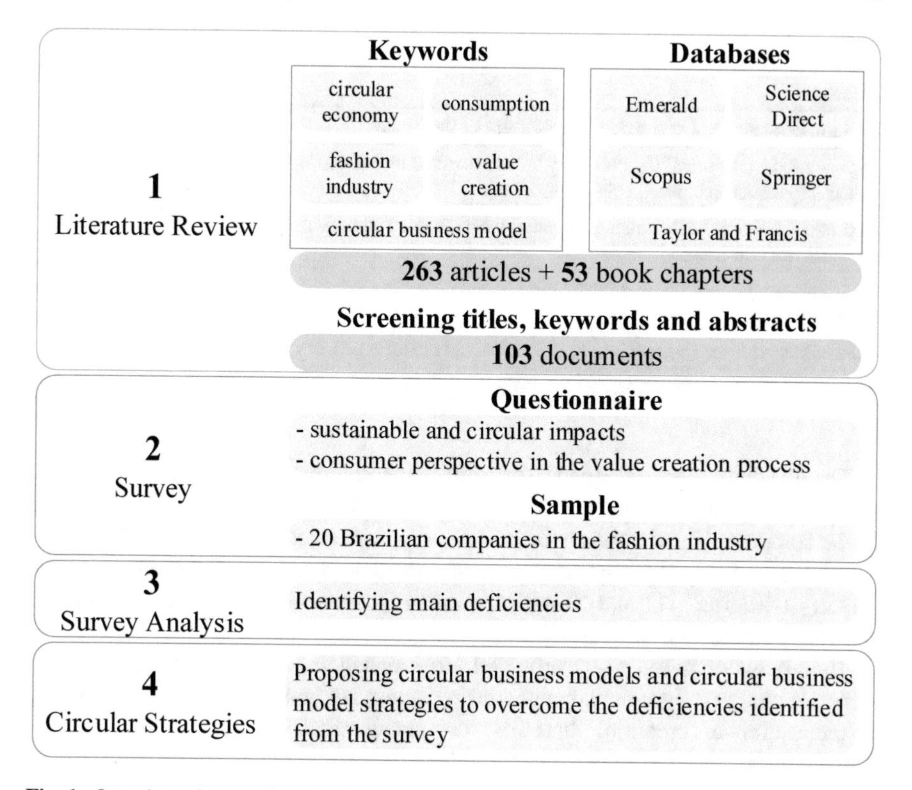

Fig. 1 Overview of research methods

For carrying out the literature review, the following databases were searched: Emerald, Science Direct, Scopus, Springer, and Taylor and Francis. The combinations of keywords comprised: "circular economy", "consumption", "fashion industry", "value creation", and "circular business model". A total of 263 articles published in journals, and 53 book chapters were identified in the raw portfolio, from the mentioned databases. After analyzing titles, keywords, and abstracts, 103 documents were left in the final portfolio.

Step 2—Survey

From the literature review, a survey was designed to investigate sustainable and circular issues in the value creation process of fashion companies. The questionnaire for the survey was composed of six questions that addressed the product development process in the company, including environmental and social impacts in the manufacture of clothing products, whether the company knew its consumers, and also considered the phases of use and end-of-life of the clothing products produced. It is understood that this information makes it possible to identify the opportunities for incorporating the consumer's perspective in value creation in the fashion industry. The questions included in the survey can be seen in Table 3.

Table 3 Questions in the survey

Question	Question ID
Fashion designers know the consumers/users of the brand	Q1
In product development, the company takes into account the environmental and social impacts of manufacturing the products	Q2
In product development, issues associated with the use and end-of-life phases of products are considered	Q3
The company knows the final destination of the products after the use phase by the consumer/user	Q4
Consumers participate/participated in the product creation process	Q5
The company has cooperations with suppliers/consumers	Q6

The questions were designed to be affirmative statements with which the companies would indicate to what extent they agreed, depending on the practices adopted by the company. The questionnaire was answered in accordance with a Likert scale, which allowed recording the frequency of each statement considering the extremes "Totally Disagree" (1) and "Totally Agree" (7) (with an odd number of options allowing there to be a neutral point).

The questionnaire was provided to managers at the strategic level, owners/managers of garment manufacturing industries, and/or professionals working in the sector of creation, direction, or management. A total of 20 clothing manufacturers have participated in the survey.

The industries participating in the research were Brazilian and were located in the state of Paraná (15 companies), Minas Gerais (2 companies), and Rondônia (3 companies). Of those companies, 10% were classified as micro-sized, 65% of companies were classified as small, 20% were classified as medium-sized, and 5% as large. Data collection was carried out from July to September 2019.

Step 3—Survey Analysis

Qualitative data analysis took place through content analysis using NVivo software. The quantitative analysis used descriptive statistics methods that show the distribution and disposition of the data. The combination of analyses made it possible to identify the paths for the incorporation of the consumer's wishes and desires in the creation of value in the fashion industry, aiming at the promotion of circular paths.

Step 4—Circular Strategies

Based on the analysis of the survey, a few CBMs and CBM strategies were proposed (based on the literature review and the authors' expertise) in order to overcome the deficiencies in fashion businesses identified in the analysis.

4 The Importance of the Fashion Consumer in Product Value Creation

In order to have a dynamic application of value creation in clothing products, it is important to know the consumers/users, their wishes, and desires. Therefore, we examined the implications related to the perceptions of the managers of the clothing industries on the inclusion of the consumers' perspective in the value creation in the clothing sector.

Thus, in order to promote and incorporate the consumer's perspective in the product design stage, we seek to understand how companies deal with fundamental issues for this purpose.

First, in Q1 we deal with the relationship between fashion designers and the action of getting to know consumers/users of the brand; still, when dealing with product development, the company takes into consideration the environmental and social impacts of manufacturing products (Q2); and the fact of considering the issues associated with the phases of use and end-of-life of products by consumers (Q3), as well as the company knowing the final destination of the products after the use phase (Q4); the participation of consumers in the creation of products (Q5) and, finally, if there is any cooperation with suppliers/consumers (Q6).

These issues are relevant for understanding the internal scenario of the product design process, which plays a decisive role in integrating the consumer's vision and creating value in fashion products, which consequently favors companies to move toward greater circularity. Table 4 shows central tendency measures of the responses to the survey.

Questions Q1, Q2, and Q3, despite presenting higher metrics, still need to be optimized, as they do not fully integrate the product development of the surveyed companies. In addition, Fig. 2 shows the distribution of responses in a boxplot.

Initially, we emphasize that the company knowing consumers/users of their brand provides fashion designers with relevant information on consumption habits, behavioral issues, consumers' desires and wishes that favor the application of value creation in the development of clothing (Q1). In the same sense, a relevant aspect for the implementation of circularity refers to the influence exerted by consumer behavior (Elia, Gnoni, and Tornese, 2017), since these are important actors for the durability of clothing products [53].

The fact that the company knows the environmental and social impacts from the manufacture of its products is relevant because with this information it is possible

Table 4 Survey's central tendency measures

Question	Q1	Q2	Q3	Q4	Q5	Q6
Mode	7	7	7	1	1	5
Mean	6.20	4.89	4.55	2.70	3.60	3.90
Standard deviation	1.24	2.16	2.14	1.89	2.33	1.74

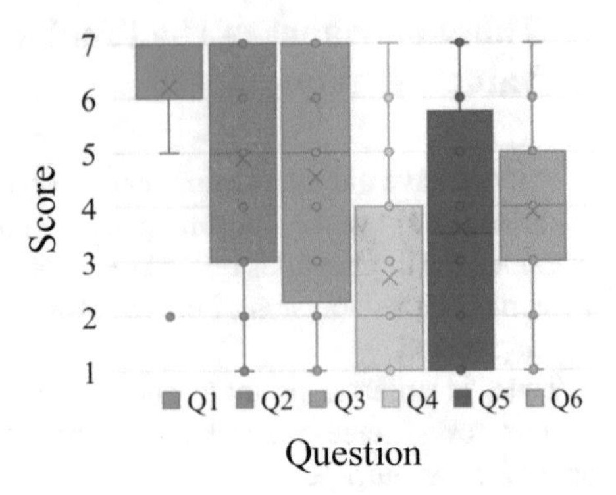

Fig. 2 Distribution of responses to the survey, considering 20 Brazilian companies from the fashion industry

to modify the paths to be followed; however, this is an aspect that still needs attention (Q2). Moreover, knowledge of environmental issues influences the purchase intention by consumers, thus revealing a way to add value to them [42], therefore, communication is signaled as a key issue and promotes consumer connection to the product.

The survey shows that there is a gap regarding the company knowing its consumers and also with regard to issues that involve the product's life cycle, from the conception, use, and disposal of post-use products (Q3). It is understood that this information is of utmost importance since it can assist in guiding a final destination that optimizes the use of the product and favors that it not be sent to landfills.

Questions Q4 and Q5 show the lowest scores obtained in the survey. The results show that there are very few actions in the responding companies aimed at knowing the final destination after the products are used by consumers (Q4). They also show the very low participation of consumers in product creation (Q5). Therefore, they signal actions to be taken when creating products.

It is evident the gap regarding the participation of consumers in product creation and the lack of cooperation between suppliers/consumers for the companies participating in the study (Q6). On that note, Lahti et al. [30] explain that one of the most complex challenges is to establish and organize activities in the reverse value chain, which encompass all activities of an organization, from the return of the product to the potential recovery of its maximum value through recycling and upcycling activities. In this context, Niinimäki [40] states that to move toward a circular economy, it is important to have a fashion perspective, where all the actors are included: designers, producers, manufacturers, suppliers, entrepreneurs, and even consumers.

Questions Q2 and Q6 show the responsibility of companies to defend the environmental and sustainable values of society, and that they must respond to a wide range of stakeholders, and not just to their closest shareholders [30]. However, we identified low adherence to these issues in the survey, and they are important for the incorporation of the consumer's vision in creating value in the fashion industry.

5 Translating Consumer Value Co-Creation into Business Modelling

A few issues that might disrupt or prevent fashion businesses from including the consumer perspective in their value creation were identified. Seeking to overcome that, this section presents a few potential solutions to include (either directly or indirectly) consumers or their perspective in the value creation process.

There are a few strategies that can be deployed from business perspectives that might enable including customer considerations in product value creation. These circular business models are presented hereafter.

Extending resource value and product life is a CBM that aims to keep materials and resources at their maximum value for as long as feasible, by promoting the production of long-lasting products and/or offering recovery options (e.g. upgradability and repair).

> Enabling multiple cycles. Fashion companies might enable multiple cycles of fashion products by offering customers' high-quality goods along with the ability of "renewing" or "upgrading" these fashion products giving them the possibility of new use cycles, allowing consumers to tailor these products according to their own style. One example of this can be seen in the service shirt (see [45]), where customers can get their shirts overprinted in store. A company could, for instance, offer different sizes of prints to be used in different renewing operations.

This would address issues related to Q1, Q5, and Q6. Fashion designers would acquire further and deeper knowledge on consumers' behavior toward the frequency with which they would want to upgrade their clothes and what kind of upgrade (taste-related) they were more inclined to. Moreover, consumers would be actively participating in the value creation process, while the company would be establishing a direct partnership for value co-creation.

Resource recovery is a CBM that aims to recover resources for further cycles before they reach final disposal, by creating value from waste, establishing take-back systems, recycling, reconditioning, refurbishing, remanufacturing, or designing out waste.

> Take-back systems along with recycling options. Fashion companies might offer value to consumers if promoting in-store TBS. Consumers could bring their old clothes in exchange of discounts or vouchers for later purchases. If accepting clothes from other brands, the company will be helping to collect garments that would otherwise go to (e.g.) landfills. By accepting garments from their own brand, the company will know the end-of-life destination of their production. This might be a value offer for consumers; since they (and the company) can (re-) capture value from pieces they would give final destinations with no value recovery.

This would address issues related to Q4, Q5, and Q6. The fashion company would be taking into account and holding responsibility for the environmental impacts of fashion garments that could have been disposed of in (e.g.) landfills, while also enabling consumers to acquire products with discounts. Moreover, the company would have control over the end-of-life of the products collected. Consumers would be part of a collaboration with the company by participating in higher value creation (addressing social and environmental values).

Delivering functionality rather than ownership is a CBM that aims to make companies offer solutions to immediate problems instead of products for consumer ownership. Companies hold ownership and are held responsible for the means used to deliver a value offer. These CBMs generally take place via product–service systems using leasing/renting-like strategies.

> Product–service systems. Fashion companies might deliver value to consumers by offering garments over lease contracts, where customers can choose the garment they want and also for how long they want to use it. After the lease ends, customers might choose to keep the garment (at a reasonable price), rent another piece, renew the lease of the same piece, or return the garment and terminate the relationship with the company altogether. An example of a PSS in the fashion industry can be seen in the company MUD Jeans (see MUD Jeans [34]).

This would address issues related to Q2, Q3, and Q4. During product development, the fashion company would be more inclined to consider (mainly) the environmental impacts of the garment they will offer since they will also be responsible for the entire life cycle of the products in their value offer. Moreover, the company will have the responsibility to assign adequate end-of-life options to each of the products, which will encourage the company to design them in a more environmentally friendly way.

6 Final Remarks, Limitations, and Call for Further Research

This chapter aimed to suggest strategies for the incorporation of the consumer's perspective into value creation in the fashion industry. These strategies were proposed based on the deficiencies of fashion businesses to incorporate sustainable and circular value propositions considering the perspective of consumers. These deficiencies were identified via a survey conducted with 20 fashion companies in Brazil.

The results of the survey allowed identifying gaps in the knowledge of fashion companies in relation to consumers and also with regard to issues that involve the product's life cycle from the conception, use, and disposal of post-use products by the consumer. This information is essential since it can promote the optimization of product use and help prevent that it be sent to landfills.

Furthermore, we show that there is an absence of consumer participation in product creation and a lack of cooperation between suppliers/consumers and the companies participating in the study. Based on that, we emphasize the challenges to establish and organize reverse value chain activities, which cover all activities of the company, from the return of the product to the potential recovery of their maximum value.

Considering the results of the survey, circular strategies were proposed for fashion businesses in Brazil to incorporate customer perspectives in the process of value creation. They included creating long-lasting products with the possibility of customization as a means to "renewing" that piece to promote new use cycles; establishing take-back systems in order to draw consumer attention, be responsible and hold control over the end-of-life of textile products, and; establishing product-service-systems, in order to capture constant consumer interest and decouple revenue from the sale of textile goods.

The CBMs and respective CBM strategies presented comprise a few examples of the plethora of strategies that can be used in circular fashion businesses. They were proposed in order to address the main issues identified via the results of the survey presented in Sect. 4. All the CBM strategies sought to incorporate the customer perspective into the value offers of fashion businesses. Nonetheless, further strategies could be proposed, within the CBMs presented or via other CBMs, in order to pursue more circular practices.

Limitations to this study should be noted. The companies that took part in the survey participated voluntarily, thus, there might be certain similarities among the companies that chose to answer the survey, which might have acted as potential biases, of which the researchers might not have been aware. Moreover, the geographical distribution of the respondents covered three states in Brazil, therefore, the results cannot be said to represent the whole country or all the companies within the fashion industry in the country.

It is expected that future research covers the implementation of the proposed strategies in fashion businesses in Brazil and that other strategies be proposed in order to increase sustainability and circularity of the fashion business in Brazil.

References

1. Andersen KR (2017) Stabilizing sustainability: in the textile and fashion industry. Copenhagen Business School (CBS), Frederiksberg
2. Armstrong CM, Niinimäki K, Kujala S et al (2015) Sustainable product-service systems for clothing: exploring consumer perceptions of consumption alternatives in Finland. J Cleaner Prod 97:30–39
3. Bartlett C, McGill I, Willis P (2012) Textile flow and market development opportunities in the UK. Waste and Resources Action Programme, Banbury, UK
4. Birtwistle G, Moore C (2007) Fashion clothing–where does it all end up? Int J Retail Distrib Manag 35(3):210–216
5. Bocken N, Miller K, Weissbrod I, Holgado M, Evans S (2018) Slowing resource loops in the Circular Economy: an experimentation approach in fashion retail. International Conference on Sustainable Design and Manufacturing. Springer, Cham, pp 164–173
6. BSI (2017) Framework for implementing the principles of the circular economy in organizations—Guide
7. Carlsson J, Torstensson H, Pal R, Paras MK (2015) Re:textile—planning a swedish collection and sorting plant for used textiles, feasibility study. University of Borås, Västra Götaland
8. Comin LC, Aguiar CC, Sehnem S, Yusliza MY, Cazella CF, Julkovski DJ (2019) Sustainable business models: a literature review. Benchmarking in press
9. Diddi S, Yan R-N, Bloodhart B et al (2019) Exploring young adult consumers' sustainable clothing consumption intention-behavior gap: a behavioral reasoning theory perspective. Sustainable Prod Consumption 18:200–209
10. Ekström KM, Salomonson N (2014) Reuse and recycling of clothing and textiles—a network approach. J Macromarketing 34(3):383–399

11. European Commission (2014) Communication from the commission to the European parliament, the Council, the European economic and social committee and the committee of the regions. Towards a circular economy: a zero waste programme for Europe. European Commission, Brussels
12. Fletcher K (2013) Sustainable fashion and textiles: design journeys. Routledge
13. Fonseca LM, Domingues JP, Pereira MT et al (2018) Assessment of circular economy within Portuguese organizations. Sustainability (Switzerland) 10(7):2521
14. Franco MA (2017) Circular economy at the micro level: a dynamic view of incumbents' struggles and challenges in the textile industry. J Cleaner Prod 168:833–845
15. Frederiksen DL, Brem A (2017) How do entrepreneurs think they create value? A scientific reflection of Eric Ries' Lean Startup approach. Int Entrepreneurship Manag J 13(1):169–189
16. Ghisellini P, Cialani C, Ulgiati S (2016) A review on circular economy: the expected transition to a balanced interplay of environmental and economic systems. J Cleaner Prod 114:11–32
17. Hansen EG, Schaltegger S (2013) 100 per cent organic? A sustainable entrepreneurship perspective on the diffusion of organic clothing. Corporate Governance (Bingley) 13:583–598. https://doi.org/10.1108/CG-06-2013-0074
18. Harris F, Roby H, Dibb S (2016) Sustainable clothing: challenges, barriers and interventions for encouraging more sustainable consumer behaviour. Int J Consum Stud 40:309–318
19. Henninger CE, Alevizou PJ, Oates CJ (2016) What is sustainable fashion? J Fashion Mark Mgmt: Int J 20:400–416
20. Hirscher A-L, Niinimäki K, Armstrong CMJ (2018) Social manufacturing in the fashion sector: new value creation through alternative design strategies? J Cleaner Prod 172:4544–4554
21. Holtström J, Bjellerup C, Eriksson J (2019) Business model development for sustainable apparel consumption: The case of Houdini Sportswear. J Strategy Manage 12:481–504
22. H&M (2020) Circularity and our value chain. https://hmgroup.com/sustainability/circular-and-climate-positive/circularity-and-our-value-chain.html. Accessed 23 May 2020
23. Hvass KK, Pedersen ERG (2019) Toward circular economy of fashion. J Fash Mark Manag 23(3):345–365
24. James AM, Reitsma L, Aftab M (2019) Bridging the double-gap in circularity. Addressing the intention-behaviour disparity in fashion. Des J 22(1):901–914
25. Khan S, Maqbool A, Haleem A, Khan MI (2020) Analyzing critical success factors for a successful transition towards circular economy through DANP approach. Manag Environ Qual 31(3):505–529
26. Kirchherr J, Piscicelli L, Bour R, Kostense-Smit E, Muller J, Huibrechtse-Truijens A, Hekkert M (2018) Barriers to the circular economy: evidence from the European Union (EU). Ecol Econ 150:264–272
27. Korhonen J, Honkasalo A, Seppälä J (2018) Circular economy: the concept and its limitations. Ecol Econ 143:37–46
28. Lacy P, Long J, Spindler W (2020) Fashion industry profile. The Circular Economy Handbook. Palgrave Macmillan, London, pp 185–195
29. Lacy P, Rutqvist J (2015) Five circular capabilities for driving value. Waste to Wealth. Palgrave Macmillan, London, pp 148–167
30. Lahti T, Wincent J, Parida V (2018) A definition and theoretical review of the circular economy, value creation, and sustainable business models: Where are we now and where should research move in the future? Sustainability (Switzerland) 10
31. Larsson E (2018) Filippa K's fashion manifest: long lasting simplicity. Sustainable Fashion. Springer, Cham, pp 83–95
32. Leonas KK (2017) The use of recycled fibers in fashion and home products. Springer, Singapore
33. Lewandowski M (2016) Designing the business models for circular economy—towards the conceptual framework. Sustainability 8:43
34. MUD Jeans (2020a) It's time to rethink what we consume and how we produce. https://mudjeans.eu/lease-a-jeans/. Accessed 23 May 2020
35. MUD Jeans (2020b) Our impact. https://mudjeans.eu/sustainability-our-impact/. Accessed 23 May 2020

36. Machado MAD, de Almeida SO, Bollick LC, Bragagnolo G (2019) Second-hand fashion market: consumer role in circular economy. J Fash Mark Manag 23(3):382–395
37. McNeill L, Moore R (2015) Sustainable fashion consumption and the fast fashion conundrum: fashionable consumers and attitudes to sustainability in clothing choice. Int J Consum Stud 39:212–222
38. Moktadir MA, Rahman T, Rahman MH et al (2018) Drivers to sustainable manufacturing practices and circular economy: a perspective of leather industries in Bangladesh. J Cleaner Prod 174:1366–1380
39. Mura M, Longo M, Zanni S (2020) Circular economy in Italian SMEs: a multi-method study. J Cleaner Prod 245:118821
40. Niinimäki K (2017) Fashion in a circular economy. In: Sustainability in Fashion. Springer, pp 151–169
41. Nußholz JLK (2018) A circular business model mapping tool for creating value from prolonged product lifetime and closed material loops. J Cleaner Prod 197:185–194
42. Okur N, Saricam C (2019) The impact of knowledge on consumer behaviour towards sustainable apparel consumption. Consumer behaviour and sustainable fashion consumption. Springer, Singapore, pp 69–96
43. Pal R (2017) Sustainable design and business models in textile and fashion industry. Sustainability in the Textile Industry. Springer, Singapore, pp 109–138
44. Pal R, Aneja AP (2017) Ambidexterity drivers of value-creation and appropriation in business models: an explorative study from DuPont. Text Apparel 21:2–26
45. Pedersen ERG, Earley R, Andersen KR (2019) From singular to plural: exploring organisational complexities and circular business model design. JFMM 23:308–326
46. Peronard J-P, Ballantyne AG (2019) Broadening the understanding of the role of consumer services in the circular economy: toward a conceptualization of value creation processes. J Cleaner Prod 239:118010
47. Planing P (2015) Business model innovation in a circular economy: reasons for non-acceptance of circular business models. Open J Bus Model Innov 1–11
48. Salvador R, Barros MV, da Luz LM, Piekarski CM, de Francisco AC (2019) Circular business models: current aspects that influence implementation and unaddressed subjects. J Clean Prod 250:119555
49. Sandvik IM, Stubbs W (2019) Circular fashion supply chain through textile-to-textile recycling. J Fash Mark Manag 23(3):366–381
50. Schwab K (2017) The fourth industrial revolution. Crown Publishing Group, New York, NY
51. Silva DS, Ghezzi A, de Aguiar RB, Cortimiglia MN, ten Caten CS (2019) Lean startup. Int J Entrepreneurial Behav Res in press, Agile Methodologies and Customer Development for business model innovation
52. Stål HI, Jansson J (2017) Sustainable consumption and value propositions: exploring product-service system practices among Swedish fashion firms. Sustainable Dev 25:546–558
53. Todeschini BV, Cortimiglia MN, Callegaro-de-Menezes D, Ghezzi A (2017) Innovative and sustainable business models in the fashion industry: entrepreneurial drivers, opportunities, and challenges. Bus Horiz 60(6):759–770
54. Vehmas K, Raudaskoski A, Heikkilä P, Harlin A, Mensonen A (2018) Consumer attitudes and communication in circular fashion. J Fash Mark Manag 22(3):286–300
55. Weber S (2019) A circular economy approach in the luxury fashion industry: a case study of Eileen fisher. Sustainable Luxury. Springer, Singapore, pp 127–160
56. Wells PE (2013) Business models for sustainability. Edward Elgar Publishing, Cheltenham
57. Zacho KO, Mosgaard M, Riisgaard H (2018) Capturing uncaptured values—a Danish case study on municipal preparation for reuse and recycling of waste. Resour Conserv Recycl 136:297–305
58. Zink T, Geyer R (2017) Circular economy rebound. J Ind Ecol 21:593–602

Woven Through Trust and Affect: Four Cases of Fashion Sustainability in Brazil

Julia Valle-Noronha and Namkyu Chun

Abstract This study investigates how fashion practitioners have approached sustainability in Brazil. Through the lens of culture—a recently emerging pillar of sustainability—we look into practices that hint at plural approaches in the dominant western perspectives, especially in terms of their symbolic dimension and value systems. We will briefly present and explore cases of agroforestry, clothing upcycling, alternative leather production, and collaborative spaces and workshops. The notions of 'trust-based relationship' and 'affect' emerge as key elements in the development of more sustainable practices in the field of fashion. This chapter contributes to the sustainability discussion by presenting Brazilian cases from a cultural perspective. It further expands the discussion on decolonizing design and proposes a possible direction for decolonizing fashion. It concludes with reflections on how western and non-western societies can benefit from the addition of these dimensions to the sustainability discourse.

Keywords Sustainable fashion practices · Plural fashion sustainability · Cultural sustainability · Brazilian fashion · Fashion design culture · Decolonizing fashion · Case study

1 Introduction: (Fashion) Sustainability As We Know It

For over 50 years, researchers and practitioners in the north–west have debated the implications of current practices and the ways in which humankind relates to the Earth. The discussions started emerging in the 1960s as the levels of pollution

J. Valle-Noronha (✉) · N. Chun
School of Arts, Design and Architecture, Aalto University, Espoo, Finland
e-mail: julia.valle@artun.ee

N. Chun
e-mail: namkyu.chun@aalto.fi

J. Valle-Noronha
Estonian Academy of Arts, Tallinn, Estonia

M. Á. Gardetti and R. P. Larios-Francia (eds.), *Sustainable Fashion and Textiles in Latin America*, Textile Science and Clothing Technology,
https://doi.org/10.1007/978-981-16-1850-5_13

started increasing and in the 1970s with the oil crisis. In 1972, the initiation of global dialogues from the United Nations (UN) reinforced the gradual establishment of the concept of 'sustainable development'. In the 1980s and the 1990s, these discussions reached social and industrial levels. The most frequently used sustainable development terms were introduced by the Brundtland Commission, also known as the World Commission on Environment and Development (WCED), in the UN-commissioned report 'Our Common Future' [48]. The report not only bridged the ecological aspect of sustainable development and the social and economic aspects but also what is generally considered valuable in life. The latest version of the globally agreed idea of sustainability can be found in the UN's 17 Sustainable Development Goals (SGDs). The SDGs [43] are described as 'an urgent call for action by all countries—developed and developing—in a global partnership'.

In the field of fashion academia, the discourse gained a body in the early 2000s (e.g., [10, 18, 47]). Through a critical examination of contextual fashion practices at the beginning of the 2000s, a number of authors offered alternatives to making and producing in fashion. Kate Fletcher, via a series of both single and collectively authored publications, has been a key figure in disseminating the voice of fashion within the sustainability discussion, constantly adding dimensions and perspectives to her seminal 'design journeys' (2008). Since these initial efforts, the discussion has widely expanded within and beyond fashion itself. Today, fashion and sustainability is undoubtedly one of the most prolific directions in the field of fashion, with contributions from a diversity of disciplines and geographies. However, a clear dominance of western perspectives prevails [39], creating a considerable gap in approaches, especially in regard to value systems and structures. This predominant western discourse is anchored in a centuries-long tradition guided by mechanist, dualist and objectivist values, deemed inappropriate for addressing planetary boundaries [41]. As a response to this discourse, this chapter looks into cases from Brazil in order to broaden perspectives, welcoming other consolidated value systems that have long been overshadowed and forgotten.

The Brundtland Commission report (WCED 1987), which established the groundwork of the currently dominant view to sustainability, emphasized the balance of three pillars of sustainable development. These pillars include environmental, social and economic demands for progress. Meanwhile, the notion of culture has recently emerged as the fourth pillar of sustainability (see [3, 40, 42]). Despite the growing number of researchers looking into the culture as a means to achieve sustainability, a wide array of perspectives exist, meaning that there is no single understanding of what culture entails in the development of sustainability practices.

Here, we focus on diversity and eco-cultural resilience as aspects of culture that may support and enable sustainability from the perspective of diverse practitioners in the contemporary fashion system. The research questions that we seek to answer through this study are the following:

- What are the characteristics of sustainable fashion in Brazil?
- How do the characteristics differ from sustainable approaches in western societies?

By answering these questions, the primary aim of this chapter is to emphasize the culture of Brazilian sustainable fashion approaches. Through the lens of culture, we look into practices and models that hint at plural approaches to fashion sustainability in the dominant western discussions. The contribution of this chapter is fourfold. First, it deepens the understanding of the cultural pillar for sustainable development in fashion. Second, it introduces lesser-known yet innovative cases from Brazilian practitioners related to the field of fashion to broader audiences beyond the region. Third, it bridges the cultural perspective from fashion design to fashion sustainability. Finally, the findings contribute to the recently emerged discussion on decolonizing design and suggesting a possible path to also decolonizing fashion.

The structure of this study is as follows. It begins with the theoretical framework on the culture of fashion design. Next, it briefly presents the research methods and the context of the Brazilian fashion ecosystem. Four cases presenting the different characteristics of Brazilian approaches to fashion sustainability are introduced and reviewed through the lens of fashion design culture. Finally, after discussing the key findings from the cases, the chapter concludes with suggestions for future studies.

2 Theoretical Framework: Fashion (Design) and Culture[1]

Both together and separately, the notions of fashion and culture have been widely explored. Previous studies from the social psychology of clothing, in particular, have approached the notion of culture through an anthropological lens (e.g., [16, 21, 25]). These studies tend to explore clothing as cultural representations, traditional costumes and social phenomena, rather than associating it with a field of practice with domain-specific knowledge and skills. Due to the limited competence of the authors on the topic as well as the broad context to be covered, this anthropological view of clothing and fashion is not the main focus of this study.

This study aims to explore the practice of Brazilian sustainable fashion by highlighting discussions from the intersection of fashion and design research. Accordingly, the notion of culture is conceptualized as symbolic material constraints that are directly tied to fashion practice and shared among actors in the fashion system, especially fashion practitioners [13]. Fashion design as a profession comprises encultured practices situated within local and global contexts [9]. In this study, fashion practitioners refer to designers and entrepreneurs who reside within the ecosystem of fashion. A previous ethnographic study in Finland, looking into the practices of Finnish fashion designers, has revealed key concepts of the fashion design culture [13], and we used these as a model for this study. These concepts supported constructing two main categories: 'objectives of designing' from the meaning dimension and 'production system' from the material dimension.

[1] This section was partially adopted from Chun's research on fashion design thinking (2018).

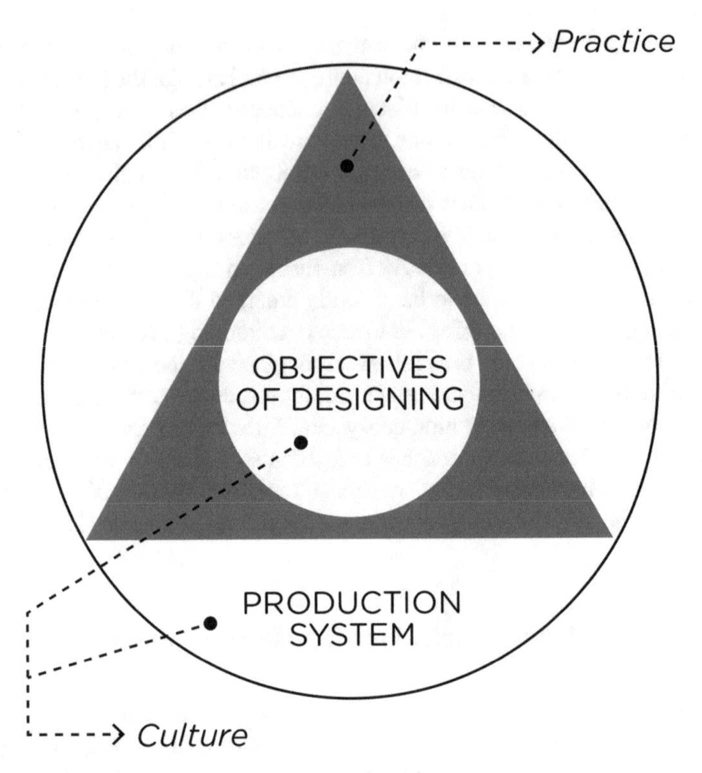

Fig. 1 Model for visualizing how the internal and external aspects of fashion design culture are situated in the practice (adopted from [13])

These categories are similar to Lawson's [28] internal and external constraints for design but require further understanding of fashion's complexity in terms of its material (clothing) and immaterial (fashion) dimensions (see Fig. 1). This complexity of spectrum allows the fashion system to produce both material and meaning [26, 29] and challenges the dualistic view by enmeshing these two dimensions. Whereas the practice of fashion practitioners may have a direct connection to clothing construction, the production of fashion requires understanding meaning-making elements. In other words, the individual practice of fashion practitioners is intertwined not only with the material production system of clothes but also with the symbolic and social production system of fashion. The following sections further discuss these aspects of the fashion design culture in relation to the meaning and material dimensions.

2.1 Meaning Dimension: Objectives of Designing

Regarding the objectives of designing being the internal aspect of fashion design culture, one point needs to be made: the practitioner is not a machine that works with an order placed by someone. A fashion practitioner makes their own choices in a situated manner [22] and is deeply affected by contexts. In other words, an action of fashion design takes place within the set of intentions of the situated designer to turn clothes into fashion, as Loscheck (2009) observed. In this respect, despite a clear limitation from the binary approach, Manzini ([30], pp. 33–37) offered a useful distinction to understand two types of design: problem-solving and sense-making. Design as problem-solving is associated with a simple daily issue in an entangled global matter in the physical and biological world. Meanwhile, design as sense-making emerges from the social and cultural world, constructing meanings and conversations for producing certain values. Manzini [30] argued that these two types co-exist. In this work, we go one step further and argue that rather than co-existing, these two types of design are entangled in such a way that a clear distinction is rare. Meanwhile, the shared culture of fashion practitioners is associated with the idea of fashion into which they intend to transform. Thus, keeping in mind the entangled and non-dualistic view of these two dimensions, it is more relevant to consider the culture of fashion design as sense-making, as clothes can only become fashion through social dialogues and experiences [29].

2.2 *Material Dimension: Production System*

Related to the external aspect of fashion design culture, we need to consider a spectrum of factors that surround the practice of individual designers in the material dimension. Different from the objectives of design that rely on the social and symbolic production of fashion, the production system is relevant for the materiality of clothes, as fashion practitioners need to join social communication in order to turn clothes into fashion [29]. However, the production of clothes is as important as the meaning production of fashion, because this aspect situates the practice of individual fashion practitioners in the physical world. In order for them to produce many pieces of clothes and other items as a collection,and for the clothes to be worn and finally become the fashion, a series of conscious efforts are required due to the complexity of the clothing industry. This complexity in the system of fashion, supported by the production and consumption of clothes, was described by Aspers and Skov [2] using the notion of 'encounters' (p. 803):

> The concept of encounters shifts focus away from individuals and entities and allows us to zoom in on interaction, negotiation and mediation between people and products, buyers and vendors, but also between different professions and different nationalities, and ultimately also between economy and aesthetics.

This view helps us to understand the entanglement of the practice of individual fashion practitioners with their shared culture. This study explored these encounters from the practitioners' perspective. However, the enmeshing of relationships is especially visible through their practices.

2.3 *Key Concepts of the Fashion Design Culture*

In addition to these categories of the fashion design culture, the corresponding concepts were identified as more detailed features [13]. Table 1 summarizes these concepts under two categories.

Despite its representation of western societies, these categories and concepts shed light on the culture of fashion design as a way to explore sustainable fashion practice. The two categories and corresponding concepts are equally important for understanding the shared culture of fashion practitioners in terms of both the symbolic and social dimension and the material dimension. The concepts of usefulness, everyday life, wearer, practitioner him/herself, and temporality are all vital for understanding the social construction of fashion through constant dialogues between individual fashion practitioners and wearers. From the perspective of fashion practitioners, especially designers, a piece of clothing needs to be both wearable and attractive for potential wearers. However, due to experience and wearing factors, pieces of clothing are not stable objects [45]. As the makers in creative practices, fashion practitioners also enjoy the development of clothes that can become a space for experimentations, experiences and meaning construction. In the meantime, the concepts related to the production system of clothes reflect the complexity in the physical and material world. The intensified biannual structure for showcasing fashion collections, also known as fashion week, triggers rapid production and cooperation among fashion practitioners with other actors on the global and local scale. At the same time, diversified feedback loops support the development of new collections. How can this model of the fashion design culture thus be used to examine fashion and sustainability in the specific context of Brazil? In order to answer this question, the following section presents the methods used for data collection and interpretation, as well as an overview of the Brazilian fashion ecosystem and its features.

3 Research Methods

Using a purposive sampling approach [19], we chose four cases that show the potential of the Brazilian approach to fashion sustainability to explore deeper. The sampling was conducted through the mobile application WhatsApp due to its wide use in Brazil. A group chat, called Grupo Transparência ('Transparency Group' in English), with a shared interest in the topic of sustainable fashion was used (see 'transparencia.me'). Different actors in the Brazilian fashion industry have been actively engaged in discussions on the topic and some of the cases for this study were invited through the application. The selected cases are FarFarm, C(+)mas, beLEAF and Ateliê Vivo. Each case will be presented in detail later.

For a broader contextual understanding of the Brazilian fashion ecosystem, we conducted semi-structured interviews with local experts in Brazil. In addition, one of the authors in this study had experiential knowledge as a native practitioner and

Table 1 Summary of concepts for fashion design culture (adopted from [13])

Category/Dimension	Concept	Description	Keywords
Objectives of designing/Meaning	Usefulness	Creating something wearable, functional and comfortable for potential wearers	Functional, comfortable, practical
	Everyday life	Making sense for potential uses in daily lives while representing wearers' personality	Experiences, personal, daily life
	Wearer	Dressing the body of wearers who reinterpret the design based on their needs and experiences	Interpretation, needs, values
	Practitioner him/herself	Deriving enjoyment and self-satisfaction from the process of designing, which conveys certain values	Process-oriented, pleasure, proposal
	Temporality	Paying attention to time changes constantly while speculating on the future	Topical issues, trend, speculation
Production system/Material	Coexistence of globalism and localism	Engaging on both global and local scales while producing fashion and textiles	Selling and sourcing globally, producing locally, local techniques and crafts
	Multiplicity of actors	Cooperating with various actors in the process of producing fashion and textiles	Building trust, mutual respect, cooperation
	Speed	Maintaining the choreographed production of fashion and textiles within a limited timeframe	Vicious rhythm, efficient planning, prioritizing
	Seasonality	Creating a new meaning or revisiting previous works while reflecting seasonal changes	Seasons in nature, fashion calendar, newness
	Plural feedback	Collecting direct and/or indirect responses from wearers and clients for sensing future production directions	Multi-channelled, indirect and direct feedback, satisfaction

this provided useful points to reflect upon. Until recently, for more than a decade, Julia Valle-Noronha worked in the Brazilian fashion industry as a fashion design practitioner on different scales; from a fast-fashion retailer to a large ready-to-wear brand, an haute couture house and her own fashion label. Her reflections on the experiences were supported by desktop research and literature on Brazilian fashion and were complemented by the interviews.

This study followed the main case study principle [49]. Multiple sources of evidence were gathered to characterize the Brazilian approaches to fashion sustainability in four distinctive cases. The collected data include semi-structured interviews with the founder or owner of each case via the video communication application Zoom, as well as secondary sources such as magazine articles, official websites and video interviews. The interviews were conducted mostly in English because one author was not a native Portuguese speaker. However, the interview participants were able to clarify their intentions in Portuguese when describing complex processes and specific terms. The fact that one author was not a native Brazilian or a Portuguese speaker allowed this study to employ triangulation of the authors for increasing reliability [37]. The findings that emerged from the collected data were reviewed constantly by two authors while interpreting. The authors discussed any possible bias and misinterpretations to present balanced perspectives of the findings.

4 Context: Fashion and Sustainability in Brazil

One obvious, distinctively unique feature of Brazil also reflected in the nation's fashion is the Amazon rainforest, a unique ecosystem, currently under threat, abundant in biodiversity and indigenous knowledge and the practices related to these [27]. This diversity is also perceived in its demographics. The Brazilian population is largely miscegenated, with racial heritage from Europe (including Portugal, Spain, Italy, Germany, Netherlands, Poland), Asia (especially Japan and Korea), the Middle East (especially Lebanon and Syria), and indigenous people, at present, are a minority in their original land [24]. This mixture of different races being involved in the local fashion ecosystem is another specific feature of the country (see [34]). Brazil is a land of disparities, especially social disparity, with massive economic inequality. The carnival and football seasons have huge impacts on Brazilian society, which embraces the production and consumption of fashion.

Before entering the discussion on fashion and sustainability in Brazil, it is important to introduce how 'fashion'—as a western conceptualization of 'what people wear' ([5], 3)—has been discussed across the country. Historically speaking, the adoption of this western conception of fashion in Brazil started in the 1920s, when the local textile industry sought to level itself with North American and European industries [35]. But it was not until the 1950s that the discourse gained a body in society, especially in the economically prosperous Central South, through the dissemination of lifestyle magazines that portrayed western styles produced in Brazil as a frequent disconnection between figure and context [46]. Together with the growth of this discourse, the integration of indigenous knowledge and practices lost space to a stronger presence of 'prêt-a-porter' clothing, especially after the 1990s when the common practice of made-to-measure was replaced by more affordable mass-manufactured clothes. Aligned with this, it was in the mid-1980s and 1990s that high fashion achieved projection [11], through the activities of, for example, the Grupo Mineiro de Moda in the 1980s [12] and Morumbi Fashion in the 1990s, later establishing itself as São Paulo Fashion Week.

After having lost much of its textile industry to the growth of the same industry in Asia, the Brazilian clothing and textile industry now holds fifth place in the world in terms of volume [7] and currently concentrates on clothing manufacturing. Recently, a number of Brazilian fashion brands have gained global recognition or/and have reached the international market, such as Osklen, Havaianas, Melissa. However, the industry is still rather dedicated to the internal market, with exports being as low as 0.05% [44]. The large Brazilian population (roughly 211 million in 2019) allows self-sufficiency without the need for external markets. Some of the difficulties in internationalizing Brazilian fashion are due to the extremely high taxes that fall upon produced goods as well as the lack of a clear standardized measurement system, which directly reflects the plurality of the bodies of the nation [33].

Until very recently, the high fashion industry and the events related to it mimicked western traditions through seasonal fashion weeks and showrooms, trends and values, with few representatives of what could be called essentially Brazilian fashion. For example, although organizing fashion weeks as Autumn/Winter and Spring/Summer seasons could somehow express the climate of the southern regions of the country (where most economic growth also resides), this does not express the climate of the northern regions, where seasons may be more related to water precipitation and tides than temperatures. Some contemporary examples that have essentially worked and valued Brazilian culture throughout the years in their designs are Auá, Ronaldo Fraga, Rosa Chá, Helen Rödel and Flávia Aranha. More recently, however, with the growth of sustainability and post-colonialist discourses, brands that evoke local practices, knowledge and values have grown in number and acknowledgement, playing an active part in the current conceptualization of fashion in Brazil. Events such as the late Prêmio Rio Moda Hype (for young designers) and the more recent Brazil Eco Fashion Week (exclusively showcasing sustainability brands) have helped in this shift.

Fashion and sustainability in the twenty-first century Brazil is a recent phenomenon. In academic environments, the discussion has spread quickly through the much consolidated and acknowledged Colóquio de Moda, currently in its 15th edition, with hundreds of papers published on the topic. The newly emerging Rio Ethical Fashion [38], with its content freely available online, has stimulated discussions on fashion and ecology. In addition to this, the work of Lilyan Berlim [8] has been seminal in the country and helped build a discourse from the Brazilian perspective, as discourses have previously been tuned to western aspects through the works of, e.g., Fletcher [18] and Black [10]. As for the commercial dimension, despite the breadth and depth of efforts to develop fashion for sustainability in the country, its reach is limited to a narrow number of individuals financially able to engage in its consumption and wear practices. However, as the number of practitioners and researchers in the field grow, dissemination and cost become more efficient, allowing a wider reach.

5 Four Cases of Brazilian Sustainable Fashion

From this context of Brazilian fashion, in this section, we present four cases that characterize certain aspects of Brazilian approaches to sustainable fashion. The cases represent four different approaches, namely agroforestry (FarFarm), clothing upcycling (C(+)mas), alternative leather production (beLEAF), and collaborative spaces (Ateliê Vivo).

5.1 FarFarm: Agroforestry for the Fashion Industry

> The SDGs [chosen for the project] are 100% on the environmental side, but when you get to the Amazon it is impossible to ignore the people, the culture (B Bina 2020, interview, 19 May)

FarFarm [17] emerged from its founder Beto Bina's desire to foster positive development in his motherland, Brazil, after a successful career abroad, aligning with the SDGs 13, 14, and 15 (Climate Action, Life under Water, and Life on Land). Inspired by his father's work with indigenous communities in the Amazon, he decided to bring agroforestry to the region by following syntrophic foundations [20]. Such foundations follow ancient knowledge, taking into consideration heritage, rituals, self-knowledge and the development of the local community to regenerate the land rather than deplete it. Though seemingly unlikely, cotton was Bina's fibre of choice after becoming acquainted with the cotton fibres Rim de Boi (*Gossypium barbadense brasiliense*) and Mocó (*Gossypium hirsutum latifolium*), as well as the successful Pima cotton plantations in the Peruvian Amazon. The chosen area for beginning the project was the Santa Bárbara community, in Pará state, Brazil, an area already familiar to Bina, where the Amazonian biome thrives.

Bina describes the community as being formed by extremely knowledgeable and sophisticated people, who hold an immense understanding of the forest—much of which is still unfamiliar to researchers in the field. He adds that 'relationship' is central to their being among themselves and with the land, with strong importance given to emotions and bonds as opposed to financial compensations. For example, the well-consolidated Maslow hierarchy of needs [31] is invalid among these communities, as the emotional dimension (described by Maslow as Love and Belonging) sits at the bottom of the pyramid. However, the contemporary western regime has impeded most male indigenous people from carrying out their traditional activities, as they became illegal under the Brazilian Republic. These activities included hunting in the

Fig. 2 FarFarm community in Pará on regenerated land. *Image Credit* renature.co / Cecilia Rechden

forest and rivers or performing indigenous rituals, which directly connected them to the land. Due to this, women have become central to the development of the project, and they have taken on core roles (Fig. 2).

The work with cotton, although small, has helped develop a connection between the community and the land and provided a financial means for people to improve their livelihoods. Community members see in this both the recognition of their knowledge and an opportunity to put it into practice, sustaining their indigenous culture. For them, the connection to the fashion and textile industry hardly affects how they perceive their work, but for the fashion industry, this is a unique opportunity to start to reverse the damage that monoculture and its consequences have caused to the environment.

Despite its small scale, FarFarm has already established some important collaboration projects with large-scale enterprises, such as Renner (the largest national fast-fashion company) and Nespresso (the international coffee processing company). In addition, the project has recently expanded beyond the Amazonian region and has now settled in the Cerrado biome in Minas Gerais, for example, and as a small organic cotton plantation in the São Paulo urban environment.

5.2 *C(+)mas: Upcycling Specialist*

Sustainability in Brazil emerges from the people ([14], interview, 21 April).

Founded by Agustina Comas (Brazil/Uruguay) in 2008 and consolidated as a brand in 2015, C(+)mas [14] has been at the forefront of upcycling in Brazil. Comas' interest in the process started during her bachelor studies when she reflected on

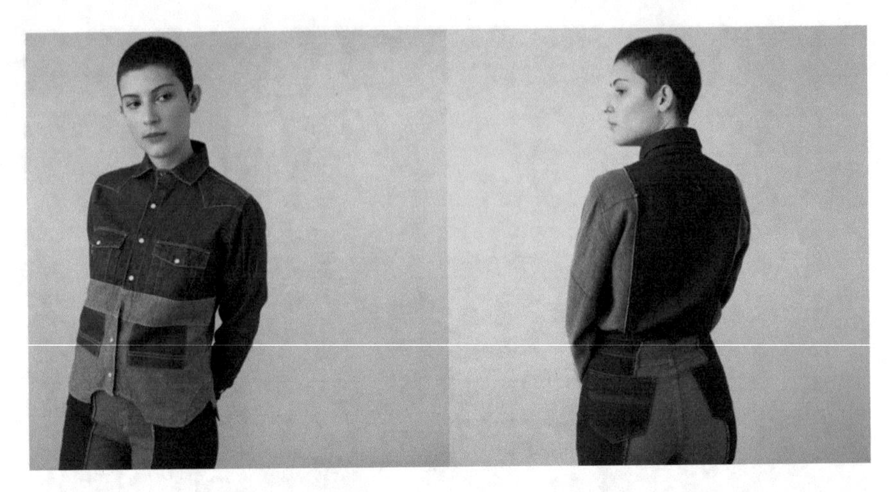

Fig. 3 Example of upcycled clothing on sale at C(+)mas online shop. *Image credit* C(+)mas

how discarded clothing could be used as 'raw material' for new clothes—a process she calls 'rescue'. Working with large-scale shirting industries in Brazil after her graduation has been both shocking and motivating to the young designer: the number of discarded shirts due to poor fit or small defects was enormous. At the same time, ready-made pieces offered a great opportunity to be redesigned into other garments. By becoming closely acquainted with factory owners, she was able to buy the discarded pieces at a low price and rework them into new, exciting pieces.

Comas notes that much in Brazil, the largest country in South America, reflects the dimensions of the country: 'it is common to find brands you have never heard of, and when you get to know about them you realize there are over 300 stores spread across the country' ([14], interview, 21 April). In 2018, the country produced 5.7 billion items of clothing, with an export rate of 0.05% [44]. To her, these numbers confirm the need to act on unsold, unused garments and circulate them back into the market and use. Her response to these worrying numbers was the development of a systematized method for upcycling clothes, especially shirt (Fig. 3), in collaboration with Ana Inés, a friend based in Uruguay, via Skype. Over the years, their explorations have expanded and the brand C(+)mas now carries a large repertoire of upcycling pieces added to fabrics made of textile salvages, such as Oricla (see Fig. 4).

According to Comas, sustainability is in its early stage in Brazil, but great individual efforts have started to gain attention and recognition, such as the Brazil Eco Fashion Week, the Grupo Transparência and the local Fashion Revolution. However, she sees that much of the initiatives focus on the social dimension of sustainability or are very experimental and difficult to scale. For the future of upcycling in fashion, she imagines a parallel system working together with the fashion industry, making use of discarded textiles in a symbiotic relationship.

Fig. 4 Oricla textile developed by C(+)mas. *Image credit* C(+)mas

5.3 beLEAF: Plant-based Textile Innovation

> Our work today is the result of a combination of almost magical coincidences (P Amaury 2020, interview, 19 May)

beLEAF is the trademark of a material innovation project developed by Paulo Amaury and Eduardo Filgueiras under the Nova Kaeru company (Nova [36], a pioneer in sustainable leather alternatives based in the town of Bemposta in Brazil. The idea for the company started when Filgueiras ran a frog farm for the food industry and was frustrated by a large amount of skin being discarded. However, frog skins were too small to use. After years of research to develop organic tanning and attempts with different food industry by-products, he was challenged by a pirarucu fish specialist (*Arapaima Gigas*) in the Amazon to tan the species. The giant fish skin leather became a significant cause of social and environmental sustainability in the Brazilian Amazon.[2] More recently, looking at the plants in the backyard of Nova Kaeru's factory, the duo decided to take on a new challenge: tanning plants.

Thus, beLEAF development [6] targeted yet another giant species, this time the *Alocasia Macrorrhizos* plant, native to Asia, but which also thrives in South American ground (nicknamed elephant's ear). This development is set to change how humankind relates to leather-like surfaces. With its 120 cm long and 100 cm wide

[2] Pirarucu is a fish native to the Amazon Rivers and one of the largest freshwater fish in the world, weighing up to 200 kg and measuring over 2 m in adulthood. More about Nova Kaeru's work on pirarucu as a sustainable approach to leather can be read in D'Itria and Valle-Noronha's [15] case study.

Fig. 5 Application of treated plant Elephant ear (*Alocasia Macrorrhizos*) in a bag by designer João Maraschin. *Image Credits* João Maraschin

leaves, the plant allows enough surfaces for the production of both accessories and clothing and requires little resources for growth (Fig. 5). The project was officially launched after 5 years in 2020 and was carried out by the company's technicians and staff. The unique network created through supplying pirarucu leather to key fashion companies in the world offered them a great testbed and high-quality feedback from the early stages of development.

Amaury assigns much of the success of their developments to a feeling of trust among the collaborators and to Filgueiras's true love and passion for sustainable material innovation. To the local community, the company has brought training and job opportunities. Bemposta is a small town in the state of Rio de Janeiro and has a high unemployment rate. Its efforts focus on working with and providing jobs for the local population, with many individuals working on the same project with a sense of shared ownership. They do not see the lack of financial backing from research institutes or governmental bodies—frequent in material development projects in western countries—as an issue; they see it as a quality that grants them the independence to develop projects at their own pace and according to their own values. For the world, beLEAF has the potential to create a permanent shift in the aesthetic orientation of leather-like products as well as radically reduce environmental impacts as the material requires minimum processing, unlike other current developments in plant-based textiles (e.g., cellulosic fibres). In this way, it is taking a step away from human-centred perspectives towards material innovation, relying on close engagement with vegetable and social entities.

5.4 Ateliê Vivo: Collaborative Space for the People

> We are motivated by the desire of encounters and transformation [...] and believe in the power of the collective (G Cherubini 2020, interview, 7 June)

Ateliê Vivo ('Living Studio' in English) started in 2015 as a space to rethink and motivate new forms of making, experiencing, and consuming fashion. It was initiated by multidisciplinary artist–researcher Karla Girotto and had its first home in Casa do Povo—a cultural centre in the city of São Paulo, in Brazil (Ateliê [4]. Currently, the project is led by a collective formed by Andrea Guerra, Carolina Cherubini, Fábio Lima Malheiros, Flávia Lobo de Felicio, and Gabriela Cherubini, as well as a number of collaborators and has recently moved to a new space in the city. In addition to working on the same project, the collective shares a background in the fashion and creative industries and restlessness to promote change. The space houses a public pattern-cutting library and a lab for fashion and textile practices, cross-disciplinary study and research groups in aesthetics (Fig. 6). Through the practices of making and reflecting, Ateliê Vivo seeks to explore the symbolic dimension of making through a constant state of sharing knowledge. Participants are particularly interested in raising awareness of the importance of time in developing relationships with clothing and fashion. They also emphasize how making together—when different bodies engage in performing similar tasks in the same physical space in silence—creates a special opportunity for deep reflection .

Fig. 6 Ateliê Vivo's physical space at Casa do Povo. Sewing machines, pattern cutting library and cutting tables. *Photo* Mariane Lima

For the participants, coming to the workshop is more than merely a means to make new clothes or learn a new skill. Social interaction is just as important for the exchange of ideas, personal growth and increasing one's knowledge of the self. The participants and workshop facilitators (both internal and external) become essential partners in consolidating the roles of the space. To the fast-paced, consumption-driven city of São Paulo, the space promotes resistance as it invites slow making, and material and knowledge sharing (Fig. 7).

For the first 2 years, the space was funded by public initiatives and held open cutting and introductory sewing workshops every first Saturday of the month, where volunteers guided the participants. As the project evolved and the demand for extra workshop hours grew, classes began charging a fee, though the library remains open to

Fig. 7 Facilitator and participants in a hand-stitching dressmaking workshop. *Image Credits* Ateliê Vivo (Instagram)

Table 2 Summary of the relationship between fashion design culture and cases of Brazilian sustainable fashion

Fashion design culture		Cases of Brazilian sustainable fashion			
Category/dimension	Concept	FarFarm	C(+)mas	beLEAF	Ateliê Vivo
Objectives of designing/Meaning	Usefulness		x	x	x
	Everyday life		x		x
	Wearer		x	x	x
	Practitioner him/herself	x	x	x	x
	Temporality	x	x	x	x
Production system/Material	Coexistence of globalism and localism		x	x	
	Multiplicity of actors	x	x	x	x
	Speed		x		x
	Seasonality	x		x	
	Plural feedback	x	x	x	x

the public and the introductory workshops continue. A number of fully-funded positions are also available for those unable to cover the costs. The instructors promote the sharing of materials (such as paper or fabric) within the classes. In addition to the courses offered at Ateliê Vivo, members also facilitate workshops in other institutions, raising funds to sustain the project. In the first semester of 2020, the project's business model was reviewed. Despite the financial struggle, the members find their 'everyday, love and life' (G Cherubini 2020, interview, 7 June) in the space and are truly motivated to continue affecting how people and clothes relate.

6 Discussion and Conclusion

In this study, we asked what are the characteristics of sustainable fashion in Brazil, and how these characteristics are different from the approaches in western societies. The questions are addressed by the following Table 2, on the basis of the individual description of the cases. The table summarizes the analysis of the cases of Brazilian sustainable fashion via diverse concepts introduced as the features of fashion design culture.

Overall, an added emphasis on the notions of 'affect'[3] and 'relationship' can be related to different concepts. Regarding the notion of affect, through the concepts

[3] In this work, affect is understood as 'an ability to affect and be affected' ([32], p. xvi) and includes both emotional and non-emotional changes in a body's ability to relate to other bodies (c.f., [45]).

of 'practitioner him/herself' and 'temporality', the ways in which Brazilian fashion practitioners intend to conduct their sustainable practices while fully embracing their personal emotions and emphasizing topical issues of society to which they are emotionally attached are clear. In relation to the concepts of 'multiplicity of actors' and 'plural feedback', we found that the Brazilian fashion practitioners had a relationship-centred approach. They heavily rely on local actors and resources and the community-driven bottom-up ecosystem, rather than on centralized support from governmental or regional funds. This encourages them to be innovative and to continually revisit their resources. Overall, it can be argued that Brazilian fashion practitioners demonstrate their particular culture for actualizing fashion sustainability through their affective and relationship-centred approaches. Our findings illustrate distinct differences from the western approaches to fashion sustainability. Whereas western approaches are heavily motivated by official bodies, such as political and economic organizations and units, the cases from Brazil show approaches centred on people and practices, starting from smaller scale actions and reaching wider audiences through time. Brazilian indigenous philosopher, Ailton Krenak, conceptualizes such efforts as those of '[…] "collective people", cells able to transmit through time their visions about the world' ([27], p. 12).

However, the purpose of this study was not to segregate different approaches. In order to address complex issues on sustainability, conjoined efforts are needed that complement each other. The findings from the cases invite further reflections to seek their possible applications in other contexts. In sum, this study looked into the ways in which fashion design practitioners have approached sustainability in Brazil. Through the lens of culture, we reviewed practices and models that both potentially advance and challenge the traditional western perspectives to fashion design. More specifically, we explored the cultural pillar of sustainability [23, 40], to view where its symbolic and material dimensions situate the local fashion practitioners.

Emerging from the four cases, the notions of affect and relationship and how these may support sustainability practices became the focus of this study. These findings need to be further studied to make a meaningful impact on the mainstream, western-centric, academic discourses on fashion sustainability. Related to this point, a clear shortcoming of this study is that it employed a European perspective to culture while exploring Brazilian cases of sustainable fashion (e.g., [30]). Thus, more systematic and in-depth reviews of knowledge that speak to the notion of culture in South America, beyond Brazil, are required. The recent development of the discussion on 'decolonizing design' can provide useful guidance for navigating the underexplored indigenous knowledge on fashion design and culture (see Abdulla et al. [1]). This chapter introduces four cases of Brazilian sustainable fashion, mostly in the Central South region, and depicts only a piece of a comprehensive and complex puzzle. The richness of Brazil as a country and a culture requires further studies on fashion sustainability by not only the local academic community but also the global community.

In the cases described here, however, affect is tightly connected to the emotional dimension of engagements.

The open-endedness and limitation of this study are due to the emerging stage of the discussions on the pluralities of sustainable development. Instead of isolating the world, the future of sustainable development should fully acknowledge and embrace different approaches. Accordingly, we suggest that both western and non-western societies become inspired by the cases from Brazil and that Brazilian fashion practitioners, with their innovative approaches, become more active in the global conversations on fashion sustainability.

Acknowledgements We are grateful for the support from the ENCORE research group at Aalto University and the constructive feedback that we received during the review process.

Appendix: List of Interviewees

- FarmFarm—Beto Bina (founder/owner)
- C(+)mas—Agustina Comas (founder/owner)
- beLEAF—Paulo Amaury (co-founder/owner)
- Ateliê Vivo—Gabriela Cherubini, Flávia Lobo, and Andrea Guerra (co-founders).

References

1. Abdulla D, Ansari A, Canli E et al (2019) The decolonising design manifesto. J Future Stud 23:129–132. https://doi.org/10.6531/JFS.201903_23(3).0012
2. Aspers P, Skov L (2006) Encounters in the global fashion business: Afterword. Curr Sociol 54:802–813. https://doi.org/10.1177/0011392106066817
3. Astra OH (2014) Culture as the fourth pillar of sustainable development. Sustain Dev Cult Tradit 1a:93–102
4. Ateliê Vivo (2020) Ateliê Vivo. https://atelievivo.com.br. Accessed 1 Jun 2020
5. Barnard M (ed) (2007) Fashion theory: a reader. Routledge, London and New York
6. BeLEAF (2020) beLEAF. https://www.novakaeru.com.br/beleaf. Accessed 8 May 2020
7. Berlim L (2014) A indústria têxtil brasileira e suas adequações na implementação do desenvolvimento sustentável. Modapalavra E-periódico 7:15–45
8. Berlim L (2012) Moda e sustentabilidade: uma reflexão necessária. Estação das Letras e Cores, São Paulo
9. Bertola P, Vacca F, Colombi C et al (2016) The cultural dimension of design driven innovation: a perspective from the fashion industry. Des J 19:237–251. https://doi.org/10.1080/14606925.2016.1129174
10. Black S (2008) Eco-Chic: The Fashion Paradox. Black Dog Publishing, London
11. Brandini V (2008) South American Style, Culture and Industry. In: Paulicelli E, Clark H (eds) The Fabric of Cultures: Fashion, Identity, and Globalization. Taylor & Francis, Abingdon
12. de Queiroz C, Braga CA (2015) A moda como instituição social no contexto belo-horizontino na década de 1980: a contribuição do grupo mineiro de moda na promoção de identidades e subjetividades. Universidade de Minas Gerais, Belo Horizonte
13. Chun N (2018) Re(dis)covering fashion designers: Interweaving dressmaking and place-making. Aalto University School of Arts, Design and Architecture

14. Comas A (2020) Comas. https://wwww.comas.com.br. Accessed 15 May 2020
15. D'Itria E, Valle-Noronha J (2020) Nurturing environmental and socio-cultural sustainable practices through materials innovation: The Osklen's case. In: Bertola P (ed) Fashioning social innovation: Design empowering communities to foster sustainability in culture intensive industries. Mandragora, Milano
16. Eicher JB, Evenson SL, Lutz HA (2008) The visible self: Global perspectives on dress, culture, and society. Fairchild Publications, New York
17. FarFarm (2020) Farfarm. https://www.farfarm.co/ Accessed 22 Jun 2020
18. Fletcher K (2008) Sustainable Fashion and Textiles: Design Journeys. Earthscan, London
19. Flick U (2009) An introduction to qualitative research, 4th edn. Sage, Thousand Oaks, CA
20. Götsch E (2020) Agenda Götsch. https://agendagotsch.com/en/agenda-gotsch-collection/. Accessed 1 Jun 2020
21. Hamilton JA (1987) Dress as a cultural sub-system: A unifying metatheory for clothing and textiles. Clothing Text Res J 6:1–7. https://doi.org/10.1177/0887302X8700600101
22. Haraway D (1988) Situated Knowledges: The Science Question in Feminism and the Privilege of Partial Perspective. Feminist Stud 14:575–599
23. Hawkes J (2001) The fourth pillar of sustainability: Culture's essential role in public planning. Common Ground
24. IBGE (2020) População. Instituto Brasileiro de Geografia e Estatística. https://www.ibge.gov.br/estatisticas/sociais/populacao.html Acessed 20 Jun 2020
25. Kaiser SB (1996) The social psychology of clothing: Symbolic appearances in context. London
26. Kawamura Y (2005) Fashion-ology: An introduction to fashion studies. Berg Publishers, Oxford
27. Krenak A (2019) Idéias para adiar o fim do mundo. Companhia das Letras, São Paulo
28. Lawson B (2005) How designers think: The design process demystified, 4th edn. Routledge, London
29. Loschek I (2009) When clothes become fashion: Design and innovation systems. Berg Publishers, Oxford
30. Manzini E (2015) Design, when everybody designs: An introduction to design for social innovation. MIT Press, Cambridge, MA
31. Maslow A (1954) Motivation and personality. Harpers, New York
32. Massumi B (1987) Preface in Deleuze G, Guattari F 1987. University of Minnesota, A thousand plateaus. Capitalism and Schizophrenia
33. Monteiro Schnaid G, Schemes C (2010) O processo de difusão da moda brasileira no mercado internacional. Gestão e Desenvolv 7:69–81
34. Moon CH (2011) Material intimacies: The labor of creativity in the global fashion industry. Yale University
35. Neira LG (2008) A invenção da moda no brasil. Caligrama 4:1–11. https://doi.org/10.11606/issn.1808-0820.cali.2008.68123
36. Nova Kaeru (2020) Nova Kaeru. https://www.novakaeru.com.br. Accessed 8 May 2020
37. Patton MQ (2002) Qualitative research and evaluation methods. Sage, Thousand Oaks, CA
38. REF (2019) Rio Ethical Fashion. https://www.rioethicalfashion.com/. Accessed 20 Jun 2020
39. Savelyeva T (2020) Vernadsky meets Yulgok: a non-western dialogue on sustainability. In: Jackson L (ed) Asian perspectives on education for sustainable development. Routledge
40. Soini K, Birkeland I (2014) Exploring the scientific discourse on cultural sustainability. Geoforum 51:213–223
41. Sterling S (2012) Let's face the music and dance? In: Wals AEJ, Corcoran PB (eds) Learning for sustainability in times of accelerating change. Wageningen Academic Publishers, Wageningen
42. United Cities and Local Governments (2010) Culture: Fourth pillar of sustainable development. Barcelona
43. United Nations (2015) Sustainable development goals. https://sustainabledevelopment.un.org/. Accessed 20 Jun 2020
44. Valente F, Santos L (2018) Relatório 2018 ABIT. Associação Brasileira da Indústria Têxtil e de Confecção, São Paulo

45. Valle-Noronha J (2019) Becoming with Clothes. Aalto ARTS Books, Espoo, Activating wearer-worn engagements through design
46. Vilaça N (2007) Brasil: Da identidade à marca. FAMECOS mídia, Cult e Tecnol 33:61–65
47. von Busch O (2009) Fashion-able: Hacktivism and engaged fashion design. University of Gothenburg, Gothenburg
48. World Commission on Environment and Development (WCED) (1987) in the UN-commifssioned report 'Our Common Future'
49. Yin RK (2014) Case study research: design and methods, 5th edn. Sage, Thousand Oaks, CA

Brazilian Organic Cotton Network: Sustainable Driver for the Textile and Clothing Sector

Larissa Oliveira Duarte, Marenilson Batista da Silva, Maria Amalia da Silva Marques, Barbara Contin, Homero Fonseca Filho, and Julia Baruque-Ramos

Abstract The influence of stakeholders' pressures on adopting better environmental practices is well reported in the supply chain management literature. In this context, product development policies focused on sustainability require integration between economic, social and environmental issues that cover the entire production chain. Clothing and fashion are highly visible elements of society; therefore, the textile industry serves a manner to promote a sustainable and eco-friendly mindset. The incorporation of eco-friendly and fair-trade fibers can be a starting point for changing the existing industrial paradigm within the textile industry. At the same time, cotton fiber is the most commonly utilized natural fiber in the textile industry. Cotton is a soft, staple fiber that grows around the seeds of the cotton plant, native to tropical and subtropical regions around the world, such as the Americas, South Asia and Africa. It is mainly used in spinning to produce ring or open-end yarn for weaving and knitting applications. The annual global production of cotton fiber is about 26 million tons. However, cotton production worldwide uses more than 20% of all insecticides employed in agriculture. In many areas, irrigated cotton cultivation has led to the depletion of ground and surface water sources. Many conventional cotton farmers in developing countries are in a crisis due to decreasing soil fertility, increasing production costs, resistant pests, or low cotton prices. In this scenario, an increasing number of cultivators turn to organic cultivation in order to restore soil fertility, reduce production costs, or to get a better price for their certified organic harvest. Organic cotton appears as an environmentally preferable product,

L. O. Duarte · B. Contin · H. F. Filho · J. Baruque-Ramos (✉)
School of Arts, Sciences and Humanities, University of São Paulo, São Paulo, SP, Brazil
e-mail: jbaruque@usp.br

H. F. Filho
e-mail: homeroff@usp.br

M. B. da Silva
Brazilian Agricultural Research Corporation (EMBRAPA Cotton), Campina Grande, PB, Brazil
e-mail: marenilson.silva@embrapa.br

M. A. da Silva Marques
Sítio Uruçu, Gurinhém, PB, Brazil

M. Á. Gardetti and R. P. Larios-Francia (eds.), *Sustainable Fashion and Textiles in Latin America*, Textile Science and Clothing Technology,
https://doi.org/10.1007/978-981-16-1850-5_14

of added benefit to the environment, farmers and consumers. Organic farming is slowly gaining ground in the global cotton market. It is often promoted to address the economic, environmental and health risks of conventional cotton production. Moving from the language of commodity chains to commodity networks, helps portray the complex network of material and nonmaterial relationships connecting the social, environmental, political, and economic actors. Understanding how individuals, firms, government authorities, and nongovernmental organizations (NGOs) are involved in economic and social transactions and how these different actors both shape and are shaped by network relations. In Brazil, organic and fair-trade cotton are widely seen as opportunities for smallholder farmers to improve their livelihoods. The cotton crop, due to its agronomic characteristics on climate adaptation, historical, cultural and economic value, established and gained prominence in family agriculture in the semi-arid region of Brazil. In this production network, trust was a critical factor in recruiting farmers and ensuring their continued participation in the organic and agroecological cotton production system and securing European customers' organic profile. Farmers' organizations as well as national and international environmental NGOs are instrumental in mediating and (re)building social networks among organic farmers and with the other actors in the supply chain. Linking small producers to markets, as integrating them into value chains is widely recognized as a valuable way to increase community development and benefit sustainable fashion brands.

Keywords Organic cotton · Sustainability · Sustainable networks · Agroecological · Textile and apparel · Biodegradable · Cellulosic fiber · Design · Fashion

1 Introduction

Cotton (*Gossypium* spp.) is cultivated for over 7,000 years, mainly for the production of its fiber [1]. Cotton and flax are the oldest natural plant fibers grown by man [2, 3]. In Europe, cotton played an important role in industrial development, especially in England [3, 4]. In America, this fiber was one of the most important sources of pre-Columbian civilizations, such as the Mayans in Guatemala, and the people of Chimu, Nazca and Paracas in Peru. In Brazil, cotton was already cultivated by natives [2, 3, 5].

Global cotton industry includes more than 100 million farm families across 75 countries [6]. It is one of the most widespread crops in the world in terms of land area [7]. Its production systems vary globally, ranging from labor-intensive systems in Africa and Asia, to highly mechanized systems in Australia, Brazil and the United States [7]. Cotton supply network actors range from small and large farmers, intermediates, traders and ginners, to sophisticated mills, textile processors, brands, exporters, global manufactures and retailers, transnational NGOs and consumers [8–11].

The market is expected to witness significant growth in demand for environment-friendly, versatile, biodegradable, and cost-effective fibers globally [12].

The sustainable raw materials contemplate the development and adoption of different types of environmentally friendly raw fibers, such as organic cotton, hemp, bamboo, lyocell, reused textiles and recycled fibers [13]. As a driver of innovation and sustainability in business models, the use of sustainable raw material means reliable access to a source of materials, technological development and communication of brand commitment to sustainable practices. Impacts on cost structure are not uncommon, as these technologies often have to be internally developed or externally acquired [14].

In this way, both agricultural and textile industries have experienced social-environmental issues linked to unhealthy workers conditions, water and soil pollution and biodiversity loss, characterized by the intense use of chemical products and natural resources [15]. Furthermore, the textile supply chain is long, complex, fragmented and opaque. Successfully translating demand for more sustainable cotton into actual sourcing depends on the entire supply chain taking part. However, many brands and retailers still have a limited overview of their own supply chain. They are therefore denied the opportunity to engage with key suppliers. Without clear indications of market demand, more sustainable cotton producers may return to conventional cotton farming [16].

Organic cotton appears as an environmentally preferable product, of added benefit to the environment, farmers and consumers [17, 18]. It is produced without the use of synthetically compounded chemicals [11, 19]. It emphasizes a system of production that seeks to maintain and replenish soil fertility and the ecological environment of the crop [17, 20].

In addition to these issues, this study is composed of this introduction (**1.0** subitem) and four following main topics. The **2.0** outlines data on cotton production in the world, Latin America and Brazil, and also aspects on fiber quality, colored cotton, certifications, characterizing production systems and producers. The **3.0** is devoted to organic cotton aspects, such as sustainable agriculture, a panorama of organic cotton (worldwide, in Latin America and Brazil), motivations and constraints, market potential and communicating organic cotton in retail. The **4.0** explores sustainable network management, including trustability and transparency, small scale and better quality process and local materials and local design. In the end, **5.0** presents the last considerations in conclusion.

2 Cotton Production

The cotton plant includes 52 species in the genus *Gossypium* (family Malvaceae). Species of cotton grown for commercial purposes are G. *hirsutum,* G. *barbadense,* G. *arboreum* and G. *herbaceum* [21, 22]. G. *hirsutum* is the main cultivated species and has medium fiber length. G. *barbadense* is the most appreciated as it presents long and extra-long fibers. G. *arboreum* and G. *herbaceum* have short fibers [21].

Cotton exhibits a certain degree of tolerance to salinity and drought, and it is grown in arid and semi-arid regions. However, higher and consistent yield and fiber quality levels are generally obtained with irrigation or enough rainfall [6]. There are several species of wild cotton in the world. They are found in Australia, Africa, Arizona, Central America, Lower California, Brazil, Mexico and other tropical countries and islands. Because of problems related to their refinement, however, they are not economically feasible to use [6].

Cotton is naturally a perennial plant that is now commercially cultivated as an annual plant in many parts of the world [1]. There are five major types of cotton being grown commercially around the world [23, 24]:

(i) **American Upland** (*G. hirsutum*): the most commonly planted type of cotton in the world, making up about 90% of the world's cotton crop. The multi-branched shrub-like plant may grow 1 to 7 feet (30.5–213 cm) tall, has creamy-white flowers, and produces white fibers up to 1¼ inches (3.2 cm) long. It can be made into many kinds of fabrics, and is used both for heavy canvas and for expensive shirts. It is grown as an annual;
(ii) **Egyptian** (*G. barbadense*): *Menoufi*, the most widely used variety, has exceptionally strong fibers about 1½ inches (3.8 cm) long. It has lemon-colored flowers and long, silky, light-tan fibers. It is made into clothing, balloon cloth, typewriter ribbons, and other fine fabrics;
(iii) **Sea-Island** (*G. barbadense*): It is now grown primarily in the West Indies. One of the most valuable and costly kinds of cotton is silky fibers about 1¾ inches (4.5 cm) long that can be made into very high-quality textiles. The plant has brilliant yellow flowers and white lint. However, it is expensive to raise because it grows slowly and has a low yield and small bolls. Technically, Sea-Island is closely related to Egyptian cotton, but growers consider it a separate kind of cotton because of its different fiber characteristics;
(iv) **Asiatic** (*G. arboreum*): grown mainly in China, India, and Pakistan. It has short, coarse, harsh fibers, and low yields. It is used for blankets, padding, filters, and coarse cloth.
(v) **American Pima**: a hybrid derived from Egyptian and American Upland cottons. It is the only variety of long-fiber cotton grown in commercially significant quantities in the American continent (especially in the United States and Peru).

'Mocó' cotton (*Gossypium hirsutum* r. marie galante Hutch) is grown in Northeastern Brazil, but its origin is unknown. The hypothesis is the mocó cotton lineage consists of several lineages rather than one, and that the 'marie galante' variety is one of them [25]. It presents great adaptability to semi-arid conditions [26]. It is also known as **'Seridó' cotton**. In this case, the name derives from the region of Rio Grande do Norte (state of Northeastern Brazil), which is the natural habitat of the mocó. This name can be used to identify the plant or even the long fiber that is obtained in this micro-region with the cultivation of mocó [27, 28].

The various kinds of cotton plants resemble each other in most ways; however, they differ in such characteristics as the color of flowers, character of fibers, and time of blooming. In addition, some varieties grow best on irrigated land. Some have lint 1¾ inches (4.5 cm) long, and others have lint only ½ inch (1.3 cm) long. Some varieties have stronger fibers than others. And some are easier to harvest by machine than others [6].

The diameter of the elementary fiber is about 15–25 μm and the length of cotton fiber is about 10–50 mm. Waxes and fats present on the surface of the fiber make it smooth and flexible. Cotton fibers are characterized by small elongation what ensures that the clothing will not change its shape during use. Cotton is also used for technical applications (industrial, medical, etc.) and interiors. In this last case, the best examples are highly absorbent bath towels and robes, upholstery material, household textiles, such as dishcloths, dusters, bed linen, tablecloth, handicraft, roller blinds, curtains, pillows, etc. Also, cotton is widely used in blends with other natural and man-made fibers [29]. Cotton has many positive characteristics such as versatility, comfort, color retention, absorbency, strength and durability [1].

In cotton cultivation, as the plant grows, flowers are formed in a vertical as well as horizontal direction. Shedding of fruit forms, particularly buds, could occur due to many complex factors, including meteorological, physiological, entomological, and nutritional. The cotton plant cannot retain all flower buds and convert them into bolls and retains only as many bolls it can afford to feed (Fig. 1a). By weight, seed cotton is composed of roughly one-third cotton lint and two-thirds cottonseed [20].

Then it splits open, showing four or five *locks* (groups of 8 to 10 seeds with fibers attached). The open-dried boll, which holds the fluffed-out cotton, is called the *bur*. An average boll will contain nearly 500,000 fibers of cotton, and each plant may bear up to 100 bolls. When fully matured, cotton bolls are picked and transported for processing, leaving the remaining plant as field trash. During the refining process or ginning of the harvested cotton, impurities are removed from the cotton fibers and are recovered as a processing by-product (CGT). From germination and emergence

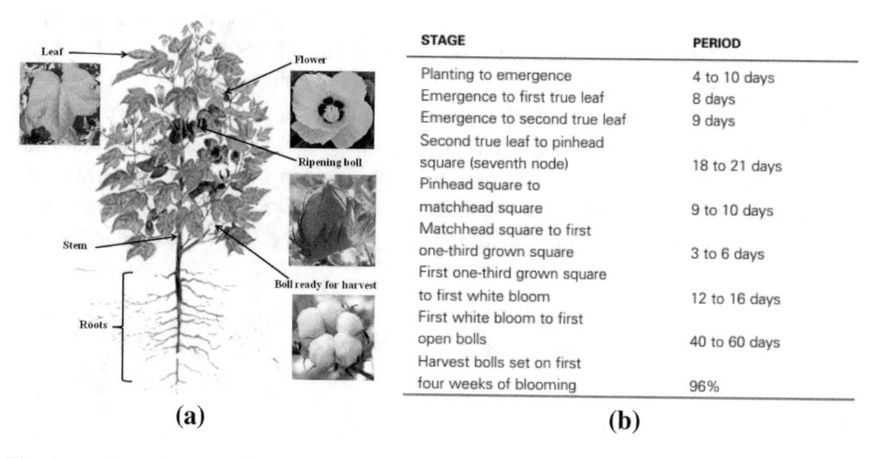

STAGE	PERIOD
Planting to emergence	4 to 10 days
Emergence to first true leaf	8 days
Emergence to second true leaf	9 days
Second true leaf to pinhead square (seventh node)	18 to 21 days
Pinhead square to matchhead square	9 to 10 days
Matchhead square to first one-third grown square	3 to 6 days
First one-third grown square to first white bloom	12 to 16 days
First white bloom to first open bolls	40 to 60 days
Harvest bolls set on first four weeks of blooming	96%

Fig. 1 **a** Cotton plant, the flower, boll and fiber [30]; **b** Average growth and fruiting period of the cotton plant [30]

of shoots to flower bud and peak flowering to boll development and bursting (Fig. 1b) [23, 24, 30].

Requirements of the cotton crop [31] (i) High temperature (ideally 30 degrees); (ii) Long vegetation period; (iii) Ample sunshine; (iv) Dry climate; (v) Min. 500 mm rainfall or irrigation; (vi) Deep soils; (vii) Heavy clay soils, ideally black soils; (viii) No waterlogging; (ix) Strong root growth in first two weeks; and (x) Natural bud shedding (only approx. 1/3 of flowers develop bolls).

During the refining process or ginning of the harvested cotton, impurities are removed from the cotton fibers and are recovered as a processing by-product. By-products of cotton include the edible oil gained from the seeds, and the seed cakes and husks are used as fodder and manure [32].

The production begins with the farmer, who grows cotton and harvests the lint (fiber) from the bolls of the plant [9]. The lint is separated from the seed using a cotton gin, a ginning process, and it is sold to spinners, who produce yarn (ICAC, 2010). Textile manufacturers transform yarns into the fabric, by knitting or weaving, and applying dyes and finishes [10]. In the final stage, end products (garments, home textiles, etc.) are made from fabrics [33, 34]. Cotton has many positive characteristics such as versatility, comfort, color retention, absorbency, strength and durability [1]. A summary of the complete process is presented in Fig. 2.

Cotton fiber is the most commonly utilized natural fiber in the textile industry. It is mainly used in spinning to produce ring or open-end yarn for weaving and knitting applications [36]. The annual production of cotton fiber is about 26 million tons [37]. At the same time, cotton consumes more pesticide than any other crop; it is estimated that 25% of the worldwide use of insecticide and 10% of pesticide use is accounted for by cotton cultivation [38].

The excessive use of agrochemicals is one of the main environmental problems related to the cultivation of cotton globally, the reason to seek to modify cultivation patterns and alternative methods of pest control that propitiate the sustainability of the agricultural system [39, 40]. The use of toxic substances to target and dye the cotton in the finishing stage is another problem of the cotton chain [40, 41].

The cotton textile and clothing industry is a complex and multi-tiered system consisting of cotton cultivation and harvesting, fiber production, yarn manufacturing, fabric preparation, fabric processing that includes bleaching and dying sub-processes, and fabrication of the final fabric product. An array of environmental concerns is associated with this sector, the most significant of which are issues related to the

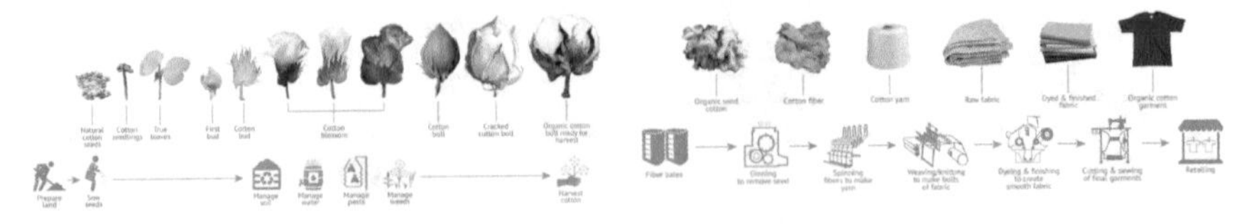

Fig. 2 Field to Fashion: cotton development and general process of transforming seed cotton from the farm into final garment [35]

use of agrochemicals in the cultivation of cotton and water, energy and chemical consumption in the fabric processing stage [42, 43].

The situation is further complicated since some of the leading cotton-producing governments have begun to reduce the size of cultivated cotton due to the severe shortage of water resources—the cultivation of cotton requires huge amounts of water, consuming 2.6% of the global water supply [42]. Moreover, the strain on agricultural land is further increased due to the terrible increase in the population size [44].

2.1 Cotton in the World

The world cotton trade generates around USD 12 billion annually and involves more than 350 million people in its production, from farms to logistics, ginning, processing and packaging [45]. According to the *Instituto de Economia Agrícola* ("Institute of Agricultural Economics") (2020), in 2019, 35 million hectares were planted, with a production of 25 million tons of lint cotton in the world. Cotton is planted in more than 60 countries, being present on five continents [45]. The forecast for the 2019/20 harvest is 26.23 million tons of cotton [46].

Many conventional cotton farmers in developing countries are in a crisis due to decreasing soil fertility, increasing production costs, resistant pests, or low cotton prices. In this scenario, an increasing number of farmers turn to organic cultivation to restore soil fertility, reduce production costs, or get a better price for their certified organic harvest [31].

The livelihood of 17 million people in India depends on cotton farming. The Indian cotton textile industry contributes 38% of the country's export earnings. In some African countries like Burkina Faso, Mali, and Benin, cotton plays an even more dominant role in agricultural exports. World market cotton prices fluctuate to a great degree and have come down considerably over the last two decades. According to Oxfam and other NGOs, this is partly due to high farm subsidies in the US [21]. This scenario mostly considers the main countries in cotton production and exports 2018–2019 and 2020 (Table 1).

Commodity prices are primarily driven by supply and demand. Aspects such as fiber quality (staple length, strength, color, leaf grade, trash content, etc.) also play a part. Other price influencers and considerations include stocks and subsidies, logistics, transportation and warehousing, trader costs, currency conversions and insurance. Agricultural policies and strategies applied by some of the main producer countries (China, India, Brazil and the USA) influence the market, as have environmental factors and competition from other commodities. The prices of competing crops influence farmers' decisions about what to grow. Higher prices for crops such as corn and soybean obviously make those crops more attractive to farmers—and, as a result, can displace cotton production and drive up prices. Additionally, of course, there is competition between fibers and, with polyester being so competitive in price, for example, the price of cotton is impacted [48, 49].

Table 1 Main countries in cotton production and exports 2018/19, May 2019/20 and May 2020/21 [47]

World cotton production				World cotton exports			
Million 480 lb bales		2019/20	2020/21	Million 480 lb bales		2019/20	2020/21
	2018/19	May	May		2018/19	May	May
India	25.8	30.5	28.5	US	14.8	15.0	16.0
China	27.8	27.3	26.5	Brazil	6.0	8.6	9.0
US	18.4	19.9	19.5	India	3.5	3.2	4.5
Brazil	13.0	13.2	12.0	Greece	1.4	1.4	1.5
Pakistan	7.6	6.2	6.3	Benin	1.3	1.2	1.3
Rest of World	26.1	25.6	26.2	Rest of World	14.2	10.6	10.7
World	118.7	122.7	119.0	World	41.1	40.0	42.9

The international market trends for cotton in 2020 reflect upward factors: the possible commercial agreement between the USA and China; good volume exported by US respect the gathered from the 2019/20 harvest; and diminishing in the area to be planted in the US. And the downward factors: estimated global productive surplus for the 2019/20 and the 2020/21 harvests; trade war between the USA and China; coronavirus and its effects on the economy; and oil prices at very low levels. In this way, considering the decrease in global consumption, the expectation is that the prices will remain under pressure along 2020 year [47, 50].

2.2 *Cotton in Latin America*

In Latin America, the three main cotton-producing countries are Brazil, Mexico and Argentina, followed by Venezuela, Peru, Colombia and others [51].

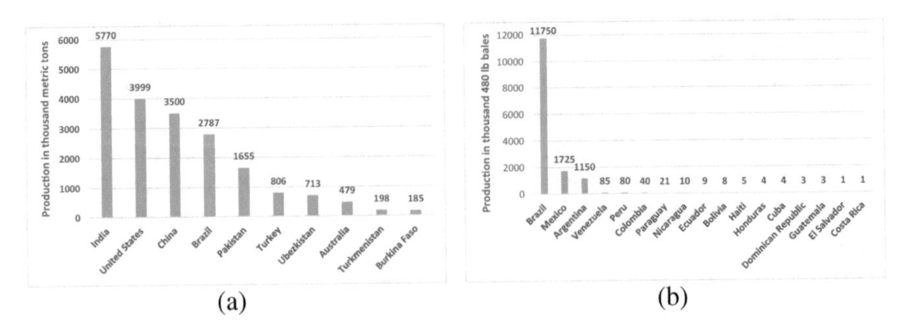

Fig. 3 **a** Cotton production by country worldwide in 2018/2019 (in 1,000 metric tons) [52], **b** Cotton production in selected Latin American and Caribbean countries in 2018 (in 1,000 480-lb bales) [51]

Complementing the data in Table 1 (2017–2018), Fig. 3a shows the world's leading cotton-producing countries in the period of 2018/2019.

Meanwhile, taking into account the cotton production in Latin America in the same period (Fig. 3b), Brazil stands out as 1st in the region, far surpassing other countries in the region [51, 52].

However, these data are about cotton production in general, which neither highlight the historical importance of cotton production in Latin America or the importance of specific cotton cultivations (Pima, organic, natural colored, etc.) in this region, for which a more detailed approach will be made forward in this text.

2.3 *Cotton in Brazil*

The second half of the 90 s marked the migration of the cotton crop from the traditional production areas in the semi-arid to the Brazilian Cerrado biome [53]. In 1997, the first producers' association was created in Mato Grosso state, AMPA (Matogrossense Cotton Producers' Association). Later in 1999, it was created the Brazilian Association of Cotton Producers (ABRAPA). In the same year, they were also created important states associations, in São Paulo, Minas Gerais, Goiás and Mato Grosso do Sul. And in the next year, in the states of Bahia, Parana and Maranhão. In 2001, the Brazilian cotton market was self-sufficient, supplying 100% of the textile industry demands, producing around 900 thousand tons per crop [5]. Brazilian fiber production is now characterized by high agroindustry scale, mechanization, analyses and fiber classification considering international standards, traceability system, environmental certification (**BCI—Better Cotton Initiative**).[1]

Brazil has achieved and maintains the productivity in cotton crops under the rainfed system, which is called cultivation without irrigation [32]. In the 2016/2017 harvest, of the 940 thousand hectares planted, only 40 thousand (4.3%) were irrigated. In the general ranking of productivity, the country ranks fourth, with 1,600 kg of feather per hectare, behind Israel, which harvested 1.76 thousand kilos, but irrigates 100% of its areas, Australia (1.74 kg/ha), where irrigation reaches 95% and China (1.66 kg/ha), which has 80% of the crops dependent on the artificial addition of water [56]. Mato Grosso is also the largest cotton producer in the country, accounting for about 70% of Brazilian cotton production. The national average is around 1.5 tons per hectare, while the United States produces an average of 0.8 tons per hectare. However, the country has great potential to be improved [57].

---

[1] According BCI, the Better Cotton Standard System is a holistic approach to sustainable cotton production which covers all three pillars of sustainability: environmental, social and economic. Each of the elements—from the Principles and Criteria to the monitoring mechanisms which show results and impact—work together to support the Better Cotton Standard System, and the credibility of Better Cotton and BCI. The system is designed to ensure the exchange of good practices, and to encourage the scaling up of collective action to establish Better Cotton as a sustainable mainstream commodity [54, 55].

The success of the corporate system implemented in the Cerrado biome is based largely on the intensive use of modern agricultural inputs, mechanized operations, use of skilled labor, and access to large buyer markets in Brazil and abroad. This model involves higher production costs and the need for scale, encouraging cotton production on large farms. As a result, around 90% of Brazilian cotton is produced on properties with an area equal to or greater than 1,000 hectares [32]. Cotton farmers in the Cerrado, in general, have good access to domestic and foreign markets, a result of the high quality of the fiber produced, the creation of sales coops, the strong performance of agricultural commodity trading companies, and the professionalism with which they meet deadlines and comply with legislation [57, 58].

Due to its agronomic characteristics of adaptation to the region's climatic conditions, the small-scale familiar cotton crop, its historical-cultural value and, mainly, economic, established and gained prominence in family agriculture in the semi-arid region of Brazil. However, throughout history, cotton has gone through ascension, crises and declining production and productivity. In summary, one can attribute the reasons that led to the decline of cotton production: (i) non-conservative crop management (ii) government policies of low and price variations; (iii) occurrence of extreme droughts; and (v) the advent and spread of the boll pest (*Anthonomus grandis* Boheman) [39, 40].

There are alternative production systems practiced by family farmers, or even by small- and medium-scale growers, aimed at exploiting niche markets, among which are: production of colored cotton, organic cotton, and agroecological cotton. These systems are of major social importance and have received growing support from government policies and companies in the textile and apparel sectors that operate in niche markets. These types of products are highly valued [32]. However, market access is a major challenge for small- and medium-scale growers.

Presenting both cotton scale productions, Fig. 4a indicates the states and its production volume in Brazilian states in 2017 for *G. arboreum* and Fig. 4b, for *G. herbaceum*.

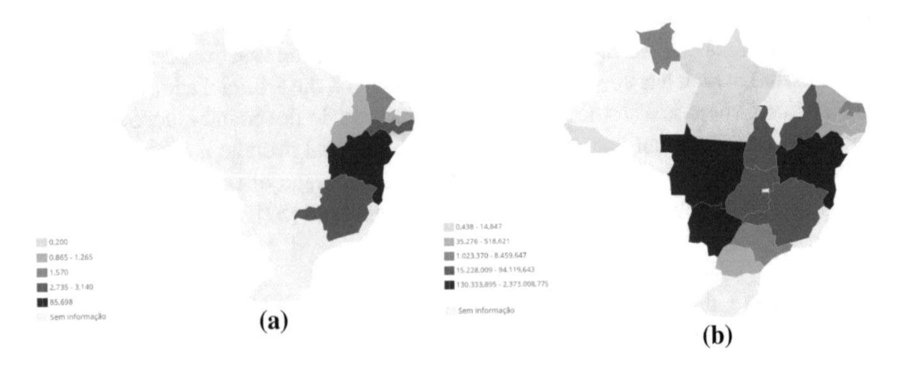

Fig. 4 Volume of Cotton Produced in Brazilian States in 2017: **a** G. *arboretum* (96,225 tons and 135 establishments) [59]; **b** G. *herbaceum* (3,664,808,060 tons and 3,081 establishments). On the left side of each map, the tons amount scale [59]

The cotton fiber obtained in Brazil is marketed in more than 40 countries and can be considered one of the best in the world. Despite the great national production, Brazil is still an importer of cotton fiber. This is because most of the fiber produced in the country is medium in size, and the textile industry still lacks reasonable amounts of longer fibers. The longer fibers allow manufacturing of lighter fabrics, which according to the analyst, are a trend. The new challenge of the Brazilian cotton sector is to produce enough fine or extra-long fibers, thinner and more resistant, that generate lighter fabrics that please consumers [32].

There is a growing demand for products and services generated without aggression to the environment and respect to the worker's dignity. In 2010, ABRAPA implemented BCI in Brazil. BCI is an international organization aimed at improving good production practices, fair working relationships and the transparency and traceability of cotton in the market. Following the same line, ABRAPA created the Brazilian Responsible Cotton (ABR) program in 2012, –81% of Brazilian producers are ABR certified nationwide, and 71% also have the international seal BCI. Brazil supplies 30% of all the BCI cotton in the world. Brands such as Adidas, Nike, Levi Strauss and Co, and C&A are some of BCI's most influential partners [54].

BCI cotton has been used as an answer to the sustainability challenges of the traditional cotton culture. The model represents an advance, especially in terms of social and labor rights aspects. However, it deserves criticism regarding the maintenance of the use of agrochemicals and transgenic seeds, in a similar way to conventional cotton [60].

2.4 *Fiber Characteristics and Quality*

2.4.1 Fiber Characteristics

A large number of literatures are available on the characterization of cotton fibers. Cotton fiber is composed mostly by cellulose (82.7%), and minor fractions of hemicellulose (5.7%) and wax (0.6%), among other constituents [19]. The cellulose in cotton fibers is also of the highest molecular weight among all plant fibers and highest structural order, i.e., highly crystalline, oriented and fibrillar. Cotton, with this high quantity and structural order of the most abundant natural polymer, is, not surprisingly, viewed as a premier fiber and biomass [61].

The cross-sectional view of cotton fiber is kidney-shaped (Fig. 5a, b). The cross-section tends to indicate the relative dimensions of the lumen and fiber walls [62]. The cotton walls are constituted by various layers with different cellulose structural orders [63], as presented in Fig. 5c.

The fibers are classified according to their length. Lower if smaller than 22 mm. Averages, if measured from 28 to 34 mm. Long above 34 mm. The good length of the cotton fiber helps in the easier spinning into a smoother, stronger yarn. Yarn quality parameters such as evenness, strength, elongation and hairiness are correlated to the length of cotton fibers. Spinning parameters depend on the length of cotton

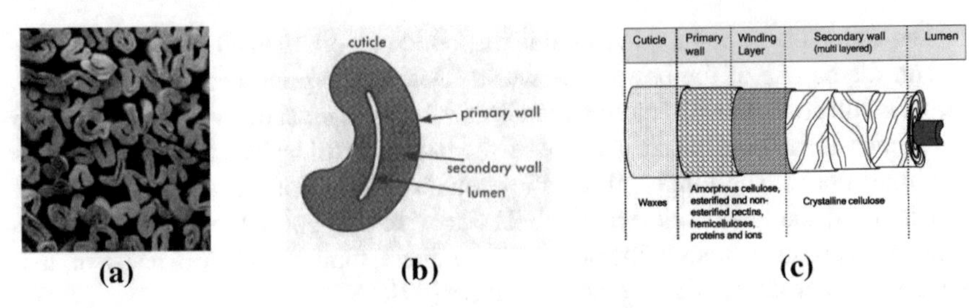

Fig. 5 Cotton fiber: **a** cross-section of raw cotton [64]; **b** macro structure [64]; **c** schematic representation of mature cotton fiber showing its various layers [63]

fibers, which results in a more comfortable, more durable, more attractive fabric and garments [62].

Murugesh et al. [19] research characterized organic and conventional cotton fiber properties, such as the surface morphology, surface chemical and elemental composition, internal chemical composition and architecture. The main properties pointed by them are: fiber length (10–50 mm); diameter (10–27 μm); moisture regain at 65% R.H. (8.5%); elongation strain to failure (7%); dry tenacity (0.3–0.5 N/tex) and flammability limiting oxygen index (LOI) (17–19%).

2.4.2 Fiber Quality

The quality of the cotton is evaluated according to the length of its fiber, following the fineness, color and purity. The classification of fiber quality is also made according to the resistance test, uniformity of length, and the relationship between maturity [62].

The mix of perennial crop and annual growth habits contribute to the variability in cotton lint quality. The amount of sunlight, day and night temperatures during growth, variety and agronomic inputs are responsible for year-to-year variations in quality. Fiber properties have been studied since the early 1900s, but electronic and physical sciences have been employed in measuring quality parameters only since the 1950s. High Volume Instrument (HVI) is a machine for measuring quality characteristics in cotton, to reduce the time required to measure fiber properties [20]. HVI was adopted in Brazilian labs following an international standard, in 2003, supporting production and chain rastreability and fiber quality control [5].

The main HVI determinations include [65]:

Fiber Length—the average length of the longer one-half of the fibers (upper half mean length). It is reported in both 100ths and 32nds of an inch.

Length Uniformity—the ratio between the mean length and the upper half mean length of the fibers and is expressed as a percentage.

Fiber Strength—reported in terms of grams per tex. A tex unit is equal to the weight in grams of 1,000 m of fiber. Therefore, the strength reported is the force in grams required to break a bundle of fibers one tex unit in size.

Micronaire—a measure of fiber fineness and maturity (μg/in). An airflow instrument is used to measure the air permeability of a constant mass of cotton fibers compressed to a fixed volume.

Color Grade—The color grade is determined by the degree of reflectance (Rd) and yellowness (+b) as established by the official standards and measured by the HVI. Reflectance indicates how bright or dull a sample is and yellowness indicates the degree of color pigmentation.

Trash—a measure of the amount of non-lint materials in the cotton, such as leaf and bark from the cotton plant. The surface of the cotton sample is scanned by a video camera and the percentage of the surface area occupied by trash particles is calculated.

The complementary classer determinations include [65]:

Leaf Grade—a visual estimate of the amount of cotton plant leaf particles in the cotton. There are seven leaf grades, designated as leaf grade "1" through "7", and physical standards represent all. Also, there is a "below grade" designation which is descriptive.

Preparation—describe the degree of smoothness or roughness of the ginned cotton lint. Various methods of harvesting, handling, and ginning cotton produce differences in roughness or smoothness of preparation that sometimes are very apparent.

Extraneous Matter—any substance in the cotton other than fiber or leaf. Examples of extraneous matter are the bark, grass, spindle twist, seedcoat fragments, dust, and oil.

The measurement units, values and classifications for all the determinations expressed above are provided by USDA and replied by many other institutions related to cotton production and trade [65–68].

Specifically, about the color grade, it is determined by the degree of reflectance (Rd) and yellowness (+b) as established by the official standards and measured by the HVI. Reflectance indicates how bright or dull a sample is and yellowness indicates the degree of color pigmentation. A three-digit color code is used. The color code is determined by locating the point at which the Rd and +b values intersect on the Nickerson-Hunter cotton colorimeter diagram for Upland cotton. The color of cotton fibers can be affected by rainfall, freezes, insects and fungi, and staining through contact with soil, grass, or the cotton plant's leaf. Color also can be affected by excessive moisture and temperature levels while cotton is being stored, both before and after ginning. As the color of cotton deteriorates due to environmental conditions, the probability of reduced processing efficiency is increased. Color deterioration also affects the ability of fibers to absorb and hold dyes and finishes. There are 25 official color grades for American Upland cotton, plus five categories of below grade color, as shown in the tabulation below. USDA maintains physical standards for 15 of the color grades. The others are descriptive standards (Table 2) [65–67].

Classification procedures for American Pima cotton are similar to those for American Upland cotton, including the use of HVI measurements. The most significant difference is that the American Upland color grade is determined by instrument measurement, while highly trained cotton classers still determine the American Pima

Table 2 Color Grades of Upland Cotton [65, 67]

	White	Light spotted	Spotted	Tinged	Yellow stained
GM (Good Middling)	11.1[b]	12	13	–	–
SM (Strict Middling)	21.2[b]	22	23[a]	24	25
M (Middling)	31.3[b]	32	33[a]	34[a]	35
SLM (Strict Low Middling)	41.4[b]	42	43[a]	44[a]	–
LM (Low Middling)	51.5[b]	52	53[a]	54[a]	–
SGO (Strict Good Ordinary)	61.6[b]	62	63[a]	–	–
GO (Good Ordinary)	71.7[b]	–	–	–	–
BG (Below Grade)	81	82	83	84	85

[a]Physical standards for color grade only;
[b]Physical standards for color grade and leaf grade; All others are descriptive

color grade. Different grade standards are used because the color of American Pima cotton is a deeper yellow than that of Upland. Also, the ginning process for American Pima cotton (roller ginned) is not the same as for Upland (saw ginned). The roller gin process results in an appearance that is not as smooth as obtained with the saw ginned process. There are six official grades (grades "1" through "6") for American Pima color and six for leaf. All are represented by physical standards. There is a descriptive standard for cotton, which is below grade for color or leaf. A different chart is used to convert American Pima fiber length from 100ths of an inch to 32nds of an inch [65, 68].

The quality of fibers and their properties correlate with the textile process and product development. The length characteristics, for example, mean the longer the fiber is, it will enhance the machinery production. Resistant fibers will influence the textile softness. Also, the fiber color uniformity and cleaning will influence processes quality and costs. The micronaire index measuring fiber maturity and density, will indicate that the fibers with low value and high maturity will be more resistant and longer. The more the yarn is long and homogeneous, the high its quality [64].

2.5 Colored Cotton

Naturally colored cotton was cultivated by the Incas from 4500 B.C., then later by other peoples in the Americas, Asia, Africa and Australia [3, 69, 70]. The growing demand for organic products generates interest in this crop because it does not require dyeing [21, 62]. More than 39 wild cotton species with colored fibers have already been identified. In most of these primitive species, cotton has colored fibers, mainly in brown. However, colored cottons in shades of green, yellow, blue and gray have already been described. These cottons, for long periods, were discarded by the global textile industry and their exploitation in several countries was even banned because they are considered as undesirable contamination of normal white cottons. These colorful types have been preserved by native peoples and in cotton collections in several countries [69].

Naturally colored cottons are considered environmentally friendly because they have many insect and disease-resistant qualities, are drought and salt tolerant and do not have to be dyed artificially. Colored cottons have been grown successfully with organic farming methods [62].

Although the organically grown naturally colored cotton is more expensive than traditional varieties, it is being exploited in relatively small but potentially lucrative market niches, similar to the 'organic food' and 'fair trade' markets. One niche is the sales of clothing in the local tourism sector; another one is children's clothing which parents do not want their children to be exposed to any chemical process-based clothing; and generally, consumers who desire to support healthy and ethical new-growth businesses [70].

In Brazil, cotton plants, in cream and brown tones, were collected in mixtures with cultivated white cottons, of the species *G. barbadense* L. and *G. hirsutum* L., Marie Galante Hutch variety, known as arboreal cottons. Since 1989, EMBRAPA (Brazilian Company of Agricultural Research) has begun to study the formation of genetic crosses to improve the raw material. The investments in production meant the resumption of the cotton crop in the northeastern semi-arid region, which had practically disappeared due to the beetle. Colored fiber has a market value of 30% to 50% higher than white cotton fibers. The colored cotton developed by EMBRAPA is considered ecologically correct, since it eliminates the process of dyeing, one of the most polluting of the textile industry [71].

To date, five cultivars BRS 200, BRS Verde, BRS Rubi, BRS Safira ("sapphire") and Topázio ("topaz") have been presented commercially (Fig. 6a, b).

Also, clothes made with this material can be used without problems by people allergic to dyes. Another quality of these seeds lies in the fact that they can be planted in dry regions and use less pesticides, that there are not many pests that attack this variety [62]. Cotton occurs naturally in white, brown, green colors, varying in a range of shades. The colored lint is shorter in length. Lack of sufficient sunlight affects the ability of a genotype to express its color. Colored cotton lint varieties have the same agronomic management requirements as white ones (Table 3) [20].

The production of colored cotton on a commercial scale began only in the first half of the 2000s [21, 73].

2.6 Cotton Certifications

With the growth in demand for organic products, it was necessary to create mechanisms for the certification of such products, based on standards that regulate production, processing and marketing [74]. This socio-environmental certification is fundamental for sustainable agricultural promotion, and it generates product differentiation with sustainable management, thus ensuring that the product consumed comes from ecological management and that there is no incentive to degrade natural resources

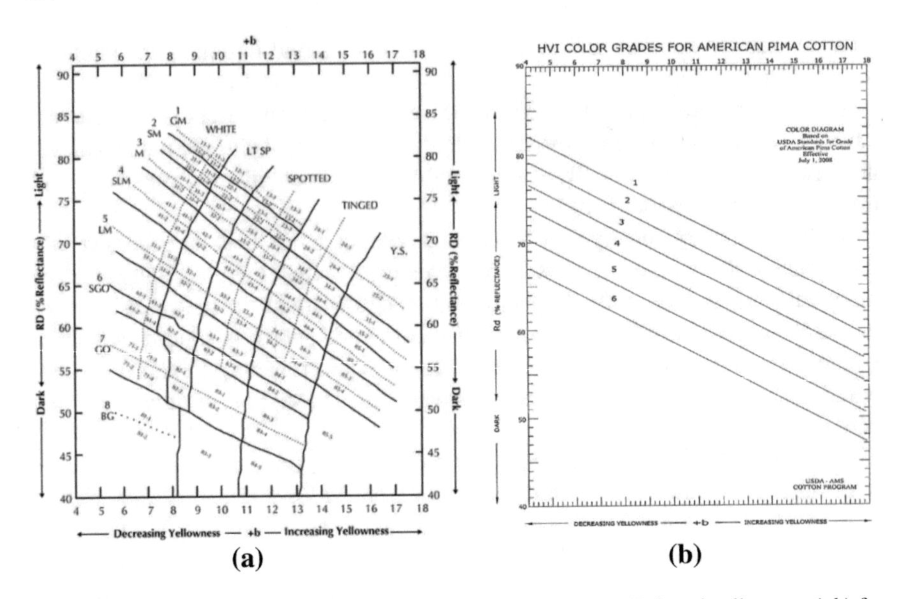

Fig. 6 **a** HVI color chart plotting of the distribution of reflectance (Rd) and yellowness (+b) for American Upland cotton. It includes color grades for white, light spotted, spotted, tinged and yellow stained cotton [65]; **b** HVI Color Chart for American Pima Cotton [67]

Table 3 Color inheritance of cotton fiber and geographical origin [21, 67, 72]

Gene symbol	Fiber color	*Gossypium* species	Region
Ld^{1k}	Khaki	Arboreum and herbaceum	Africa and Asia
Lc^{2b}	Light brown	Arboreum and herbaceum	Africa and Asia
Lc^{2k}	Khaki	Arboreum and herbaceum	Africa and Asia
Lc^{2M}	Medium brown	Arboreum and herbaceum	Africa and Asia
Lc^{2v}	Slight brown	Arboreum and herbaceum	Africa and Asia
Lc^{3B}	Light brown	Arboreum and herbaceum	Africa and Asia
Lc^{4k}	Khaki	Arboreum	Asia
D^{W}	Off white	Raimondii	America
Lg_1	Green	Hirsutum	America
Lc_2	Brown	Hirsutum	America
Lc	Brown	Barbadense, darwinii and tormentosum	America

[75]. Furthermore, the main aim of the certification processes is to maintain the integrity of the organic nature of the fiber as much as possible. Regulations are important because they standardize criteria for organic production and post-harvest handling/processing that facilitate domestic and international trade [8].

In 2018, the global cultivation of organic cotton grew by 10% compared to 2017 [7]. Although organic cotton production worldwide is still small the demand for the product only grows [76]. To produce and sell organic cotton, it is necessary to follow the principles of organic production, certifying the production systems. The organic certification is the guarantee that the producer has followed norms/standards for organic production, and in this way, it can be called an organic product. Products labeled as green, agroecological, biodynamic, ecological or sustainable, which have not undergone an organic certification process, cannot be called an organic product [77].

The organic conformity assessment is a systematic process, composed of pre-established rules and standards, which undergo monitoring and evaluations, promoting security and guarantee that a product, process or service, meets the requirements determined in the rules and regulations. The conformity assessment involves the following activities: selection of standards or regulations, collection of samples, carrying out laboratory analyzes, carrying out inspections, carrying out audits and product traceability [78]. Furthermore, a three-year transitional period from conventional to organic cotton production is required for certification [42].

These certifications are made by public or private agencies, and are important to discourage possible opportunistic actions [79], in addition to ensuring the integrity and origin of the product, so only certified products contain a guarantee of organic property.

According to the IFOAM—International Federation of Organic Agriculture Movements [80], "there are two different kinds of organic farms in the world (1) Certified organic farms producing for a premium price market and (2) Non-certified organic farms producing for their own households and for local markets". Organic production takes place in more than 180 countries; of this total, only 87 countries have their own regulations [81], with Brazil among these countries. In each country or region,, organic cotton is subject to organic production laws, such as Organic Regulation No 834/2007 in Europe, USDA National Organic Program (NOP) in the United States, and the National Organic Production Program (NPOP) in India [60]. Also, in organic agriculture, international standards and certificates exist, such as specific standards such as ISO 9000, ISO 14,000 [82].

The Brazilian Organic Law, Law No. 10,831/2003, was sanctioned in 2003, is regulated in 2007, through the publication of Decree No. 6,323/2007. The Ministry of Agriculture, Livestock and Supply is the supervisory body. The Brazilian legislation regulates two certification mechanisms, which are: Audit Certification or Third-Party Certification and the Participatory Guarantee System or Participatory Certification or **Participative Guarantee System (PGS)**.[2] Brazil was the pioneer country in participatory certification, being a world reference [77].

The Participative Guarantee System, together with the certification by external audit, composes the Brazilian System of Evaluation of Organic Conformity (SisOrg) [84].

[2] Participatory Guarantee Systems (PGS) are an alternative to third party certification. As per IFOAM—Organic International's definition, PGS are locally focused quality assurance systems that certify producers based on active participation of stakeholders and are built on a foundation of trust, social networks and knowledge exchange. IFOAM—Organics International has a list of recognized PGS programs [83].

Participatory certification is a process that involves knowing how to produce, managing production and marketing and registering all information, a process that, for many, is complex and demands support [85]. It is a system composed of commissions and councils, with the mandatory effective and direct participation of producers interested in certification, with the participation of technicians and employees. In order for the Participatory Guarantee System to carry out certification activities in the production systems of its members, it is necessary to accredit a **Participatory Organism for the Evaluation of Organic Conformity** (OPAC) in the Ministry of Agriculture, Livestock and Supply [84]. (MAPA). The mechanisms of evaluation of organic conformity have the as main objective to certify the quality of organic products to consumers [77]. The OPAC is a legally constituted association with an ethics council, which assumes the formal responsibility for the set of activities, attesting to whether the products and producers meet the requirements of the organic production regulation. After the MAPA accreditation, OPAC may authorize the producers controlled by it to use the Standard of the Brazilian Organic Conformity Assessment System and become responsible for launching and keeping up to date all the data of the production units it controls, in the National Register of Organic Producers and in the National Register of Productive Activities, so that information is available to society [84].

Certification by external auditing is carried out by independent third-party organizations, which ensure compliance with organic production procedures—provided for in international standards and legislation. Despite the importance of certification in expanding the production and marketing of organic products in Brazil attesting to the credibility of the product, certification by external auditing has high costs; it is often not feasible for the family farmer [84].

Audit certification is a private service carried out by contracting services from certifying companies. The certifying companies follow the standards of the countries in which they operate, utilizing the standards from IFOAM and of ISO Guide 65, which are internationally recognized references [77].

According to Ferraz [60], there are various others standards and certifications for classified cotton called by Textile Exchange as "Preferred Cotton" (cotton that is ecologically and or socially progressive because it has more sustainable properties compared to other conventional options). The organization, "Preferred Cotton" (pCotton) includes: Recycled, Organic, Fair Trade, CmiA, Better Cotton (BCI) and its equivalences.

Also, there are initiatives aimed at certifying recycled cotton. In order to ensure the production model, origin and traceability, there are some types of standards and certifications, legally defined as external audit models.

In Brazil, organic cotton is attested by auditory certification (carried out by private companies: IBD Certifications and ECOCERT Brazil). The PGSs in the Northeastern Semi-Arid are: *Rede Borborema de Agroecologia* ("Borborema Network of Agroecology"), ACEPA, ACEPI, APASPI, ACOPASA and ECOARARIPE.

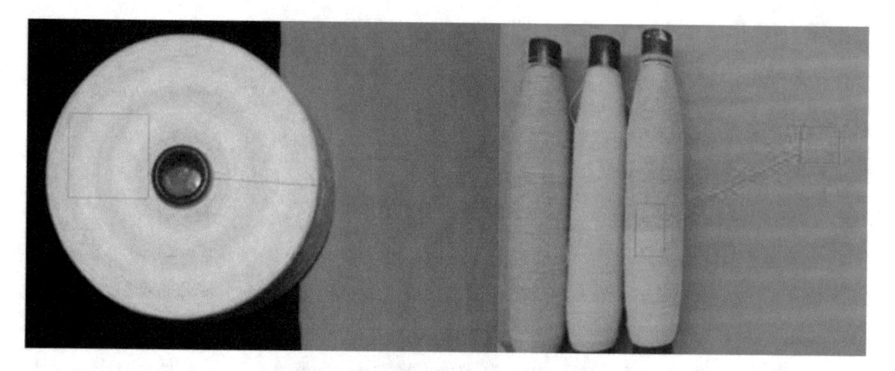

Fig. 7 Example of yarns and textiles with color variation [63]

In addition, some companies and farmers' associations also employ **GOTS**[3] and **Fair Trade**[4] certifications. For example, the family farmers of the Association for Educational and Cultural Development of Tauá (ADEC), located in Ceará state (Brazil), certifies cotton by IBD Certification and has the Fair Trade certification. On the other hand, the OCC (Organic Cotton Colors) company, does not require organic cotton certification, but certifies its manufactured products by GOTS certification.

2.7 Characterizing Production Systems

In Brazil, the conventional crop is characterized as using machinery and agroindustry systems for scale production in extensive areas (Fig. 7).

The organic crop is characterized as small properties (Fig. 8a), family farming and manual activities such as planting and harvest [88]. Since the organic system was labor-intensive, it was more attractive for producers with smaller areas of cultivation, especially in a context of production with the predominance of family farming, where the hiring of temporary or permanent labor is very scarce [60].

[3] GOTS Certification is an internationally accredited certification by the Global Organic Textile Standard (GOTS). It guarantees the organic quality of textiles, starting from the harvest of the raw material, passing through all stages of textile processing, from fiber to finished product. It aims to guarantee the process traceability, evaluating the reduction in the use of chemicals, energy and social and economic relations [86].

[4] Fair Trade Certification: it is an international certification, accredited by the International Federation of Alternative Trade (IFAT) with the mission "to improve the livelihoods and well being of disadvantaged producers by linking and promoting Fair Trade Organizations, and speaking out for greater justice in world trade" [87].

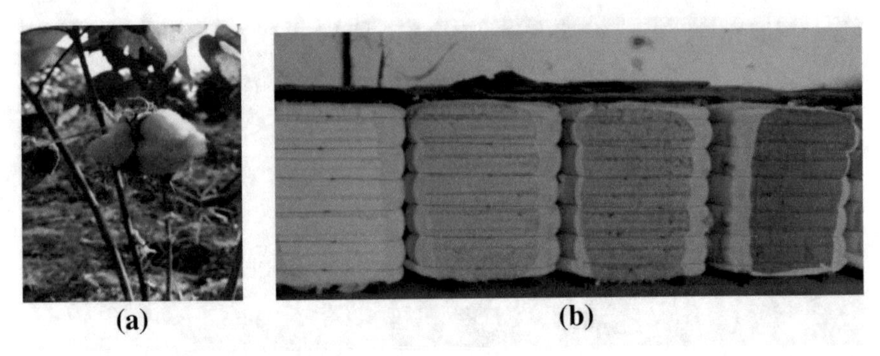

Fig. 8 **a** Colored Cotton BSB Rubi from EMBRAPA Cotton (Authorship); **b** From left to right, white color lint, BRS 200, BRS Green, BRS Rubi from EMBRAPA Cotton [70]

2.8 Characterizing Producers

According to Ref. [89], the size of the area, the rural properties are classified in:

- *Minifundio*—is the rural property with an area less than 1 (one) fiscal module;
- *Small Property*—the property of area between 1 (one) and 4 (four) fiscal modules;
- *Average Property*—the rural property of area greater than four (4) and up to fifteen (15) fiscal modules;
- *Large Property*—the rural property of the upper area 15 (fifteen) fiscal modules.

To the Brazilian Ministry of Agriculture and Supply [90], small holders and family agriculture (Fig. 8b) in Brazil is characterized as using the family's own labor force in rural economic activities in an area of maximum 4 fiscal modules predominantly, having a minimum family income originating from rural economic activities in their establishment and/or enterprise; and driving the establishment with the family. Family farming is an important segment for the development of Brazil. There are approximately 4.4 million farm families, representing 84% of Brazilian rural establishments. For the economy, it represents 38% of the gross value of agricultural production and the sector accounts for seven out of ten jobs in the field. It is productive since it is responsible for producing more than 50% of the food supplies of Brazilian "cesta básica" (set consisting of basic food products consumed by a family per month), being an important instrument of inflation control [91].

Enhancing smallholders' capabilities [10, 92]:

- *Training*: The provision of farmer training enables farmers to improve their production capacity and productivity and their abilities to meet the quality standards demanded by international supply chains.
- *Information Systems*: Providing smallholders with access to information and communications can help them make decisions and reach new or more beneficial markets. Market information is crucial for good decision-making.

- *Financial Services*: Access to finance has been identified as a major issue for small farmer inclusion. There is a growing need to facilitate and adapt financial products for small farmers, such as access to loans, crop finance, and crop insurance advances.
- *Social Entrepreneurship*: Social entrepreneurship aims to improve smallholders' inclusion by providing entrepreneurial opportunities within the supply chain. Several authors identify the relevance of entrepreneurs within farmer organizations who might bridge the gap and coordinate small farmers and market actors [93, 94]. Entrepreneurs are more likely to try new technologies and methods and can be triggers for innovation.

According to DataSebrae [95], referring to the first quarter of 2018, the highest proportion of rural producers is between 45 and 55 years of age, representing 26.3% of the total. Next are those who are between 55 and 65 years old (20.5%). On the other hand, the younger rural business owners, who are up to 25 years of age, are the minority. They represent only 6.7% of the total. Also, they affirm [95]:

- Personnel employed in agricultural establishments: 15,036,978 people.
- In the case of agricultural establishments: 5,072,152 establishments.

3 Organic Cotton

Organic cotton is grown without the use of any synthetically compounded chemicals (i.e., pesticides, plant growth regulators, defoliants, etc.) and fertilizers are considered 'organic' cotton [96]. The production of cotton using organic farming techniques (Fig. 9) seeks to maintain soil fertility and to use materials and practices that enhance the ecological balance of natural systems and integrate the parts of the farming system into an ecological whole [97].

Organic cotton cultivation is reported in the following countries: Africa: Benin, Burkina Faso, Egypt, Mali, Mozambique, Senegal, Tanzania, Togo, Uganda, Zambia, Zimbabwe. Asia: China, India, Kyrgyzstan, Pakistan. South America: Argentina, Brazil, Nicaragua, Paraguay, Peru. Middle East: Turkey, Israel. Europe: Greece. The USA and Australia [98]. The main producers of organic cotton are India, China, Kyrgyzstan, with India alone accounting for 56% of the global production [99]. While African organic cotton Sourced from six different nations holds a 4% share [99].

Surface morphology and surface chemical composition of both the organic and conventional cotton fibers are similar [19].

Production support comes considerably from financial institutions, local banks, donors and governments and NGOs [11]. Intermediate stakeholders, as transnational and local environmental NGO networks 'Solidaridad' and 'Helvetas', are important instruments in the construction, maintenance and transformation of the organic cotton network [10]. They enable services such as training, storage facilities, logistics, insurance and financial services, marketing, technical support and the supply of

Fig. 9 Organic color cotton clothing from "Flávia Aranha" Brazilian fashion brand, employing cotton cultivated in Paraiba state—Brazil [73]

seeds and inputs [10]. International institutions, such as Textile Exchange, also play an important role in financing capacity building in farmers' groups, acting as an agent in cotton marketing, promoting international events and publishing information [9].

Proponents of organic cotton and those who market organic cotton products promote the perception that conventional cotton is not an environmentally responsibly produced crop [99]. Some of the reasons used to support they contend that traditional cotton production greatly overuses and misuses pesticides/crop protection products that have an adverse effect on the environment and agricultural workers and conventionally grown cotton fiber/fabrics/apparel has chemical residues on the cotton that can cause skin irritation, and other health-related problems to consumers. Organic cotton production is not equivalent to sustainable—either organic or conventional cotton production practices may be sustainable [100].

With its tiny market share, organic cotton represents a viable option and a lucrative niche for many small-scale farmers in developing countries, in particular, due to attractive price premiums. However, these premiums may encourage more and larger producers to enter the market [101]. The aim is not to compare conventional and organic cotton value chains but to provide the necessary reference to understanding the context of emergence and the dynamics within the organic cotton network [9].

3.1 *Sustainable Agriculture*

Growing demand is forcing the conversation on reconciling economic growth with environmental sustainability. Local scaled decisions include but are not limited to the following: farm design, crop allocation, adoption of equipment and infrastructure, landscape planning, and groundwater management [43].

Analyzing the many definitions of Sustainable Agriculture, it can be found two that are most frequent, maybe the most accepted internationally, the one elaborated by FAO (United Nations Food and Agriculture Organization) [102] and the other by NCR (National Research Council) [103]. These two, complement each other, probably because there is still no consensus about Sustainable Agriculture definition.

The concept of Sustainable agriculture in farming means meeting society's present food and textile needs, without compromising the ability of current or future generations to meet their needs. Sustainable agriculture is not a set of unique practices, but rather an objective: to achieve a productive system of food and fibers that; increase the productivity of natural resources and agricultural systems, allowing producers to respond to demand levels engineered by population growth and economic development; produce healthy, wholesome and nutritious foods that enable human well-being; ensure a sufficient net income for farmers to have an acceptable standard of living; to invest in increasing the productivity of soil, water and other resources; and meets community standards and expectations [104].

Practitioners of sustainable agriculture seek to integrate three main objectives into their work: a healthy environment, economic profitability, and social and economic equity. Every person involved in the food system—growers, food processors, distributors, retailers, consumers, and waste managers—can play a role in ensuring a sustainable agricultural system [104].

Wide-scale transformation promoting sustainable agricultural production in the tropics will be crucial to global sustainability and development. Although contemporary agricultural production has increased alongside international demand, it has resulted in extensive changes in land cover, often at the expense of tropical forests and other native habitats. Conservation and development professionals from civil society, private foundations, multilateral and specialized international agencies, along with academic organizations and, increasingly, the private sector, have cited the urgent need to transform tropical agricultural production to meet current and future food needs without compromising environmental, economic, and sociocultural outcomes for present and future generations [105].

3.1.1 Agroecology

The term 'Agroecology' was first time coined in the scientific publication by Bensin [106, 107] and recently reaffirmed by Gliessman [108] and Warner [109]. The scientific discipline uses ecological theory to study, design, manage, and evaluate sustainable agriculture systems that are productive and resources conserving. Drawing on the natural social sciences, agroecology provides a framework for assessing four keys (productivity, resilience, sustainability and equity). Hence its importance is greatly realized by the dominant food policy and agricultural research bodies around the world [108, 110–112].

3.1.2 Organic Agriculture

The concepts of organic agriculture were developed in the early 1900s by Sir Albert Howard, F. H. King, Rudolf Steiner, and others who believed that the use of animal manures (often made into compost), cover crops, crop rotation, and biologically based pest controls resulted in a better farming system [113].

"Organic agriculture is a holistic production management system that promotes and enhances agroecosystem health, including biodiversity, biological cycles, and soil biological activity. It emphasizes the use of management practices in preference to the use of off-farm inputs, considering that regional conditions require locally adapted systems. This is accomplished by using, where possible, agronomic, biological, and mechanical methods, as opposed to using synthetic materials, to fulfill any specific function within the system [114].

The latest statistics reveal that Australia now has more certified organic agriculture hectares than the rest of the world put together. Organic agriculture is reported from 181 countries. Australia reported 35,645,038 certified organic hectares and the world total is 69,845,243 hectares. The development and growth of organics in Australia have always been driven by two factors, ideology and the market. Australian organics has received scant support from government and institutions, often being ignored and sometimes derogated [115, 116].

It is possible to say that all definitions express the need to establish a new productive pattern or system, that uses natural resources more rationally and maintains productive capacity in the long term. The word "sustainability" is known worldwide and its use in many sectors of the economy has been increasing. However, there is no consensus about its real concept. Therefore, a single, globally agreed definition of the terms "sustainability", "sustainable development" or "sustainable agriculture" is inadequate [117].

3.2 Organic Cotton in the World

In the cultivation of organic cotton, chemical fertilizers, synthetic pesticides, genetically modified seeds, chemical growth regulators, and chemical defoliants are prohibited [60, 118, 119].

Although 19 countries produce organic cotton, only seven of them (India, China, Turkey, Kyrgyzstan, Tajikistan, USA and Tanzania) represent 98% of the total production. India has the largest share, with 47% of global production. Evidence from India, the USA and Turkey shows that, if supported by good science, high yields of more than 1,000 kg per hectare can be obtained. Current global average yields are low, 375 kg of feather per hectare. Property yields may actually be higher due to intercropping with other crops produced organically, different from monoculture on conventional properties [35].

The demand for organic cotton is constantly growing, as more and more buyers are looking for high-quality cotton, produced according to strict environmental and social standards. The increase in responsible consumption, involving social, environmental and economic issues, led to strong market growth, increasing the demand for organic cotton. By using organic cotton in their products, buyers are promoting a more ecologically sustainable social production cycle, contributing to the Sustainable Development Goals (SDGs) achievement, giving greater credibility to the sustainable activities carried out by your company. The continuous expansion will depend on overcoming challenges: technologies appropriate to organic production systems, the need for progressive increments in areas of arable land under organic management to meet growing consumption, the standardization of certification criteria and the great concentration of world demand for consumption [35, 60, 118–120].

3.3 *Organic Cotton in Latin America*

In Latin America, the production of organic cotton takes place in Peru, Brazil and Argentina, mainly being carried out by family farmers. According to the report of the Organic Cotton Market of the Textile Exchange (2019), the 2017/18 crop represented 0.29% of the global production of organic cotton. It counted with the participation of 1,172 farmers, and the planted area was 1,300 hectares of land, where 526 tons of feathered organic cotton were produced [35].

Peru is the largest producer of organic cotton in Latin America and its production is responsible for 2.3% of the overall cotton production in the country. Brazil is the second, and Argentina occupies third place in the production of organic cotton [35].

The organic certification adopted by farmers is done through certification by auditory and through the Participatory Guarantee System. Some companies and farmers' associations access other types of certification, such as Global Organic Textile Standard (GOTS) and Fair Trade.

Production is supported and monitored by the following institutions:

- Bergman Rivera Service Company—located in Peru;
- SoCiLa Non-profit—located in Colombia;
- Stay True Social Enterprise—located in Argentina;
- EMBRAPA Cotton, ESPLAR, EMPAER, Textile Exchange, Renner Stores Institute, Food and Agriculture Organization of the United Nations (FAO/WHO), Laudes Foundation, DIACONIA, ASPTA and Sustainable Fashion Lab—all located in Brazil.

In Latin America, cotton production is being encouraged through the actions of the *Projeto + Algodão* ("Project + Cotton"), which has been developing actions since

2013. The project is a partnership between FAO, the Brazilian government and seven other countries in Latin America and the Caribbean (Argentina, Bolivia, Ecuador, Colombia, Paraguay, Peru and Haiti), which were organized through Trilateral South-South Cooperation. One of the project's objectives is to achieve SDG targets, which could be a strong ally in organic cotton production in the near future [121].

3.4 Organic Cotton in Brazil

Organic production began around the 1940s, arriving in Brazil around 1989 [74]. Cotton is a plant that resists drought, so it is an option for planting in semi-arid regions, which generally do not have many alternatives for the population in rural areas of income generation. Besides its resistance, in cotton, almost everything is used; the seed and fiber are the most relevant, representing around 65% and 35% of production weight, respectively [122].

In Brazil, the cultivation of organic cotton, whether white or naturally colored, is produced through agricultural cooperatives and/or small farmers due to the needs of the management itself. The technical limitations end up including small farmers in the production chain, thus generating gains and social inclusion [123]. In addition to economic gains, there is also an improvement in life quality, health and well-being. In the environment, it is important to highlight the value of biodiversity, in addition to the non-use of water, since the water coming from the rains is sufficient to irrigate the plantation and since there is no use of pesticides, there is also no environmental contamination [82].

The production of organic cotton in Brazil is made by small producers in a scheme of family labor. The production of organic cotton in Brazil, carried out by family farmers, is concentrated in the semi-arid Northeast (Paraiba, Piaui, Pernambuco, Rio Grande do Norte and Ceará). The states of Alagoas and Sergipe are also producing; however, their areas are transitioning. Among the Northeast states, only Paraiba and Ceará produce white and colored organic cotton. In the state of Mato Grosso Sul, family farmers work with the production of colored organic cotton.

The data presented below are from the Organic Cotton Market Report—Textile Exchange [83]:

- Planted area: 619 ha, of this total amount, 358 ha PGS certified;
- Number of family farmers: 930, out of this total, 700 PGS certified;
- Organic cotton production: 22 tons, of this total value, 13 tons PGS certified;
- Area planted in the process of organic transition: 318 ha; of this total, 30 ha are under the responsibility of the PGS.

In this way, in the agricultural aspect, the management adopted for cotton crops in large areas is practically established. Still, the main challenge for organic cotton production is to put into action more researches with cotton-organic/agroecological management [53].

Table 4 Advantages of growing organic cotton compared with conventional farming [85]

	Conventional cotton	Organic cotton
Environment	Pesticides kill beneficial insects	Increased bio-diversity
	Pollution of soil and water	Eco-balance between pests and beneficial insects
	Resistance of pests	No pollution
Health	Accidents with pesticides	No health risks from pesticides
	Chronic diseases (cancer, infertility, weakness)	Healthy organic food crops
Soil fertility	Risk of declining soil fertility due to use of chemical fertilizers and poor crop rotation	Soil fertility is maintained or improved by organic manures and crop rotation
Market	Open market with no loyalty of the buyer to the farmer	Closer relationship with the market partner
	Open market with no loyalty of the buyer to the farmer	Option to sell products as 'organic' at higher price
	Dependency on general market rates Usually individual farmers	Farmers usually organized in groups
Economy	High production costs	Lower costs for inputs
	High financial risk	Lower financial risk
	High yields only in good years	Satisfying yields once soil fertility has improved

Besides that one related to the agricultural activity, many other actors are involved in the organic cotton chain in Brazil, constituting a network. The main actors and respective roles of this organic cotton network in Brazil are summarized in Table 4.

One of Brazil's firsts organic cotton network dynamics includes the French shoes' brand named "Veja". Since its beginning, Veja has been signing one-year contracts with farmers' associations and setting a market price per kilo of organic cotton. Veja also pre-finances the harvest up to 50% and pays a premium per kilo of cotton produced that associations must use to develop community projects. The agricultural approach farms use based on mixed farming provides food independence and maintains nutrient balance in cultivated land (agroecological practices). These families are in rural communities and cultivate cotton together with other crops such as beans, corn, sesame, manioc (cassava), sunflower, and pumpkin in areas of maximum of two hectares. They benefit from regenerating the land, producing food and commercializing diverse products. In 2018, more than 23 tons of agroecological cotton was bought by Veja directly from seven different associations in northeast Brazil. This cotton was cultivated by 259 families [124].

The second dynamic and very important one is this one initiated by "Laudes Foundation", "Embrapa Cotton" and "Diaconia". The objective of the Program "Agroecological cotton cultivated in consortium with food crops" is the expansion of organic cotton supply network. At the beginning of this project, in 2017, 2 tons of organic cotton was produced and 80 smallholder farmers were involved. This number has significantly increased with the expectation of 50 tons to be produced in 2019 and more than 900 farmers producing cotton. This increased productivity has a lot to do with the intensive capacity-building for agroecological intercropping management and processing. Another objective is to guarantee women's effective participation and recognition in both agricultural activities and political organization. More than 160 women have taken on leadership positions in their communities in these three years of the project. This program has also strengthened OPACs to obtain the certificate of organic compliance and consolidate the producers' network and autonomy. There is an important role undertaken by "Laudes Foundation", which relates to the participation in negotiation between representatives of OPACs and potential buyers of cotton. Besides strengthening local farmers' collectives, the Program also aims to create a regional network to promote collective actions. With a focus on income generation, development of family-based social organizations, conservation of natural resources, and introduction of smallholders into fair trade and the organic market, the initiative will benefit at least 1,000 families by the final of 2020. The main buyers of the organic cotton produced are the brands Veja and "Organic Cotton Color" [125–128]. Laudes Foundation has also been working with ESPLAR and Embrapa Cotton to expand the production of Mocó (*G. hirsutum* subsp. marie gallant) organic cotton supply (long and resistant fiber) in Ceará State [53].

The third dynamic is called "Institute Casaca de Couro" (in reference to the Caatinga singer bird *Pseudoseisura cristata*). The Institute is formed by Coopnatural, EMBRAPA Cotton, EMPAER, Norfil and other organizations. Their partnership includes actors that understand the organic cotton reality in Paraíba state (Brazil), offering technical support to the small producers and its families and straightening the dialog with the textile industry, considering the increase of organic cotton fabrics production. The main aim is to support local communities, the environment, and the development of Caatinga Biome (the driest area of Brazil), including the cultivation of native plant species for coloring the produced cotton clothes [129, 130].

The fourth is organized by company "Natural Cotton Color", which commercializes textiles and produces and commercializes clothes and bags, mainly for exportation. The company is connected with family farmers from the "Margarida Maria Alves community" in (Juarez Távora, Paraíba state, Brazil) and also works with natural colored cotton [124].

Figure 10 overlooks the process from farm to clothing in the regions cited above.

3.5 *Motivations and Constraints of Organic Cotton*

Whereas speaking about smallholder organic cotton farmers in developing countries, the following motivations are stated as the most important ones: (i) to improve the fertility of the soil (softer soil, greater absorption of water, better water holding capacity, healthy crops); (ii) to reduce the production costs and thus the financial risk; (iii) to get a better price for the cotton (organic premium); (iv) to get rid of the

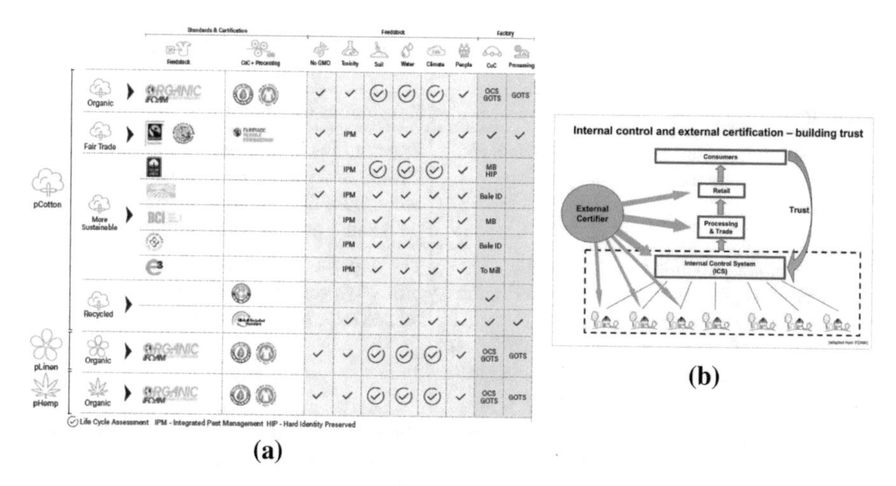

Fig. 10 **a** Cotton standards and certifications [83]; **b** Internal and external control in an organic cotton project (adapted from IFOAM) [27]

negative effects of conventional farming: declining yields, the resistance of pests and diseases, health hazards of chemicals; (v) to improve the profitability of the farm in the long run [98].

However, various issues hinder the adoption of organic cotton production. These include production problems, particularly insect and weed control, and marketing problems, particularly price variability and unstable, underdeveloped markets, crop rotation problems, lack of organic cotton marketing information, and organic certification issues. One of the most important aspects of organic cotton production and expansion requires improvement in marketing and market linkages between cotton producers and international organic cotton buyers, including access to market information distribution channels. Improvements in all these areas are needed to promote organic cotton properly [100]. Some challenges are also linked with lack of information on cost of production and production methods, lack of work force and tax incentives, unstable market, development of new markets and international certification issues [41, 60]. Table 5 summarizes the main topics.

In this way, complementing the literature information, motivations and restrictions (challenges) for planting organic cotton and its consortia observed by the present authors are highlighted in Table 6.

Furthermore, from the textile and clothing perspective, there is a lack of publicly available data on more sustainable cotton as well as a need for consistent uptake production reporting. Greater transparency and coordination between standard organizations would support by providing the sector with clear indications of market demand and understanding where the bottlenecks are in the supply chain. Thus, companies have numerous concerns about more sustainable cotton supply chains, including issues such as the disconnection between harvest and production times combined with the unavailability of stocks; the challenge of adapting to new suppliers and different business practices, and; the lack of supply of certain qualities of cotton

Table 5 Motivations and restrictions (challenges) for planting organic cotton and its consortia (Authorship)

Motivations	Restrictions (challenges)
Improving food security, family health, and the economic situation that organizes property income	The need for a closer relationship between farmers and purchasing companies to accordance the organic market related to production processes in the field
The women participation in organic cotton production in agri-food consortia in developing countries is more frequent due to the non-handling of dangerous chemicals. The access to training and financial income for women has a positive impact on their social autonomy	The improvement of the organic certification process to comply with laws and regulations related to organic products and processes
Young people have the opportunity to participate effectively in the productive actions of the property. The production of organic cotton in agri-food consortia has a positive impact on their financial independence and activates the process of rural succession	Difficulty in accessing organic inputs such as bioinsecticides and seeds with organic certification
The diversified cultivation with the use of natural fertilizers and pesticides, let the environment clean and balanced, where groundwater and water reservoirs are free of chemical contaminants. The biodiversity of the property systems and subsystems is improved	Expansion of the market for machines and implements adapted to production systems combined with field activities and processing of organic products. The prices of machines and implements are most often incompatible with the reality of small cotton producers
Enabling the organization of their participatory certification institutions, generating autonomy in the commercialization process	Access to a market with a fair price and compatible with the organic quality of the other products of the agri-food consortia

Table 6 Comparison of production cost between conventional, BCI, fair trade, organic and organic-fair trade for a T-shirt [48]

	Conventional Cotton		Better Cotton Initiative			Fair Trade Cotton			Organic Cotton			Organic-Fair Trade Cotton		
	USD/kg	B/down	USD/kg	B/down	Differential	USD/kg	B/down	Differential	USD/kg	B/down	Differential	USD/kg	B/down	Differential
Seed cotton price farm gate	0.61	6.4%	0.62	6.1%	2.0%	0.67	5.9%	8.9%	0.64	5.6%	4.0%	0.69	5.8%	12.9%
Ginning, baling, transportation, handling & seed recovery	0.84	8.8%	0.94	9.2%		1.03	9.1%		0.96	8.4%		1.08	9.2%	
Fiber price ex mill	1.45		1.56		7.7%	1.70		17.0%	1.60		10.4%	1.77		22.4%
Spinning, packing, transportation & margin	1.50	15.8%	1.55	15.2%		1.61	14.2%		1.57	13.7%		1.65	13.9%	
Yarn price ex mill	2.95		3.11		5.6%	3.31		12.3%	3.17		7.5%	3.42		16.2%
Knitting, dyeing, finishing, loss, transportation & margin	1.47	15.5%	1.50	14.7%		1.54	13.5%		1.51	13.2%		1.56	13.2%	
Fabric price	4.42		4.61		4.4%	4.85		9.6%	4.68		5.9%	4.98		12.7%
Certification & traceability		0%	0.06	0.6%		0.15	1.3%		0.15	1.3%		0.15	1.3%	
Total fabric price	4.42		4.67		5.7%	4.99		13.0%	4.83		9.3%	5.13		16.1%
Fabric usage per t-shirt @ 13% fabric usage	**0.57**		**0.61**			**0.65**			**0.63**			**0.67**		
Standard marking	0.03	2.2%	0.03	2.2%		0.03	2.0%		0.03	2.0%		0.03	1.9%	
Accessories / printing	0.28	23.0%	0.28	21.3%		0.28	19.2%		0.28	19.1%		0.28	18.4%	
Packing	0.11	8.6%	0.11	8.0%		0.11	7.2%		0.11	7.2%		0.11	6.9%	
Cutting, making & trimming	0.24	19.7%	0.24	18.3%		0.24	16.5%		0.24	16.4%		0.24	15.8%	
Integrity, certification & traceability			0.06	4.5%		0.16	11.1%		0.19	13.1%		0.21	13.6%	
FOB price per t-shirt	**1.23**	**100.0%**	**1.33**	**100.0%**	**7.7%**	**1.48**	**100.0%**	**19.6%**	**1.48**	**100.0%**	**20.3%**	**1.54**	**100.0%**	**24.6%**

or from preferred locations. Brands and retailers have a crucial role to play as they can pull the sector toward greater sustainability by demanding and sourcing more sustainable cotton [16].

3.6 Market Potential

Market demand for textiles made from organic cotton mainly exists in Europe, the USA, Canada, Japan and Australia [7]. Some large companies become involved with organic cotton textiles in order to improve their corporate image with respect to environmental and social accountability. The main reasons for consumers to buy textiles made out of organic cotton are: To reduce the risk of skin irritation and allergies; To protect the environment from toxic chemicals; To support sustainable agricultural production in the country where the cotton is grown; To ensure that the farmers in developing countries receive a fair price [98].

It is important to avoid contamination throughout the entire organic cotton processing chain and separate organic from conventional cotton. As most spinning mills and processing entities process organic and conventional cotton on the same machinery, it is important to separate the cottons and clean the equipment before processing an organic lot. Some labels and brands have certain restrictions on which dyes can be used. Initially, most organic cotton was processed into garments containing 100% organic cotton fiber. Some large garment brands have recently decided to blend a certain percentage (usually 5–10%) of organic yarn into their entire range of articles rather than selling purely organic clothes. This could increase the demand for organic cotton fiber considerably. Companies can communicate to their customers that they support organic cotton farming, which helps them to improve their corporate image [98]. The main organic cotton markets are presented in Fig. 11.

In the future, a number of changes to the cotton production industry are likely to affect the shape and scope of the value chain. It is predicted that the demand for organic cotton will grow substantially in the coming years [10]. This demand is likely to be increasingly met by producers in developing countries, who are now benefiting from better support services, know-how, and the economic and regulatory infrastructure necessary to allow them to shift to organic production [8].

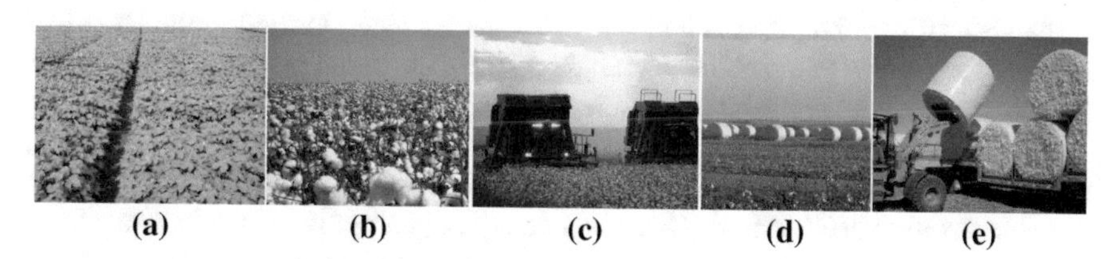

Fig. 11 **a, b** Conventional cotton farming in Brazil for scale production [63]; **c** mechanical harvesting [63]; **d** Conventional cotton crop at Brazilian Cerrado biome [63]; and **e** bales transport [63]

In the event “Brazil Eco Fashion Week” (São Paulo city—Brazil, 2019), the Natural Cotton Color company presented a new denim (“jeans”) manufactured with BRS Rubi fibers (Fig. 6b). This product had wide approval at the “Première Vision” (Paris—France), considered the main international textile fair. Besides denim, the company developed other new materials, including a biodegradable elastane and a fabric combining colored organic cotton and silk. This segment tends to grow and have a strong impact on the future market, based on customer demand and the trend toward more sustainable field practices and circular materials [53].

Engagement of stakeholders will motivate them to take ownership of initiatives and improve production and market coordination [131]. Collectivism may be conducive to loyalty and at the same time promote creativity and diversity of solutions. Organic production involves collaboration between network participants in terms of transparency, sharing information on technical and managerial practices, and fair trade and price-related contractual agreements [60].

3.7 Communicating Organic Cotton in Retail

Organic cotton cultivation has been gaining notoriety in recent years and increasing its production in planted hectares and amount of production [132, 133].

As the cultivation of organics has grown, so has the relevance of sustainable issues in production, aiming at sustainable development and practices to curb environmental and social disasters [123]. In this way, sustainable products are beginning to integrate some environmental costs into business practices. In retail, they are beginning to express values to the consumer, not just monetary issues, but also influence cultural thinking [134].

Advertising is considered to be informative and wishful thinking, seen as fundamental for placing new products on the market and increasing sales. However, this requires consumers to be educated about the impacts of clothing production and sustainability, so that attitudes and behaviors can be changed [135].

Flavia Aranha is one of the main clothing brands utilizing organic cotton in Brazil. The brand also works with natural dyes, emphasizing more sustainable processes. It started in 2009, working with natural fibers such as organic cotton, linen, silk and wool. Using fabrics from Natural Cotton Color, an organic cotton chain organized in the state of Paraíba (Brazil) and “Justa Trama”, a chain located in the Brazilian states of Ceará, Sao Paulo and Rio Grande do Sul. The brand prefers organic cotton considering the land regeneration potential, soil and water sustainable use.

The company carries the name of the owner and director, Flávia Aranha. According to Flavia “The challenge of organic cotton is not in the technique of cotton cultivation itself. The challenge concerns the perspective of the industry as an all, understanding the materials and the chains. It’s a mindset-changing challenge.”

The brand uses different communication tolls to engage and educate costumers. The product tag contains a short text explaining about Brazilian plant-based textile fibers and pigments and the sustainable process utilized in production. Also, the QR

(a) (b)

Fig. 12 **a** Manual harvest agroecological cotton in the Tiracanga rural settlement, Canindé (Ceará, Brazil) [89]; **b** C&A Foundation´s Sustainable Cotton Program in Brazil - Araripina Community Field Work (Brazil's Caatinga biome) [90, 91]

code inside the clothes has a link with videos explaining the process of the outfit manufacture, showing the fabric, the color, the plant extraction, sewing, etc., and always looking to the most affective and real way to tell these stories. Also, since the first collection, all clothes were packed with organic cotton bags. In the Flavia Aranha store, the brand promotes events and workshops. Nowadays, how to bring the brand experience to the digital platforms is the main challenge (Fig. 12).

According to Magnuson et al. [136], most customers do not know-how their clothes were manufactured. It would be up to the brand to inform them about the production processes and materials used, adding value to the product. Communication in retail and e-commerce becomes essential to prospect and inform new customers about ethics and social-environmental responsibility [137, 138]. In Araújo et al. [139], it is mainly through communication that consumers come to know the identity, value purpose of brands and their products so that they become known in fact and reach their target audience.

4 Sustainable Network Management

The so-called chain processes are the most used to analyze the macro-processes of production in order to ascertain performances, bottlenecks, productive and managerial processes, always to improve efficiency, quality, competitiveness and sustainability [140]. Moving from the language of commodity *chains* to commodity *networks,* Raynolds [141] helps portray the complex network of material and nonmaterial relationships connecting the social, environmental, political, and economic actors enmeshed in the life of a commodity, such as cotton. Understanding how individuals, firms, government authorities, and nongovernmental organizations (NGOs) are involved in economic and social transactions and how network relations shape these different actors both shape and. The network concept is increasingly used in studies of the horizontal and vertical relations among global manufacturing firms [141].

Many innovative approaches may contribute to delivering sustainability through business models but have not been collated under a unifying theme of business model innovation [142] that could include:

- A system that encourages minimizing of consumption or imposes personal and institutional caps or quotas on energy, goods, water, etc.;
- A system designed to maximize societal and environmental benefit, rather than prioritizing economic growth;
- A closed-loop system where nothing is allowed to be wasted or discarded into the environment, which reuses, repairs, and remakes in preference to recycling;
- A system that emphasizes delivery of functionality and experience, rather than product ownership;
- A system designed to provide fulfilling, rewarding work experiences for all that enhances human creativity/skills;
- A system built on collaboration and sharing, rather than aggressive competition.

Moreover, delivering environmental and social sustainability initiatives such as [142]:

- Employee welfare and living wages;
- Community development: Education, health, livelihoods;
- Sustainable growing and harvesting of food and other crops, minimizing chemical fertilizers and pesticides, water consumption, and top soil erosion;
- Environmental resource and biodiversity protection and regeneration.

In order to generate a sustainable network, improve research, gain credibility and fulfill missions of sustainable processes, several companies join NGOs that aim to promote sustainability through organic products, fair trade, and social responsibility. At this junction, both gain and generate greater confidence in the consumer and added value, since the consumer perceives this junction as beneficial and contributing to sustainability, in addition to improving the lives of the participating communities [143].

Therefore, collaboration is particularly relevant because it represents a chance to improve the chain's competitiveness and farmers' well-being. One of the principles of organic and fair trade funding, which translates into certification rules of transparency and, most pertinently, for fair trade and joint management procedures [144]. Thus, sustainable commodity systems will require participation and cooperation throughout the chain [145].

Glin et al. [9] analyzed the social dynamics that connect actors and practices within the organic cotton network, particularly flows of information and knowledge, trust-building mechanisms, and power relations among actors from production level to global market level. The research of these authors well-described network challenges in Benin, which are apparently very similar in Brazil. Linking small producers to markets and integrating them into value chains is widely recognized as a valuable way to reduce poverty. However, little is known about the precise conditions under which this is most likely to occur. Smallholders are often illiterate, under-educated, lack management and technical skills, and have poor access to information (about quality, buyer demand and standards). Larger farmers have more bargaining power and better access to capital, information, finance and technology [57].

So, initiated by intergovernmental sustainable development cooperation, a transnational organic cotton network evolved into a hybrid structure, combining private economic actors and domestic and international NGOs. National and international NGO networks opened spaces for value sharing and information exchange and played a brokering role in linking local producers to the global organic cotton market in Europe and vice versa [9].

International conferences and events provided important occasions for establishing linkages between organic cotton promoters and businesses, and they strengthened the organic movement. Trust was a critical factor in recruiting farmers and ensuring their continued participation in the organic cotton production system and in securing the organic profile for European customers (Fig. 13) [9].

As sustainability labeling is gaining momentum in the global apparel industry, opportunities for market expansion are related to the creation of farmer organizations [10, 146, 147]. This will allow small farmers to be more competitive, achieve economies of scale, reduce transaction costs, enhance their bargaining power, improve their market information, access technology, manage common pool of

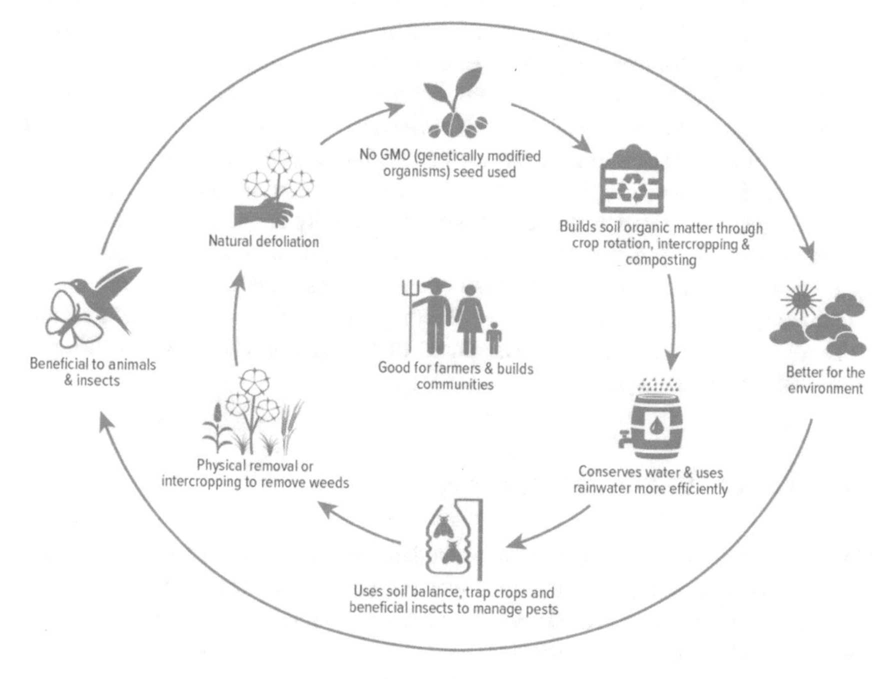

Fig. 13 Organic cotton benefits and main characteristics [31]

resources and reduce certification costs [8]. In this way, considering the various involved actors, the main advantages could be summarized as follows:

- For *farmers*, a range of perceived advantages was found to motivate converting to organic farming, including higher market prices for organic cotton (premiums), reduced costs for agricultural inputs, services provided by support organizations (e.g., access to credit, provision of seeds, marketing support, training), reduced health risks and soil improvements [101]. As many traditional farmers in developing countries are not well educated and lack the channels to share their experiences, many of these costs remain unacknowledged [148]. Potential advantages to cotton farmers, including lower expenses for farm inputs, healthier soils, diverse Sources of income, and higher prices, may be able to offer higher gross margins than conventional cotton farming [11].
- For the *textile and clothing industry*, many factors affect the growth of the organic cotton industry. These include consumer demand for organic products, a recognition by firms of the benefits in terms of sales and profits from the increasing consumer awareness of organic methods, and the institutionalization and regulation of the industry with its attendant reputational benefits [144].
- For *retail*, the market innovation in its production procedures and client's engagement can contribute to communicating to stakeholders the companies practice regarding cultural, social and environmental aspects [136–138, 149]. Opportunities for exports, particularly to the European markets, create further demand pressures on the industry, for environmental improvements including more formal certification [150], expecting to cover more than niche markets.
- For *clients*, considering the growth factors in the organic cotton industry, consumers and the various media have played probably the most important role in raising ethical expectations of business and achieving public visibility of corporate social responsibility issues. Among recent changes in consumer habits and preferences, increased awareness about sustainability is one of the most important elements [14]. Although there are a number of forces encouraging the move to organic production, other forces counteract these. Prices of organic cotton garments are still high to encourage the migration of the mass market to them. There is also, still, a lack of awareness of the consequences of the different production methods [8].

4.1 Trustability and Transparency

Since 2006, many companies have committed themselves in the long term to increase the use of organic cotton, thus enabling new brands and retailers to enter the market [143]. It is important that such companies provide sustainability reports and produce environmentally friendly collections [151].

Any large apparel brand often consists of thousands of suppliers, distributors and retailers [152]. Concerning a global supply chain that employs 58 million people

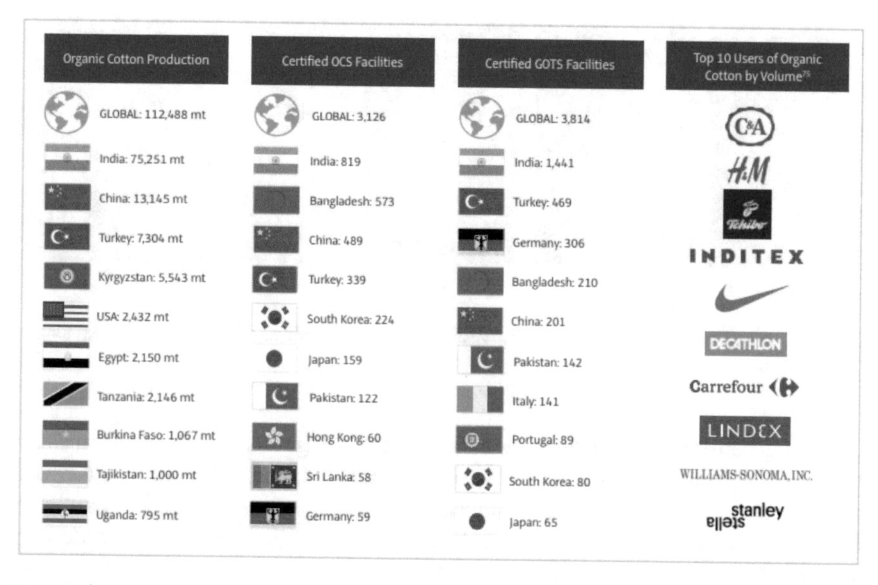

Fig. 14 Top 10 countries for organic production and certified facilities and top ten users by volume [48]

worldwide [153], this complexity introduces myriad challenges associated with monitoring, reporting, compliance and improving sustainability practices [154].

In the competitive fashion product market, the organic label can generate successful differentiation, promoting and encouraging profitability as well as consumer loyalty [155]. In this case, consumers recognize social values and seek a better quality of life [156]. The motivations of ethical consumers are their concerns about the fair trade, fair remuneration, etc. [157]. With this scenario change, many textile and clothing companies have increased their mindset transformation, toward consumers engagement [158] (Fig. 14).

Multi-stakeholder initiatives, acting beyond commercial interests, can offer guidance and promote network cohesion. The industry united around an agenda for change can drive the needed systemic change and work jointly on disruptive innovation [159]. Collaborative relationships emerge among the actors who can achieve complementary benefits by integrating their functional specializations and building a collective intelligence [160]. In other words, actors are individually intelligent, they are purpose-driven and they can mobilize resources through collaboration [161]. As a consequence, trusting relationships are often depicted as the essence of collaboration [162]. Trust is the outcome of a gradual, coherent and consistent long-term effort (Fig. 15).

As a driver of innovative and sustainable business models in fashion, collaboration refers to adopting a collaborative mindset by all stakeholders involved in a sustainable value network: suppliers, distributors, customers (who often are involved in co-creating initiatives), and even competitors. Collaboration allows the creation of a

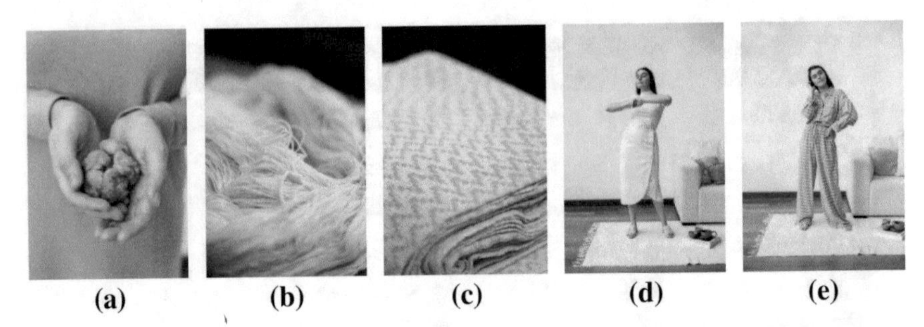

Fig. 15 Flavia Aranha Collection "The harvest", using organic cotton fabrics from Natural Cotton Color and Justa Trama [132]

supporting ecosystem that drives resource and knowledge sharing, promotes sustainable practices, and ultimately allows business model experimentation. As a result, it is critical to relate value creation (key activities, key resources, and key partners), distribution (delivery channels and customer relationship), and potential impacts on cost savings and revenue structure, as many collaboration initiatives involve revenue sharing [14] (Fig. 16).

This process is largely due to changes in consumer needs [163], therefore the transparency of the entire production cycle is essential because it is where the increase of consumer confidence lies [164].

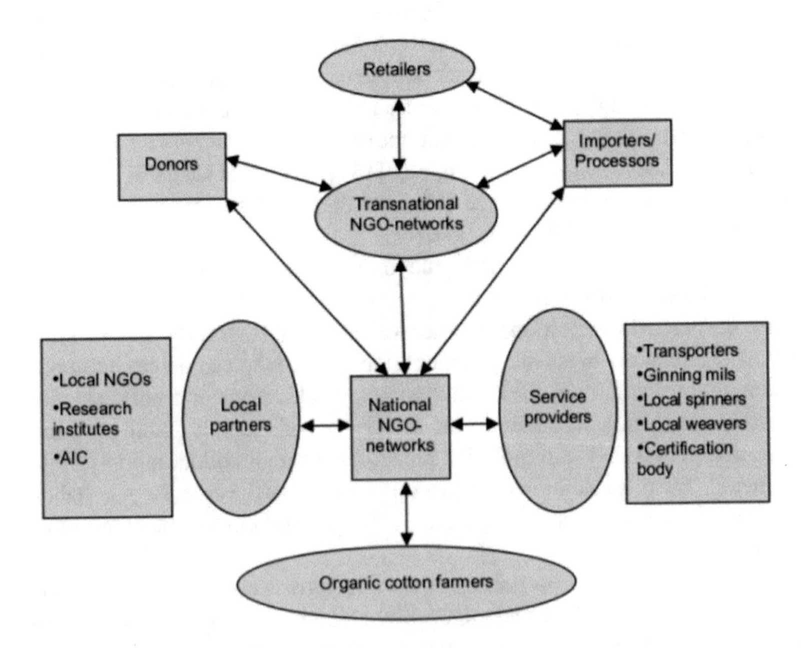

Fig. 16 The transnational organic cotton network [5]

The fashion system is dependent on flows of resources: fiber, chemicals, energy, water, human labor. The global environmental costs and consequences associated with fashion production and consumption are widely documented [165]. Therefore, it is necessary to have a vision of the whole system since several processes are related through the flows of materials, capital and information, always aiming to meet the market demand [140].

4.2 Small Scale and Better Quality Process

There is a growing need for the unique and the traditional. More customers want a one-of-a-kind, natural products, and are increasingly reverting to the idea of quality products. Old crafting techniques are gaining in value. Accordingly, small-scale companies have the ability to be flexible and innovative, which is crucial in order to take the time to create hand-crafted and exclusive designs. They are able to involve their customers, creating emotional value and belonging. Their clients are more likely to feel attached to their purchases and to keep them on the long term [166].

Traditional handicraft does not demean new technologies. They are compatible, as was pointed out by trend forecaster, Lidewij Edelkoort (one of the world's most famous trend forecasters), who believes that the two phenomena enrich one another. When technology provides new material options and smart machinery, the implementation of traditional handicraft techniques to create strong, durable and beautiful designs out of them gets accessible. Increasingly large-scale businesses are outsourcing their innovation to small, highly specialist organizations, which in turn help them to stay ahead of the crowd and remain unique while also spreading awareness and standards [57, 166].

4.3 Local Materials and Local Design

Clothing companies must face challenges posed by demand unpredictability and must adapt to a new, competitive environment [145]. But also increasingly time-based competition and mounting consumer sensibility to social-environmental issues [167], have driven fashion companies to reorganize their supply networks, searching for a new balance between local and global sourcing and production.

Fletcher and Tham [165] detailed their perspectives for the future, based on eight values: (1) Multiple centers, (2) Interdependency, (3) Diverse ways of knowing, (4) Co-creation, (5) Action research, (6) Grounded imagination, (7) Care of world, (8) Care of self.

Multiple centers include diverse ways of knowing; direct experience, practice, indigenous knowledge, artistic exploration, spirituality, and theory, among others. The promotion of multiple centers and interdependency in unison foregrounds specific skills of *collaboration*, *listening*, *dialog* and *linking*. In practice, co-creation requires a high level of collaboration between all involved which reaches beyond knowledge exchange and generates new ideas and actions. Imagination is a creative living process [165].

Motivating changes in the fashion system is a challenge to sustainability, and great potential since fashion affects the lives of almost everyone on a daily basis and has the potential to be effective in changing intentions, attitudes and behaviors [168].

As a way to add value to their products and meet consumer demand, a new generation of designers is already concerned with sustainable processes and materials. With motivations and information, these professionals seek innovative social and environmental solutions [96].

Sustainable clothing and fashion seem to be not just a trend but a movement toward changing paradigm. Fundamental in the movement, more aware consumers with their demands seem to be increasing. From supporting handcrafts and traditional local communities, using biodegradable and renewable fibers to supporting textile natural dying technics, second-hand clothes and upcycling. Furthermore, companies also began to assess the impact of not acting toward consumers choice in supporting and purchasing more sustainable products and brands [169].

In such a new textiles economy, clothes, fabric, and fibers are kept at their highest value during use, and re-enter the economy after use, never-ending up as waste. Designing and producing clothes of higher quality and providing access to them via new business models would help shift the perception of clothing from being a disposable item to being a durable product. To achieve system change, buy-in to the vision needs to be built across different actors, including industry, government and cities, civil society, and the broader public. None of these groups can do it alone. In particular, ambitious, common, time-bound commitments to the vision are required. The principles of transparency, compliance and sustainability are mandatory for all members and their activities of an organic network to the same degree of responsibility and commitment [170].

5 Conclusion

The fashion system is dependent on flows of resources: fiber, chemicals, energy, water, human labor. The global environmental costs and consequences associated with fashion production and consumption are widely documented. Cotton is one of the primary resources in many industries and with increasing demand rates. The cotton production chain is divided into production, processing, spinning, weaving, confection, and consumption. Therefore, it is necessary to have a vision of the whole system, since several processes are related through the flows of materials, capital and information, always aiming to meet the market demand. In addition, there is a need for increasingly seeking mechanisms to close the life cycle of materials, without leaving aside their nature and concept. The interest in organic cotton production has been increasing every year, yet the production still faces difficulties regarding articulation with the textile and clothing sector and the market. Multi-actors' initiatives and programs, acting beyond commercial interests, could offer guidance and promote cohesion to the network. Organic agriculture especially together with agroecological practices, has a major role to play in assisting with resource management, such as reducing water demand, diminishing soil erosion, maintaining and enhancing biodiversity. Thus, to persuade a systemic change in the current model of textile production, it is necessary to generate solid connections, educate, and involve the society to build a joint vision of a sustainable and creative economy with financial, social and environmental purpose.

Acknowledgements We gratefully acknowledge **CAPES** (Coordination for the Improvement of Higher Education Personnel of Brazilian Education Ministry).

References

1. Egbuta M, McIntosh S, Waters D, Vancov T, Liu L (2017) Biological importance of cotton by-products relative to chemical constituents of the cotton plant. Molecules 22:93
2. Beckert S (2014) Empire of cotton: a global history. Penguin Books
3. Pezzolo DB (2012) Tecidos: história, tramas, tipos e usos. Editora Senac São Paulo
4. Beltrão NE, de M, de Carvalho LP (2004) Algodão colorido no Brasil, e em particular no nordeste e no Estado da Paraíba. Embrapa Algodão -Documentos (INFOTECA-E) 128:17
5. Rodrigues JCJ (2015) Algodão no Brasil: Mudança, associativismo e crescimento. In: Algodão no cerrado do Brasil 21–37. ABRAPA (Associação Brasileira dos Produtores de Algodão
6. FAO-ICAC (2015) Measuring sustainability in cotton farming systems. FAO-ICAC - Plant Production and Protection Division
7. Textile Exchange (2018) 2018 Organic cotton market report
8. Rieple A, Singh R (2010) A value chain analysis of the organic cotton industry: the case of uk retailers and Indian suppliers. Ecol Econ 69:2292–2302
9. Glin LC, Mol APJ, Oosterveer P, Vodouhê SD (2012) Governing the transnational organic cotton network from Benin. Glob Netw 12:333–354
10. Fayet L, Vermeulen WJV (2014) Supporting smallholders to access sustainable supply chains: lessons from the Indian cotton supply chain. Sustain Dev 22:289–310
11. Lakhal SY, Sidibé H, H'Mida S (2008) Comparing conventional and certified organic cotton supply chains: the case of Mali. Int J Agric Resour Gov Ecol 7:243–255
12. GVR—Grand View Research. Cellulose Fiber Market Size, Share & Trends Analysis By Product Type (Natural, Synthetic), By Application (Textile, Hygiene, Industrial), By Regions And Segment Forecasts, 2018—2025. Cellulose Fiber Market Size & Share, Industry Report, 2018–2025 75 (2016). Available at: https://www.grandviewresearch.com/industry-analysis/cellulose-fibers-market. Accessed 10th Feb 2020
13. da Silva FM (2018) Sustainable fashion design: Social responsibility and cross-pollination! In: Montagna G, Carvalho C (eds) Textiles, Identity and Innovation: design the future proceedings of the 1st international textile design conference (D_TEX 2017). Lisbon, Portugal, CRC Press, pp 439–444
14. Todeschini BV, Cortimiglia MN, Callegaro-de-Menezes D, Ghezzi A (2017) Innovative and sustainable business models in the fashion industry: Entrepreneurial drivers, opportunities, and challenges. Bus Horiz 60:759–770
15. Oliveira Duarte L, Kohan L, Pinheiro L, Fonseca Filho H, Baruque-Ramos J (2019) Textile natural fibers production regarding the agroforestry approach. SN Appl Sci 1:914
16. Ferrigno S (2016) Mind the gap: towards a more sustainable cotton market

17. Karakan Günaydina G et al (2019) Evaluation of fiber, yarn, and woven fabric properties of naturally colored and white Turkish organic cotton. J Text Inst 1–18. https://doi.org/10.1080/00405000.2019.1702611
18. Dawson T (2012) Progress towards a greener textile industry. Color Technol 128:1–8
19. Murugesh Babu K, Selvadass M, Somashekar R (2013) Characterization of the conventional and organic cotton fibres. J Text Inst 104:1101–1112
20. Chaudhry MR, Guitchounts A (2003) Cotton facts. FAO-ICAC -International Cotton Advisory Committee
21. Lirbório LFO (2017) circuito espacial de produção do algodão naturalmente colorido na Paraíba—Brasil. Universidade de São Paulo. https://doi.org/10.11606/T.8.2017.tde-22052017-115134
22. Wendel JF, Brubaker C, Alvarez I, Cronn R, Stewart JM (2009) Evolution and natural history of the cotton genus. In: Genetics and genomics of cotton. Springer US, pp 3–22. https://doi.org/10.1007/978-0-387-70810-2_1
23. Bertoniere NR (2000) Cotton. In: Kirk-Othmer Encyclopedia of chemical technology. Wiley, New York. https://doi.org/10.1002/0471238961.0315202002051820.a01
24. The Robinson Library. Cotton. (2018). Available at: https://www.robinsonlibrary.com/agriculture/plant/field/cotton.htm. (Accessed: 14th June 2020)
25. Moreira J, de AN, Freire EC, Dossantos JW, Vieira RD (1995) Use of numerical taxonomy to compare 'Moco' cotton with other cotton species and races. Rev. Bras. Genética **18**, 99–103
26. Pinto de Menezes IP, Barroso PAV, Hoffmann LV, Lucena VS, Giband M (2010) Genetic diversity of mocó cotton implications for conservation (Gossypium hirsutum race marie-galante) from the northeast of Brazil. Botany 88:765–773
27. de Moreira JAN, Freire EC, Santos RF (1989) dos & Neto, M. B. Algodoeiro mocó: uma lavoura ameaçada de extinção
28. EMBRAPA Cotton—Empresa Brasileira de Pesquisa Agropecuária - Centro Nacional de Pesquisa do Algodão. Embrapa 112 algodão 6M : cultivar de Algodoeiro Mocó Precoc
29. Kicińska-Jakubowska A, Bogacz E, Zimniewska M (2012) Review of natural fibers. Part I—vegetable fibers. J Nat Fibers 9:150–167
30. Gaber ASE-D (2016) Ecological and toxicological studies on certain insect pests infesting cotton crop in assiut governorate. Assiut University
31. Eyhorn F, Ratter SG, Ramakrishnan M (2005) Organic cotton crop guide—a manual for practitioners in the tropics
32. Neves MF, Pinto MJA (2012) Cadeia do Algodão Brasileiro: Desafios e Estratégias. Associação Brasileira dos Produtores de Algodão (ABRAPA)
33. Desore A, Narula SA (2018) An overview on corporate response towards sustainability issues in textile industry. Environ Dev Sustain 20:1439–1459
34. Uddin F (2019) Introductory chapter: textile manufacturing processes. In: Textile manufacturing processes. IntechOpen. https://doi.org/10.5772/intechopen.87968
35. About Organic Cotton and Textile Exchange (2019) About organic cotton: organic cotton field to fashion. J Organic Cotton. Available at: https://aboutorganiccotton.org/field-to-fashion/#:~:text=FIELD-TO-FIBER&text=Ittakes approximately 60 to,for approximately 5–6 months.&text=45 days after bolls appear,open along the bolls' segments. Accessed 2nd June 2020
36. Esteve-Turrillas FA, de la Guardia M (2017) Environmental impact of Recover cotton in textile industry. Resour Conserv Recycl 116:107–115
37. Béchir W, Mohamed BH, Béchir A (2018) Industrial cotton waste: recycling, reclaimed fiber behavior and quality prediction of its blend. Tekst. ve Konfeksiyon 28:14–20
38. Malik TH, Ahsan MZ (2016) Review of the cotton market in Pakistan and its future prospects. OCL 23:D606
39. de Beltrão NEM et al (2009) Algodão agroecológico: opção de agronegócio para o semiárido do Brasil. Embrapa Algodão -Documentos (INFOTECA-E) 222:66
40. de Souza MCM (2000) Produção de algodão orgânico colorido: possibilidades e limitações. Inf Econômicas (Governo do Estado Sao Paulo - Inst Econ Agric 30:91–98

41. Cardoso NFS (2017) Algodão Agroecológico no Semiárido Brasileiro: da produção à comercialização. Universidade Federal de Viçosa
42. Baydar G, Ciliz N, Mammadov A (2015) Life cycle assessment of cotton textile products in Turkey. Resour Conserv Recycl 104:213–223
43. Accorsi R, Cholette S, Manzini R, Pini C, Penazzi S (2016) The land-network problem: ecosystem carbon balance in planning sustainable agro-food supply chains. J Clean Prod 112:158–171
44. Yousef S et al (2019) A new strategy for using textile waste as a sustainable source of recovered cotton. Resour Conserv Recycl 145:359–369
45. ABRAPA—Associação Brasileira dos Produtores de Algodão (2019) Algodão no Brasil e no Mundo. Available at: https://www.abrapa.com.br/Paginas/dados/algodao-no-mundo.aspx. Accessed 14th June 2020
46. IEA—Instituto de Economia Agrícola (2020) Algodão: conjuntura e tendências 2019/20. (2020). Available at: https://www.iea.sp.gov.br/out/TerTexto.php?codTexto=14762. (Accessed: 14th June 2020)
47. Cotton Incorporated (2020) Cotton market fundamentals, Price Outlook, Monthly Economic Letter
48. Kering and Textile Exchange (2017) Organic cotton—a fiber classification guide
49. Textile Exchange (2017) Preferred fiber and materials market report
50. Nogueira BP (2020) Algodão - Análise Mensal - Abril-Maio/2020
51. Alves B (2019) Cotton production in selected Latin American and Caribbean countries in 2018(in 1,000 480-pound bales). Latin America and Caribbean: cotton production volume 2018, by country. Available at: https://www.statista.com/statistics/1006650/latin-america-caribbean-cotton-production-volume-country/. Accessed 7th June 2020
52. Shahbandeh M (2019) Cotton production by country worldwide, 2019 Published by M. Shahbandeh, Sep 24, 2019 This statistic shows the world's leading cotton producing countries in crop year 2018/2019. In that year, cotton production in India amounted to around 5.77 million metr. Cotton production by country worldwide. Available at: https://www.statista.com/statistics/263055/cotton-production-worldwide-by-top-countries/. Accessed 7th June 2020
53. Barros MAL et al (2020) A review on evolution of cotton in Brazil: GM, White, and Colored Cultivars. J Nat Fibers 1–13. https://doi.org/10.1080/15440478.2020.1738306
54. ABRAPA—Brazilian Association of Cotton Producers. Better Cotton Initiative (BCI). Available at: https://www.abrapa.com.br/Paginas/sustentabilidade/better-cotton-initiative.aspx. Accessed 30th May 2020
55. BCI—Better Cotton Initiative. Better cotton standard system. Available at: https://bettercotton.org/better-cotton-standard-system/. Accessed 12th June 2020
56. Notícias Agrícolas (2018) Cotonicultura brasileira é campeã de produtividade sem irrigação. Notícias Agrícolas. Available at: https://www.noticiasagricolas.com.br/noticias/algodao/210393-cotonicultura-brasileira-e-campea-de-produtividade-sem-irrigacao.html#.XsrWSWij-Um. Accessed 24th January 2019
57. Silva J (2018) Cotton growers bet in agriculture 4.0 to ensure the sustainability of the crop. EMPRAPA Instrumentação—Empresa Brasileira de Pesquisa Agropecuária. Available at: https://www.embrapa.br/en/busca-de-noticias/-/noticia/35007487/cotton-growers-bet-in-agriculture-40-to-ensure-the-sustainability-of-the-crop. Accessed 14th June 2020
58. Neves MF et al (2017) A cadeia do algodão brasileiro: safra 2016/2017: desafios e estratégias. ABRAPA—Associação Brasileira dos Produtores de Algodão
59. IBGE (2017) Censo Agro. IBGE—Instituto Brasileiro de Geografia e Estatística. Available at: https://censos.ibge.gov.br/agro/2017/templates/censo_agro/resultadosagro/produtores.html. Accessed 12th February 2019
60. de Ferraz FPC (2018) Sustentabilidade na cadeia de suprimento do algodão: um estudo de caso da relação entre uma empresa de calçados esportivos e produtores de algodão orgânico. Fundação Getúlio Vargas (FGV)
61. Khadi BM, Santhy V, Yadav MS (2010) Cotton: an Introduction, pp 1–14. https://doi.org/10.1007/978-3-642-04796-1_1

62. Yu C (2015) Natural textile fibres. In: Textiles and fashion. Elsevier, pp 29–56. https://doi.org/10.1016/B978-1-84569-931-4.00002-7
63. Rocky AMKBP (2012) Comparison of effectiveness between conventional scouring and bio-scouring on cotton fabrics. Int J Sci Eng Res 3:1–5
64. Belot J-L (2018) A Indústria Têxtil e a Qualidade da Fibra. In: Manual de Qualidade da Fibra da AMPA- Safra 2018 (eds. IMAmt & Ampa) 155–191
65. USDA—United States Department of Agriculture (2001) The classification of cotton—agricultural handbook 566. USDA—United States Department of Agriculture
66. Cotton Incorporated (2019) HVI® Color Chart. Available at: https://www.cottoninc.com/cotton-production/quality/us-cotton-fiber-chart/hvi-color-chart/. Accessed 9th June 2020
67. International Trade Centre. Cotton Guide—Grade standards——Chapter 2—Cotton value addition—Classing and grading. Available at: https://www.cottonguide.org/cotton-guide/cotton-value-addition-grade-standards/. Accessed 9th June 2020
68. Cotton Incorporated (2019) Classification of American Pima Cotton. Available at: https://www.cottoninc.com/cotton-production/quality/classification-of-cotton/classification-of-american-pima-cotton/. Accessed 9th June 2020
69. Freire EC (1999) Algodão Colorido. Biotecnol. Ciência Desenvolv. 1:36–39
70. Hall J, Matos SV, Martin MJC (2014) Innovation pathways at the Base of the Pyramid: Establishing technological legitimacy through social attributes. Technovation 34:284–294
71. de Carvalho LP, de Andrade FP, da Filho JLS (2011) Cultivares de algodão colorido no Brasil. Rev Bras Ol. e Fibrosas, Camp. Gd 1:37–44
72. Endrizzi JE, Turcotte EL, Kohel RJ (1984) Qualitative genetics, cytology, and cytogenetics. In: Kohel RJ, Lewis CF (eds) Cotton, vol 24, pp 81–129. ASA, CSSA, SSSA Books - Copyright © 1984 by the American Society of Agronomy, Inc. Crop Science Society of America, Inc. Soil Science Society of America, Inc. https://doi.org/10.2134/agronmonogr24.c4
73. Maia AG, Miyamoto BCB, Silveira JMFJ (2016) A adoção de Sistemas Produtivos entre Grupos de Pequenos Produtores de Algodão no Brasil. Rev Econ e Sociol Rural 54:203–220
74. Lima PJBF (1995) Algodão orgânico: bases técnicas de produção, certificação, industrialização e mercado. In: VIII Reunião Nacional do Algodão 18. IAPAR
75. Pereira RMPG, Marinho MRM, Pereira JPG (2001) Certificação e Algodão Orgânico no Brasil. In: III Congresso Brasileiro de Algodão 884–886 (AMPA (Associação Matogrossense dos Produtores de Algodão)
76. Santos E (2017) Cadeia produtiva do algodão orgânico debate estratégias para aumentar produção. EMPRAPA - Empresa Brasileira de Pesquisa Agropecuária. Available at: https://www.embrapa.br/busca-de-noticias/-/noticia/28873222/cadeia-produtiva-do-algodao-organico-debate-estrategias-para-aumentar-producao. Accessed 22nd May 2020
77. da Marques MAS (2019) Autonomia ou Submissão? Uma Análise sobre os Mecanismos de Certificação Orgânica Adotados pelos Agricultores Familiares do Estado da Paraíba. UFRPE - Universidade Federal Rural de Pernambuco
78. Fonseca MF, de AC, de Souza C, da Silva GRR, Colnago NF, Barbosa SCA (2009) Agricultura Orgânica—Regulamentos técnicos e acesso aos mercados dos produtos orgânicos no Brasil. PESAGRO-RIO
79. de Souza MCM (2000) A produção de têxteis de algodão orgânico: uma análise comparativa entre o subsistema orgânico eo sistema agroindustrial convencional. Agric em São Paulo 47:83–104
80. Rundgren G (2006) Organic agriculture and food security. IFOAM—International Federation of Organic Agriculture Movements
81. Liu MO ano de (2019) pode ser o marco para os produtos orgânicos. Globo Rural. Available at: https://revistagloborural.globo.com/Noticias/Sustentabilidade/noticia/2019/01/o-ano-de-2019-pode-ser-o-marco-para-os-produtos-organicos.html. Accessed 5th July 2019
82. da Cunha SGC, de Oliveira AJ (2019) A adesão da fibra de algodão orgânico branco e o naturalmente colorido ao mercado da moda sustentável. in: Blucher design proceedings. Editora Blucher, pp 413–423. https://doi.org/10.5151/7dsd-2.2.038
83. Textile Exchange (2019) 2019 Organic cotton market report

84. Muñoz CMG, Gómez MGS, Soares JPG, Junqueira AMR (2016) Normativa de Produção Orgânica no Brasil: a percepção dos agricultores familiares do assentamento da Chapadinha, Sobradinho (DF). Rev Econ e Sociol Rural 54:361–376
85. Hirata AR, da Rocha LCD, Bergamasco SMPP (2020) Panorama Nacional dos Sistemas Participativos de Garantia. In: Hirata AR, da Rocha LCD (eds) Sistemas Participativos de Garantia do Brasil Histórias e Experiências. IFSULDEMINAS, pp 10–44
86. Global Organic Textile Standard (GOTS). Comprehensive rules for ecological and socially responsible textile production. Available at: https://global-standard.org/. Accessed 31st May 2020
87. Faircompanies (2007) International Federation for Alternative Trade (IFAT). Available at: https://faircompanies.com/articles/international-federation-for-alternative-trade-ifat/#:~:text=IFATisaninternationalnetwork,andfairtradesupportorganizations. Accessed 11th June 2020
88. Lima PJBF, de Souza MCM (2007) Produção brasileira de algodão orgânico e agroecológico em 2006
89. INCRA—Instituto Nacional de Colonização e Reforma Agrária (2020) Classificação do Imóveis Rurais. Available at: https://www.incra.gov.br/tamanho-propriedades-rurais. Accessed 12th February 2012
90. Ministério da Agricultura PEA (2020) A Secretaria de Agricultura Familiar e Cooperativismo Available at: https://www.gov.br/agricultura/pt-br/assuntos/agricultura-familiar/secretaria-de-agricultura-familiar-e-cooperativismo. Accessed 3rd June 2020
91. Ministério da Agricultura PEA (2019) Plano Safra 2019–2020. Available at: https://antigo.agricultura.gov.br/plano-safra. Accessed 3rd June 2020
92. Parikh TS, Patel N, Schwartzman Y (2007) A survey of information systems reaching small producers in global agricultural value chains. In: 2007 international conference on information and communication technologies and development. IEEE, pp 1–11. https://doi.org/10.1109/ICTD.2007.4937421
93. Hall J, Matos S (2010) Incorporating impoverished communities in sustainable supply chains. Int J Phys Distrib Logist Manag 40:124–147
94. Mangnus E, de Piters BS (2010) Dealing with small scale producers—linking buyers and producers. KIT Publishers
95. DataSebrae (2018) Perfil do Produtor Rural. Available at: https://datasebrae.com.br/perfil-do-produtor-rural/#onde. Accessed 19th Feb 2019
96. Alessio MA, Araujo AS, Lopes LD, Schulte NK (2014) Algodão Orgânico na Produção Sustentável. ModaPalavra e-periódico 14:136–150
97. de Oliveira CSC, Oliveira-Filho EC (2014) Agricultura ecológica e indústria têxtil: o papel da comunicação para o algodão orgânico no Brasil. Univ Arquitetura e Comun Soc 11
98. Willer H, Lernoud J (2019) The world of organic agriculture statistics and emerging trends 2019. Research Institute of Organic Agriculture (FiBL) and Organics International (IFOAM)
99. Textile Exchange (2017) 2017 organic cotton market report
100. Wakelyn PJ, Chaudhry MR (2007) Organic cotton. In: Cotton. Elsevier, pp 130–175. https://doi.org/10.1533/9781845692483.2.130
101. Bachmann F (2012) Potential and limitations of organic and fair trade cotton for improving livelihoods of smallholders: evidence from Central Asia. Renew Agric Food Syst 27:138–147
102. FAO—Food and Agriculture Organization of the United Nations. Sustainable Agriculture. Sustainable Development Goals (2020). Available at: https://www.fao.org/sustainable-development-goals/overview/fao-and-the-post-2015-development-agenda/sustainable-agriculture/en/. Accessed 11th June 2020
103. USDA—United States Department of Agriculture. Sustainable Agriculture: Definitions and Terms. Special Reference Briefs Series no. SRB 99–02 (2007). Available at: https://www.nal.usda.gov/afsic/sustainable-agriculture-definitions-and-terms#:~:text=%22Sustainableagriculture%3A A whole-,including international and intergenerational peoples. Accessed 11th June 2020

104. UCDavis—University of California—Davis. What is Sustainable Agriculture. Agricultural Sustainability Institute Available at: https://asi.ucdavis.edu/programs/ucsarep/about/what-is-sustainable-agriculture. Accessed 11th June 2020
105. Erbaugh J, Bierbaum R, Castilleja G, da Fonseca GAB, Hansen SCB (2019) Toward sustainable agriculture in the tropics. World Dev 121:158–162
106. Bensin BM (1930) Possibilities for international cooperation in agroecology investigation. Int Rev Agric Mon Bull Agric Sci Pract 21:277–284
107. Gliessman S (2013) Agroecology: growing the roots of resistance. Agroecol Sustain Food Syst 37:19–31
108. Gliessman SR (2006) Agroecology: the ecology of sustainable food systems. CRC Press, Taylor and Francis Group
109. Warner KD (2006) Agroecology in action extending alternative agriculture through social networks. MIT Press
110. Tripathi N, Singh R, Pal D, Singh R (2015) Agroecology and sustainability of agriculture in india: an overview. EC Agric 2:241–248
111. Tilman D, Balzer C, Hill J, Befort BL (2011) Global food demand and the sustainable intensification of agriculture. Proc Natl Acad Sci 108:20260–20264
112. Groundswell International. Our Approach. (2020). Available at: https://www.groundswellinternational.org/our-approach/. Accessed 11th June 2020
113. Adamchak R (2020) Organic farming. Encyclopedia Britannica. Available at: https://www.britannica.com/topic/organic-farming. Accessed 11th June 2020
114. FAO—Food and Agriculture Organization of the United Nations (1999) Press Release 99/40 - CODEX ALIMENTARIUS COMMISSION TO APPROVE INTERNATIONAL GUIDELINES FOR ORGANIC FOOD. Available at: https://www.fao.org/waicent/ois/press_ne/presseng/1999/pren9940.htm. Accessed 11th June 2020
115. Paull J (2019) Organic agriculture in australia: attaining the global majority (51%). J Environ Prot Sustain Dev 5:70–74
116. Wikipedia. Organic farming. (2020). Available at: https://en.wikipedia.org/wiki/Organic_farming. Accessed 11th June 2020
117. Kamiyama A (2011) Cadernos de Educação Ambiental 13 - Agricultura Sustentável. SMA = Secretaria de Meio Ambiente - Coordenadoria de Biodiversidade e Recursos Naturais
118. Retamiro W, da Silva JLG, Vieira ET (2013) A sustentabilidade na cadeia produtiva do algodão orgânico. Lat Am J Bus Manag 4:25–43
119. Duarte AYS, Baruque-Ramos J, Sanches RA, Mantovani W (2010) Produção de algodão orgânico no Brasil e seu potencial de uso na moda. Química Têxtil
120. Gadaleta C (2017) EcoEra: algodão orgânico no radar da moda. Vogue. Available at: https://vogue.globo.com/EcoEra-Chiara-Gadaleta/noticia/2017/07/ecoera-algodao-organico-no-radar-da-moda.html. Accessed 11th April 2019
121. FAO—Food and Agriculture Organization of the United Nations. Program of Brazil-FAO International Cooperation—Project +Cotton. Available at: https://www.fao.org/in-action/program-brazil-fao/projects/cotton-sector/en/. Accessed 11th June 2020
122. Buainain AM, Batalha MO (2007) SÉRIE AGRONEGÓCIOS—Cadeia Produtiva do Algodão - Volume 4. (Instituto Interamericano de Cooperação para a Agricultura no Brasil (IICA); Ministério da Agricultura, Pecuária e Abastecimento (MAPA); Agência Brasileira de Cooperação do Ministério das Relações Exteriores (ABC/MRE)
123. Barbieri JC, Cajazeira JER (2008) Desenvolvimento Sustentável. In: Responsabilidade Social Empresarial e Empresa Sustentável da teoria à prática 66–80. Saraiva
124. Textile Exchange (2020) 2025 sustainable cotton challenge—second annual report 2020
125. Pereira LB (2019) Manager of sustainable raw materials at Laudes Foundation—Panorama of sustainable cotton and projects happening in Brazil
126. Arriel NHC (2019) Researcher at EMBRAPA cotton—the agroecological cotton project in the northeast of Brazil
127. da Silva MB (2019) Researcher at EMBRAPA cotton—improvement of agroecological cotton systems in diversified crops

128. Pereira DF (2019) Organic cotton colour- in-country manager—perspectives in organic cotton production, partnerships and market
129. Gadelha MM (2019) Director of coopnatural—challenges of organic cotton production in Brazil
130. Cavalcante CC (2019) Planning and operations manager at EMPAER—Organic cotton production in the state of Paraiba in Brazil
131. Ghazinoory S, Sarkissian A, Farhanchi M, Saghafi F (2020) Renewing a dysfunctional innovation ecosystem: the case of the Lalejin ceramics and pottery. Technovation 96–97:102122
132. dos Santos CE (2018) Fashion does good - Movement Sou de Algodão ('I Am Cotton') tries to make final consumers aware of the benefits from the fiber, whilst stressing cotton´s sustainable footprints. Anuário Bras. do algodão 2018:104
133. dos Santos CE (2018) Production—More Color Please. Anuário Bras. do algodão 2018:104
134. Fletcher K, Grose L (2012) Moda and sustentabilidade: design para mudança. Senac São Paulo
135. Furlow NE, Knott C (2009) Who's reading the label? millennials' use of environmental product labels. J Appl Bus Econ 10:1–12
136. Magnuson, B., Reimers, V. & Chao, F. Re-visiting an old topic with a new approach: the case of ethical clothing. *J. Fash. Mark. Manag. An Int. J.* **21**, 400–418 (2017).
137. Matthews D, Rothenberg L (2017) An assessment of organic apparel, environmental beliefs and consumer preferences via fashion innovativeness. Int J Consum Stud 41:526–533
138. Rothenberg L, Matthews D (2017) Consumer decision making when purchasing eco-friendly apparel. Int J Retail Distrib Manag 45:404–418
139. de Araújo MBM, Mota-Ribeiro S, Broega AC (2016) Marcas de moda sustentável: a importância das mídias sociais na aproximação com o público. In: Congresso Internacional Negócios da Moda 14. Instituto Brasileiro de Moda (IBModa)
140. de Castro AMG, Lima SMV, Cristo CMPN (2002) Cadeia Produtiva: Marco Conceitual para Apoiar a Prospecção Tecnológica. In: XXII Simpósio de Gestão da Inovação Tecnológica 14
141. Raynolds LT (2004) The globalization of organic agro-food networks. World Dev 32:725–743
142. Bocken NMP, Short SW, Rana P, Evans S (2014) A literature and practice review to develop sustainable business model archetypes. J Clean Prod 65:42–56
143. Berlin L (2012) Moda e sustentabilidade: uma reflexão necessária. Estação das Letras e Cores
144. Rota C, Pugliese P, Hashem S, Zanasi C (2018) Assessing the level of collaboration in the Egyptian organic and fair trade cotton chain. J Clean Prod 170:1665–1676
145. Fernandez-Stark K, Gereffi G (2019) Global value chain analysis: a primer (second edition). In: Handbook on global value chains 54–76. Edward Elgar Publishing. https://doi.org/10.4337/9781788113779.00008
146. Ozturk E et al (2016) Sustainable textile production: cleaner production assessment/eco-efficiency analysis study in a textile mill. J Clean Prod 138:248–263
147. Pal R, Gander J (2018) Modelling environmental value: An examination of sustainable business models within the fashion industry. J Clean Prod 184:251–263
148. Wilson C, Tisdell C (2001) Why farmers continue to use pesticides despite environmental, health and sustainability costs. Ecol Econ 39:449–462
149. Wagner M et al (2017) Fashion design solutions for environmentally conscious consumers. IOP Conf Ser Mater Sci Eng 254:192017
150. Foure P, Mlauli T (2007) Eco initiatives in the textile pipeline—a south african experience. In: Ecotextiles 96–106. Elsevier 2007. doi:https://doi.org/10.1533/9781845693039.2.96
151. Vehmas K, Raudaskoski A, Heikkilä P, Harlin A, Mensonen A (2018) Consumer attitudes and communication in circular fashion. J Fash Mark Manag An Int J 22:286–300
152. Fletcher K (2010) Slow fashion: an invitation for systems change. Fash Pract 2:259–265
153. Moorhouse D, Moorhouse D (2018) Designing a sustainable brand strategy for the fashion industry. Cloth Cult 5:7–18
154. Kozlowski A, Searcy C, Bardecki M (2015) Corporate sustainability reporting in the apparel industry. Int J Product Perform Manag 64:377–397

155. Tong X, Su J (2018) Exploring young consumers' trust and purchase intention of organic cotton apparel. J Consum Mark 35:522–532
156. dos Santos EF, Silva CE (2012) A influência das estratégias de marketing na captação de recursos para o Terceiro Setor. Rev Bras Adm Científica 3:94–106
157. Bray J, Johns N, Kilburn D (2011) An exploratory study into the factors impeding ethical consumption. J Bus Ethics 98:597–608
158. Joergens C (2006) Ethical fashion: myth or future trend? J Fash Mark Manag An Int J 10:360–371
159. Kerr J, Landry J (2017) Pulse of the fashion industry—global fashion agenda—executive summary
160. Stam E, van de Ven A (2019) Entrepreneurial ecosystem elements. Small Bus Econ. https://doi.org/10.1007/s11187-019-00270-6
161. Oh D-S, Phillips F, Park S, Lee E (2016) Innovation ecosystems: a critical examination. Technovation 54:1–6
162. Bryson JM, Crosby BC, Stone MM (2015) Designing and Implementing Cross-Sector Collaborations: Needed and Challenging. Public Adm Rev 75:647–663
163. De Oliveira JB, Severiano Filho C (2005) Considerações sobre a produção do algodão colorido e a importância do Consórcio Natural Fashion como último elo da cadeia produtiva. In: X Congresso Brasileiro de Custos 14. Associação Brasileira de Custos
164. Harris F, Roby H, Dibb S (2016) Sustainable clothing: challenges, barriers and interventions for encouraging more sustainable consumer behaviour. Int J Consum Stud 40:309–318
165. Fletcher K, Tham M (2019) Earth logic: fashion action research plan. JJ Charitable Trust
166. Global Fashion Agenda—Design (2017) Design for Longevity: Inspiration, knowledge and tools for future-proof design. Available at: https://globalfashionagenda.com/design-for-longevity-inspiration-knowledge-and-tools-for-future-proof-design/. Accessed 14th June 2020
167. Caniato F, Caridi M, Crippa L, Moretto A (2012) Environmental sustainability in fashion supply chains: An exploratory case based research. Int J Prod Econ 135:659–670
168. Fletcher K, Grase L (2012) Moda and Sustentabilidade, Design Para Mudança. Editora Senac São Paulo
169. Lee M (2009) Eco Chic - O Guia de Moda Ética Para a Consumidora Consciente. Editora Larousse
170. Ellen MacArthur Foundation (2017) A new textiles economy: redesigning fashion's future
171. Lima PJBF (2008) Algodão agroecológico no comércio justo: fazendo a diferença. Rev. Agric. 5:37–41
172. Cardeal T (2019) Photo: araripina community field work
173. Aranha F (2020) Photos: Flavia aranha collection "The harvest", using organic cotton fabrics from Natural Cotton Color and Justa Trama

Sustainable Fashion

B. Luis Chaves and A. Shirley Villalobos

Abstract One of the main problems in the fashion industry is its impact on the environment as a result of its value chain due to its energy consumption, exploitation of natural resources, waste disposal, CO_2 emission, generation of solid waste, indiscriminate use of synthetic materials, among others. Reducing, avoiding, minimizing, limiting, stopping, etc., are key points in the sustainability agenda: we must go further, considering that the best way to reduce the environmental impact is not to recycle but to produce and waste less [1]. The meaning of sustainable fashion is not very clear, particularly when this can be analyzed from different perspectives: a commercial and an anthropological one. Journalists and scholars have defined fashion in terms of history, cultural identity, personal communication, social position, lifestyle, change, speed, and even sexuality and eroticism [2]. Color is a critical element in fashion and is probably one of the principal causes of environmental problems. Substituting fossil fuels in the production of clothes is possible through renewable energy; however, these are still essential to manufacturing many textile materials, including textile colorants. The truth is, the fashion industry is facing a complicated environmental problem, also considering the accumulation of toxic and inorganic residues and water overconsumption. This situation is a reality. We only need to look at the standards of the European Chemical Agency—REACH. Currently, the term 'sustainability' feels too ambiguous; many companies claim they are sustainable when, in reality, they only carry out superficial environmental activities, so much so that a term for this practice has already been coined: Green Washing. In the medium-term, sustainability will end up being a better-defined term each time and will work as a restriction and a standard for the fashion industry and its value chain, which, in many cases, will be difficult to meet. Developing concepts related to durability and versatility through

B. L. Chaves (✉)
Industrial Engineer, Pontificia Universidad Católica del Perú, Lima, Perú
e-mail: lchaves@incalpaca.com

Master of Business Administration, Polytechnic University of Madrid, Madrid, Spain

A. S. Villalobos
Chemical Engineer, National University of San Agustín, Arequipa, Peru

Master in Textile Engineering, Polytechnic University of Valencia, Valencia, Spain

M. Á. Gardetti and R. P. Larios-Francia (eds.), *Sustainable Fashion and Textiles in Latin America*, Textile Science and Clothing Technology,
https://doi.org/10.1007/978-981-16-1850-5_15

materials and intelligent design of processes and products will be a significant way to care for and protect the environment.

Keywords Trends · Fashion · Sustainability · Environment

1 Fashion

Understanding the meaning of fashion is complicated due to its connotation and outreach. We can draw a comparison to any economic activity, be it individual or communal, that revolves around people if we study the definition of tribe presented by the writer Seth Godin [1]. Said definition could explain the meaning of fashion from different perspectives: "A tribe is a group of people interconnected to a leader or an idea." For many years, humans have belonged to one tribe or another. A group only needs two things to be a tribe: a common interest and a means to communicate.

(a) **What does fashion mean?**

Associating the term fashion with its meaning in Spanish (Moda) is interesting: Statistically, Moda (fashion) is defined as the value with the most frequency in a data distribution. This is a priority in companies of product sales and general services and clothing; the serious issue occurs when the mechanisms used to reach targets threaten the environment and, in many cases, respect for people.

Anne Hollander [2] defines fashion in the following way: "Everyone has to dress up tomorrow and go to work during the day. This has been shaped in the West in the last 700 years, and this is fashion."

A simple interpretation of this definition circumscribes fashion around western culture. It has probably been to a certain extent true over the last 100 years and is associated with the progress of communication media and advertisement. However, it has been seen throughout history in the richness of attires in many Asian, American, African, and even European cultures thousands of years; paradoxically, M. Bernard [3] synthesized it with this phrase: "When we see a movie or TV show and a person wearing an outfit is shown, we can identify a place in history."

The first time that the men from the Spanish Empire made contact with the Andean textile art was in the year 1526; art made excellence by over 4500 years of uninterrupted development. It was a place in the world where rags did not exist, where dressing up was a cult to beauty to which every individual had the right.

We do not want to forget a quote from Fernando de Szyszlo, one of the most renowned Peruvian plastic artists (1925–2017): "Undoubtedly, the most important contribution of Peru to the history of universal art are the textiles created between approximately 1400 BC and 1600 AC." Fashion had been accompanying humanity for thousands of years.

Malcolm Barnard's [3] observations referring to the relationship between history and fashion and fashion and time are enlightening. We would like to highlight three of these observations.

- History is a background to fashion. It is the stage of important tales in mankind's history that can be simply associated through an article of clothing.
- History is a context for fashion; a possible explanation for this is the French revolution and its extravagant dresses.
- In its relationship with history, fashion is a seesaw and tightly connected; due to this relationship, fashion can transform history.

In the book "Fifty Dresses that Changed the World" [4], it is mentioned that young modern women of the 1920s changed the perception of their role in a male-dominated world through morally aggressive dresses.

In the last 100 years, scholars and researchers have been attempting to understand the meaning of fashion. It is undeniable that it is intimately associated with articles of clothing, without failing to recognize that their primary function is to protect our body from weather conditions.

After analyzing its definition by several authors, what is fashion? It is art, design, cultural expression, communication, emotions, identity, image, social position, and color.

There are many ways to answer this question and we have found several theories and postulates searching for the meaning of fashion. It is evident that there are discrepancies across different interpretations of fashion and several aspects are involved when we search for a definition. We face contradicting references in personal and cultural emotions, expressions, design, brands, and especially thinking and culture that lead to representing an idea or cultural experience.

One of the most controversial topics is the acceptance of fashion as art. From a wide perspective, it is possible to consider fashion an art; its aesthetic and communicative objectives through which ideas, concepts, emotions, and a general view of the world can be expressed support this postulate.

However, different authors, including Baudrillard Jean, 1981 [5], conclude that fashion is not art through the mechanical reproduction of articles of clothing. Fashion manifests itself through an article of clothing that is fashionable. It has to be available for many people. This refers to the mechanical production of several copies; thus, fashion is not art. In the end, this discussion is about whether the articles are produced mechanically or not. We can state that if the article is created by hand, only one edition, and for personal use, then this activity could be considered art.

Particularly, the vision of fashion and art should be associated with beauty and creativity. There are creative processes, for instance, of clothes knitted in Jacquard fabrics where the combination of colors, the structure of the textile makes it possible to obtain fabric with unmatched beauty and we resist to say that this combination of processes, raw material, and creativity is not art.

The association of design with fashion is undeniable, and it is related to the necessary variety and velocity for the creation of clothes. Besides, traditionally, the designer remains anonymous, and their design can be reproduced indeterminately. Therefore, the design ends up being a tool that lets us develop concepts and reinforce the claim that fashion is not art.

Fashion and clothes can be explained as a means of communication, appearing in beliefs or ideas associated with cultural aspects. This point is considered key to understanding the meaning of fashion from the point of view of a regular person.

Belonging to a tribe—a culture associated with our beliefs and values—is often related to clothing and fashion as a differentiating element and plays a very important role in an aspiring necessity. The need for change in social status is associated with the way one dresses, translated into the need for speed and variations in design to allow the person to achieve their objectives. Change is progress. In the end, fashion requires and reflects an experience, a conception of time of change and, of course, different.

The presence of a leader or leadership in fashion can lead to a change in clothes. If the velocity decreased significantly, we would be before a concept of classical clothing.

(b) **Is it a way for us to communicate?**

It is a fact that an article of clothing can send messages. Often, it allows us to identify the behavioral characteristics of a person when they wear certain clothing. It can indeed lead us to mistaken interpretations about specific situations, but we cannot deny the fact that a message is being emitted.

The sender/receiver communication model can be inadequate for the meaning of an article of clothing, but the communication is there. The main question is whether, through fashion, it is possible to send clear messages about who we are or what we wish to be. This will be a significant challenge for fashion gurus.

In an interview with Ryan Smith for Vogue [6], actress Jessica Alba stated that fashion is your visible personality and represents who you are. These are some of the many common assumptions of the relationship between fashion and identity. Fashion can represent values and beliefs of different cultural groups and the relationship among their people, which can ultimately lead to the construction of a political position.

A clear example of this claim was the political group, the Black Panthers. Scholars mention it as the most influential organization of the black movement in the US during the late 1960s. During the times of Mao Tse Tung, the Chinese cultural revolution is another clear example of the relationship between clothing and political movements, whatever its causes were.

A controversial topic is a relationship between fashion and the large number of people that work in the textile industry. Unfortunately, a significant part of them does so in unsafe economic and ergonomic conditions. These topics have only become relevant in the fashion industry at the start of the twenty-first century. Awareness of the situation is being spread. The big question is, ethical, moral, or economic reasons?

Today, the new consumers are pushing more and more for responsible behavior from the brands, including in the fair treatment of people as well as the environment, the relationship between fashion and what we wear and the image we wish to project, specifically a physical stereotype, often associated with femininity, even able in some cases to influence and control people's behavior.

The image may or may not be associated with identity; it could be a very powerful stimulus, negative like anorexia which in some cases can reach extreme levels, or positive like online influencers who can affect the behavior of hundreds of thousands of people. It is incredible what a single article of clothing can achieve: create a new image or identity to change it.

From an anthropomorphic and social perspective, understanding the meaning of fashion is important to choose adequate routes to frame the fashion industry inside the new conditions that begin to impose social responsibility and environmental responsibility on the new consumers.

There are two topics that fashion researchers and scholars seem to place on a lower level of importance: design and color. In terms of color, we can assume it is intrinsically incorporated into the design. However, its importance, considering the current market situation, goes further.

(c) **Design versus color?**

The images we can observe through night vision goggles, whatever the design of an article of clothing may be, cannot be appreciated for their charm.

In her article published in July 2020 on WGSN [7], Jenny Clark's observations reveal the importance of color in people's daily lives, especially during these days when we are all affected by COVID-19. "Color is an important part of visual communication that can evoke a strong emotional response in all of us; in these difficult times, it is used to transmit different emotions and reactions related to different global crises."

Before examining how color is used, we must learn about our cognitive response to this and its ability to evoke physical and emotional reactions. Several studies show that certain colors evoke a subconscious physical response that is related to primal instincts.

Apart from a physical response, we also have an emotional and personal relationship with colors linked to our own memories and experiences. These factors, combined with centuries of tradition and social conditioning, have created a subjective response that is independent of color. One thing is certain: when facing the global problems of the future, the color will be used in a more conscious manner."

It is anecdotal to mention that the sense of sight utilizes 60% of all the energy consumed by all five senses. One of the greatest challenges the fashion industry faces is the production of colors, considering environmental restrictions.

The need for a change and differentiation requires the speed in processes of creation, production and commercialization of fashion, a situation that the fast fashion sector has understood well and taken advantage of.

Another aspect barely mentioned and often assumed to be intrinsic to clothing is comfort. It has been hard to find observations and/or comments about the need to wear comfortable clothing in all the analyzed data. In the near future, we foresee that it will be a primordial feature of an article of clothing, and its design will have to have meaning and necessity in its true dimension.

(d) **Concepts**

The book "The Story of Purpose," written by Joey Reiman in 2013 [8] lets us know the proposals associated with the development of concepts. In these times, purpose and meaning make better institutions and lasting legacies for everyone involved.

Brene Brown's phrase [9] is inspiring. "Maybe stories are just data with a soul." "A story is a new narrative for business. Stories shape our lives because we are creatures who search for a reason." [10].

But what does all of this have to do with fashion? Today, fashion itself does not matter to the new consumer. They need to see in it a purpose, be it social, personal, or cultural. How can we give fashion, or any other activity or business, a purpose?

To define a purpose through Concepts (master idea) is a proposal by Joey Reiman [8]. He references Theodore Roszak, a University Professor in California, and concluded that Master Ideas refer to great moral, religious, and metaphysical teachings, which are the foundations of culture.

Master ideas are not based on facts but on the conviction, that thought will shake the soul. Roszak himself presented two key characteristics of the Master Idea: they must be infectious and lasting. They are here to stay and frequently, when people hear about them, are taken into peoples' lives immediately.

Can fashion transmit this effect? We believe so. Through Concepts (Master Idea), it is possible to do so, but always seeking the common good in all its magnitude. Reiman proposes nine principles to develop a Master Idea: the nine principles of a Master Idea (Concepts):

1. The Master Idea is timeless. How can this affect fashion? It may sound confrontational with the fashion industry, but it is completely aligned with sustainability.
2. The Master Idea teaches: Storytelling is a powerful tool nowadays.
3. The Master Idea fulfills us.
4. The Master Idea is a war chant. (Steven Jobs: Think different)
5. The Master Idea is based on ethos; the character or identity of people is identified.
6. The Master Idea is transformative; it creates a holistic change around a new organization and objective.
7. The Master Idea inspires: Searching for new meanings
8. The Master Idea is born not from data processing but absolute conviction.
9. The Master Idea tells a story.

(e) **State of fashion**

- **What does fashion hold for us in the upcoming years?**

According to the publication presented by The Business and Fashion and McKinsey & Co.: The State of Fashion 2018 [11], the report mentions ten key tendencies to consider in the fashion industry in the short term; these are:

(1) Volatility and uncertainty
(2) Globalization is rebooted
(3) The Asian pioneers (e-commerce) and their influence in the market
(4) Getting personal
(5) Electronic platforms (Amazon vs Tmalls)
(6) Obsession for mobility
(7) Artificial Intelligence
(8) Sustainability
(9) Price as a strategy
(10) Startups as new players in the market.

Analyzing these ten tendencies, we can claim that eight of them are geared toward market development. In layman's terms, it can be the fashion industry or any other industry, such as the home or education industries. These are the two specific topics that must be studied and considered for future strategies in the fashion world.

Sustainability will be paramount in the future of the said industry, but it will also make it more personal through concepts and values, personalization of experiences, etc.; these will be key aspects in the future of the industry. These topics have been brought up; the first is an inherent part of this report. The second one brings us to explain the need and importance of developing concepts associated with fashion and clothing.

2 Sustainability

(a) Definition

According to the Brundtland report [12], Sustainability means that the raw materials consumed should be available in sufficient quantities for future generations. To understand what the concept of sustainable development intends to convey, we must highlight some of the conclusions drawn from the "Brundtland Report," which will be listed below.

- The globally dominant models of economic growth inevitably lead to the gradual exhaustion of the world's natural resources and a rise in poverty, reaffirming the idea of a lack of intergenerational solidarity.
- Nevertheless, with positive and conciliatory intent, the report considered that with a more equal and rationalized distribution of goods, growth is possible and poverty can be reduced, but, to achieve this, relevant actions from the political leader and effective participation from the population in integrating the objectives of economic and social development with environmental conservation are all indispensable, which are grouped in the sustainable development category.

What happens to the consumer?

Do they understand the problem with the environment? Are they conscious about the risk of the irreversibility of this problem, mostly caused by their shopping behavior?

There is no clear answer to these questions; probably environmentalists and environmental aficionados can find them.

(b) **Environmental Impact**

K. Ulrich and S. Eppinger, in their book Product Design and Development [13], present a design process based on environmental problems that can be applied to the design process of any product or service and can lead different interest groups within the fashion industry to understand the imperative need to take care of the environment throughout the different inherent stages of said industry.

A complete analysis of several industrial sectors is performed, and the necessity to explain the environmental performance of products and processes is established while considering the following factors:

- Environmental impact of materials/raw material used in the manufacturing of the product; regarding fashion, we can establish the difference between using cotton fibers, celluloid fibers, protein fibers, fibers derived from fossil fuels, among others.
- It is clear that the behavior of each of these fibers, from their extraction and/or creation to their final use, will have a different impact on the environment.
- Recycling: In general terms, any textile fiber can be recycled, but we have to consider the implications of said process for the environment, for example, the consumption of water and energy.
- The amount of material that can be incorporated into an article of clothing without affecting its features and performance has a limit, either due to market or environmental requirements.
- **Clean Energy**: The possibility of using renewable energy in the fabrication, distribution, and commercialization processes are aspects to consider during clothing design activities.
- **Emissions**: Clearly establish the permitted levels of emissions.
- **Water consumption**: A critical problem in the textile industry, which is magnified when we use this resource while diminishing food production.
- **Solid waste**: Especially hazardous and/or inorganic ones, whose degradation can take not years, but centuries.

Ulrich and Eppinger [13] clearly define two cycles of life:

- The life cycle of an industrial product begins with the extraction and processing of raw material and/or natural resource, then the production, distribution, and use of a product. Finally, when the product can no longer be used, there are several options, to recover it or reuse its components, recycle the material, burn it, or bury it.
- The life cycle represents the growth and decomposition of organic products in a continuous loop.

As shown in Fig. 1, both cycles intersect with the use of natural materials and industrial materials. While the life cycle of an industrial product can last a few months or years, the natural life cycle can take various ranges. In the case of organic materials, these quickly decompose into nutrients for the extraction of similar materials. Other natural resources are created over long periods and are considered non-renewable, and their decomposition might take thousands of years and risk polluting the earth.

There are three great challenges in product design toward sustainability:

- Eliminate the use of non-renewable natural resources, including fossil fuels.
- Eliminate the availability of synthetic and inorganic materials that do not decompose quickly.
- Eliminate the availability of toxic waste that is not part of the natural life cycle.

Lewis and Gertsakis [14] explain some of the environmental impact derived from the manufacturing sector in general:

- Global warming as a result of the buildup of greenhouse gases.
- Solid waste, especially those that cannot be recycled and must be burnt or buried.
- Exhaustion of non-renewable resources such as fossil fuels, coal, minerals, etc.
- Water pollution, especially due to discharge, results from industrial processes and is poured into rivers and public sewage, including heavy metals, fertilizers, solvents, etc.

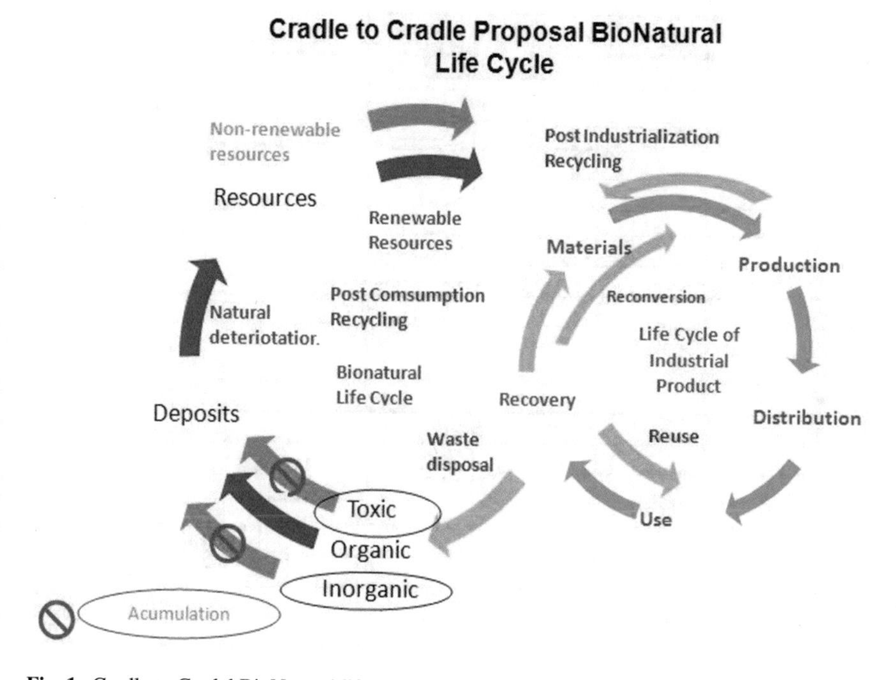

Fig. 1 Cradle to Cradel BioNatural life cycle

- Air pollution due to the emission of gases in thermal processes, from generating plants to vehicles with internal combustion engines.
- Soil degradation due to the extraction of raw materials in mining, the expansion of the agricultural borders for the production of food, the forestall industry, and even in the production of prohibited substances.
- Biodiversity: displacement of a variety of plants and animals caused by urban and industrial development.
- Exhaustion of the ozone layer due to the emission of gases that destroy it.

Ulrich and Eppinger [13] have proposed a model called Design for Environment (DFA), which should be studied by the fashion industry, which allows one to evaluate and determine whether a product and/or process is environmentally and socially sustainable.

Based on the model proposed by said researchers, we present the following table (Fig. 2), a chart to evaluate the environmental impact of the various textile fibers.

It is a challenge for the fashion and clothing industries to establish methods to measure the environmental impact of the extraction and manufacturing processes of the textile fiber, whichever its origin may be.

In a report by the House of Commons—Environment Audit Committee of the UK titles "Fixing Fashion: Clothing Consummation and Sustainability" [15], a comparison between natural and synthetic fibers is drawn.

Description	Energy consumption	Depletion of natural resources	Waste disposal	Gas emission	Solid waste production
Animal Fibers					
Alpaca - Camelids					
Wool					
Cashmere - Mohair					
Silk					
Others					
Plant fibers					
Cotton					
Linen					
Jute					
Others					
Synthetic Fibers					
Polyester					
Nylon					
Acrylic					
Polypropylene					
Others					
Viscose					

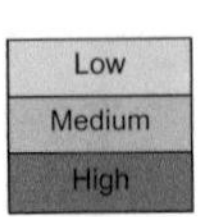

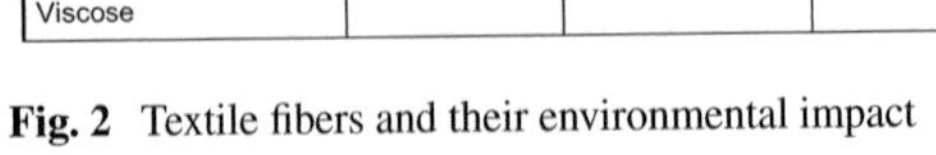

Fig. 2 Textile fibers and their environmental impact

Producing natural fibers like cotton, wool, silk, and cashmere requires the use of water, soil, animals, food, and chemicals.

Synthetic fibers like polyester are made with petroleum, a non-renewable resource that requires intensive use of energy for its production.

Cotton is the most used natural fiber. It requires a large amount of water and uses fertilizers and insecticides. The ecological cotton alternative is gaining a lot of momentum.

Animal fibers, by nature, create methane, one of the gases that contribute to the greenhouse effect. Animal mistreatment is a concern, a topic that relevant associations are trying to solve.

Synthetic fibers are manufactured from plastic, including polyester, polyamide and acrylic, or from plants (trees), ground and chemically dissolved, and then transformed in fibers such as viscose, mode, rayon, lyocell, etc.

Petroleum-based synthetic fibers have less impact on land and water, but greenhouse gas emissions are higher.

Developing the market for recycled fibers is a challenge, as the circular fashion model presents it; it is an important matter; however, it has limitations considering the characteristics and performance of the products made with recycled fibers.

A critical factor to bear in mind is plastic microparticles, which create problems in marine life and its food chain.

Synthetic textiles are the most important source of microspheres in the oceans.

It is a challenge of the fashion and garment industry to establish methods to measure the environmental impact of the production of textile fibers.

Higg Index, developed by "Sustainable Apparel Coalition" [16] (It is stated that the index allows us to measure the environmental impact of various textile fibers, specifically in the emission of greenhouse gases (methane and nitrogen dioxide, not including CO_2)).

In addition to evaluating the environmental impact of the raw material, it is important to assess it through the whole value chain in the textile industry.

In Fig. 3, we show a diagram to evaluate the environmental impact throughout the entire value chain of the textile industry from fiber made of animal fiber.

The most important aspects to consider:

Animal fibers (protean) are a renewable resource. It is possible to use renewable energy sources to process them, but often it is necessary to use non-renewable energy, especially fossil fuels.

The main problems originate from wet processes, as these cause the greatest environmental impact due to the chemical characteristics of inherent chemical materials of such processes.

The chemical industry associated with the production of these raw materials is aware of the original situation. In most cases, it meets the conditions and requirements set by regulators (Governments) of the main producing countries. Will this be enough?

The elimination of waste is one of the greatest negative environmental factors. The creation of waste in the textile industry must be very clear; identify it and evaluate its

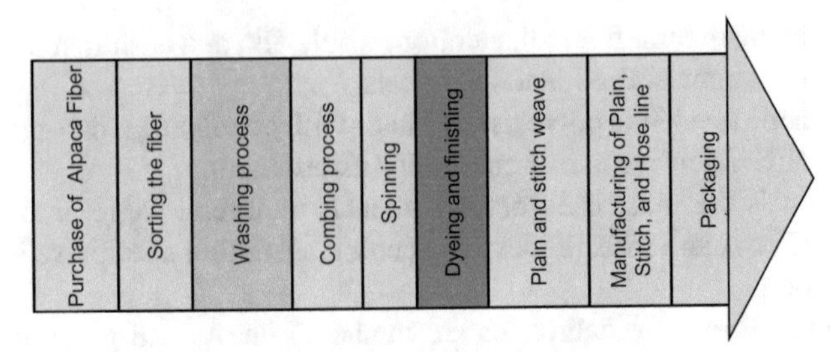

	Purchase of Alpaca Fiber	Sorting the fiber	Washing process	Combing process	Spinning	Dyeing and finishing	Plain and stitch weave	Manufacturing of Plain, Stitch, and Hose line	Packaging
Depletion of Resources									
Energy									
- Renewable									
- Solar			Panels						
- Hydraulic			Electric power						
Non-renewable									
- Of high impact			Industrial petroleum						
- Of low impact			Natural Gas						
Materials									
- Solid waste									
- Organic			Derivatives and vegetable material			Fats	Paper and cardboard derivatives		
- Inorganic				Plastic and parts					
- Toxic			Batteries, screens, integrated circuits						
- Effluents									
- Dyes / Chemicals			Detergent	Enzymes		Dyes	Softener		
- Residual water									
Gas emissions			Steam			Steam			

Fig. 3 Value chain of alpaca animal fiber

environmental impact considering the current internationally recognized regulations. REACH, implemented by the European Chemical Agency, is a good example.

Figure 4 shows a chart explaining the extent of the possible environmental impact of the different materials and waste used in the textile industry for animal fibers.

(c) Speed and buildup

The velocity of the fashion industry is a characteristic associated with their commercial and financial needs, which has been deepened with the development of fast fashion, with a negative effect on the environment.

This greater velocity creates a higher buildup of inorganic and toxic waste in the environment.

The appraisal presented in the Sustainable Fashion Handbook [17], edited by Liz Parker (UK) and Marsha A. Dickson (USA), on the topic of velocity explains the magnitude and extent of this topic.

Description	Organic	Inorganic	Toxic
Textile fibers			
- Animal fibers			
- Vegetable fibers			
- Synthetic fibers			
Containers and packaging			
- Cardboard boxes			
- Plastic bags			
- Wooden boxes			
- Paper bags			
Effluents			
- Chemicals		REACH certification	
- Dyes		REACH certification	REACH certification
- Auxiliary		Biodegradable	REACH certification
- Residual water			
Other solid waste			
- TI			Batteries and LCD
- Construction materials			
- Machinery and parts		Metal	Batteries
- Plastic			Oceans
- Wood			

Fig. 4 Elimination of waste

"They refer to the ancient Greeks where they establish two types of time: one focused at the moment and another focused in the time that has passed." Regarding nature, some ecosystems acquire long-term balance and resilience, adapting to change at different rates.

Nature combines changes that occur slowly at a large scale with small but sudden ones. The different paces in changes allow the ecosystem to survive potentially dangerous and harmful events.

They mention Stewart Brand and his book "The Clock of the Long Now," where he proposed that any transcending human civilization needs a rhythm for quick and slow activities to balance themselves out.

These layers from fast to slow: Art-Fashion, commerce, infrastructure, government, culture, and nature.

Fashion brings forth sudden and imaginative change while the slower laters maintain invariability and the long-term.

Crucially, the system works when each activity respects the rhythm of the other, but, today, the fashion industry as it is designed does not respect other activities. There is plenty of evidence that suggests fashion is disconnected from its effects on culture and nature, creating social problems, large amounts of waste, and climate change. In fact, the fashion industry's commercial agenda seems to promote the opposite to a multilayered system of variable velocity.

What is commercialized is a variety of similar products fabricated and consumed at high economic velocity instead of finding the appropriate rhythm of cultural, social, and environmental needs.

It is clear that, regarding sustainability, we must begin by being aware of the damage created by the excessive velocity in fashion toward the system in general. Which should be our response to fast fashion?

(d) **Actions**

The fashion industry is performing actions that seek to revert the environmental impact caused by their activities.

Specifically, it is worth mentioning the Sustainable Apparel Coalition [18], with participant-partners that include renowned brands, manufacturers, public and private institutions, among others. Their objective is to establish a global alliance for sustainable production.

It has developed the Higg Index, looking to standardize the protocols to measure the environmental impact of the clothing industry's value chain.

Its vision is a clothing industry that produces no unnecessary environmental harm and positively impacts people and communities associated with their activities.

Cradle to Cradle Products Innovation Institute [19] is another organization that is proposing much more aggressive solutions to eliminate environmental impacts.

Dorough and Braungart [20], authors of the book "Cradle to Cradle" (2002) are the leaders of this proposal. In their book, they explain the beginnings of the concern by industries and governments regarding the environment, which go back to the Rio Summit in 1992, where a basic strategy to deal with environmental problems is brought up. Eco Efficiency is proposed, transforming the traditional industry toward economic, environmental, and ethical concerns, doing more with less.

Stephan Schmidheiny, one of the founders of the Business Council for Sustainable Development, predicted that by 2002 it would be impossible for a business to be competitive without being ecoefficient. The bigger industries achieved impressive results, but will this be enough?

Many industries and governments are promoting the application of 4R to counter negative environmental impact: Reduce, Reuse, Recycle, and Regulate.

Dorough and Braungart [20] propose to go further and explain the problems derived from applying 4R.

Reducing the amount of toxic waste: to be more efficient in the usage of raw material and in our processes is good, but will not cease the emission of greenhouse gases, the buildup of toxic waste, the exhaustion of raw material, or the destruction of the environment. They will only be slowed and the environmental consequences will be postponed.

Very little is known about the industrial contaminants for the slowing strategy to be healthy in the short term.

Reusing waste and feeling something good for the environment is being done. If, in the end, these end up somewhere else somehow, these results are very discouraging for the environment and, thus, for people. We are postponing the problem.

Reusing mud in the sewers. These can contain residues that can pollute food-producing fields. In many cases, it would be less dangerous to bury these materials in sealed dumps.

Recycling has a problem. The quality of the material decreases with the number of cycles. Such is the case of the metallurgical industry. Recycled steel has worse properties than the original and, in many cases, cannot be used in the automotive industry.

Recycled paper releases particles that can be inhaled by people causing irritation in the respiratory system.

Recycled threads have a resistance problem that affects their performance in posterior productive processes (efficiency).

Regulations can be applied on a large scale but are traditionally applied at the end of the line, applying control and sanctions once the damage has been done.

Good intentions are present when regulations are enforced to protect the common good. However, destructive and low-intelligent designs can be regulated and can reduce immediate negative consequences. Enforcing regulations at the end is proof of an error in design. This is called license to harm.

Being less evil is to accept things the way they are, to believe that appropriately designed, dishonest, and destructive designs are the best humans can do, an imaginative flaw. A completely different model is needed.

Mc Dorough and M. Braungart [20] propose a new design objective.

- Constructions that, just like trees, produce more energy than they consume and cleanse their own residual waters.
- Factories that discharge potable water.
- Products that, once their useful life is over, do not become useless waste, but biodegrade instead.
- Recovered materials for human and natural uses.
- Means of transport that improve quality of life while also distributing products and services.
- A world of abundance.

Such a task awaits, but would it be easier to colonize Mars?

"Getting there, where no human ever has set foot." Let us not cause a great disaster here to then go to a less welcoming place. Let us use our creativity to stay here, to be once again natives of this planet.

Mc Dorough and Braungart's proposal can seem unreachable. Nevertheless, the problem exists and the challenge is clear. We have a long and arduous road to travel, and even more when we consider the havoc caused by coronavirus (COVID-19) on our society.

3 Coronavirus

The devastating effects of coronavirus can be explained with the changes in the rhythm of human activities [21]. The rhythm of the healthcare system changed, and it could not adapt to the rhythm of the coronavirus and its effects. The most amazing thing is that this scenario had been forecast by several researchers and journalists over thirty years ago.

According to an article by Robin Marantz Henig, published in National Geographic last April [22] titled "Experts warned of a pandemic decades ago. Why weren't we ready?".

She mentions a virologist named Stephen Morse who, in the 1990s, coined the term emerging viruses; then, the author wrote about the introduction of new, potentially devastating pathogens—"climate change, massive urbanization, the proximity of humans to farm and forest animals that act as viral reservoirs", accelerating their spread through war, the global economy, and international air travel. Very few people took this potential threat seriously. "The single biggest threat to man's continued dominance on the planet is the virus."

It is recommended to read the report crafted by the Board of Innovation titled Low Touch Economy [23] where they propose that, in the post-COVID-19 era, the economy will be shaped by new habits and regulations, based on the reduction of close interaction between people and things, and the tight restrictions on hygiene and health. The disruption created would change the way we feed ourselves, work, buy, exercise, manage our health, socialize, and use our free time with an unprecedented rate of change.

The world changed, and, thus, all social, cultural, and economic activities must be redesigned, including fashion.

How much? It cannot be determined. Proposals are and will be contradictory. Some people think that said changes can be very deep. Others claim or are hopeful that normalcy will return when the pandemic is overcome as a result of a vaccine or herd immunity.

It is clear that the pandemic will delay the efforts of developing sustainable industries, including fashion. We wish to highlight the report presented by the Boston Consulting Group, Sustainable Apparel Coalition, and Higg, Co.L: "Weaving a Better Future: Rebuilding a More Sustainable Fashion After COVID-19." [24]

It summarizes the actions to follow in four postulates:

- Protect critical assets to survive the crisis.
- Solve stock problems along with providers
- Integrate sustainability through a business recovery strategy
- Accelerate transparency while sustainability objectives are increased.

Sustainability is a basic necessity for society's future. However, the reason why this necessity exists must be acknowledged. Firm believers of sustainability and people who are pressured into following it due to pressure from the consumers both exist.

We must prevent that, in 30 years' time, we will be regretting situations even more severe than the pandemic because we did not know how to handle the environmental impact caused by human activity. COVID-19 is a serious warning.

4 Conclusions

Some observations on fashion and sustainability are the following: Fashion is tightly associated with people's behavior and needs. Should we not understand this, it will be very difficult for people to do so. It is their means of communication and expression and educating them on sustainability and its importance for the future of society will take time.

Sustainability can be an investment or an expense nowadays. Prioritizing sustainability in industries is not urgent. Survival and jobs are the main objective. Solution proposals to environmental impact developed by Ulrich and Eppinger in their book Product Design and Development [13] and those of Mc Dorough and Braungart in Cradle to Cradle [20] are a base to develop a path to sustainability.

The 4R proposal, reduce, reuse, recycle, and recover, is the best option today. However, it is not a solution to the problems originated by the speed set in the fashion industry, particularly in fast fashion.

The accumulation and how and where it takes place, would force us to take actions that many environmentalists may question.

The consequences of not compartmentalizing and/or isolating toxic and even inorganic waste currently have disastrous consequences on the environment. Until a solution to this problem is not found, we estimate that the best course of action is to establish high security dumps and whose responsibility should lie with the agents of the industry's value chain in general.

We must regard as important to measure this proposal's long- and short-term effects, and a very clear example is electric cars. They would be fueled with non-renewable energy and there will be no emission of contaminating gases, but the need to save this energy forces us to use batteries, which will ultimately have a negative environmental impact.

Color is a determinant element in fashion and must thus be treated in its true dimension. This is a huge challenge for the fashion industry in general to develop real colors that are completely degradable or inert toward the environment. This topic will be critical in the sustainability agenda.

What is impossible to measure is impossible to regulate. There is a need to establish protocols and procedures first to establish standards (4R) and then measure the environmental impact of the fashion industry's value chain. These must be simple so that they allow a person or agent of any group of interest to understand and measure the environmental impact of their actions, be it a farmer, stockbreeder, industrial worker, investor, and especially clients.

There are many efforts that establish indexes or parameters that allow the measurement of different forms of environmental impact. However, these focus on specific stages along the value chain or the interests of different organizations. The bias with which these results are interpreted can be unfair and cause irreparable damage to specific interest groups of the value chain in the fashion industry, and experience tells us that the weakest part will be the most affected one.

In this first document, we sought to develop an appropriate relationship between fashion and sustainability, considering the environmental impact and its negative effects.

We hope to do so again in the future, but between fashion and social responsibility with its main interest groups, especially those most exposed: workers.

Finally, and especially due to COVID-19, we recommend acting with responsibility, resilience, empathy, and gratitude.

References

1. McDonough W, Braungart M (2002) Remaking the way we make things: cradle to cradle. North Point Press, New York, 104, ISBN, 1224942886
2. Barnard M (2014) Fashion theory: an introduction. Routledge (1:Introduction)
3. Godin S (2008) Tribes: we need you to lead us. Penguin
4. Ann Hollander cited by Malcolm Barnard, 2014; "Fashion Theory" (2:17)
5. Barnard M (2014) "Fashion Theory" (5:56)
6. Design Museum (2009) "Fifty dresses that change the world" (30:)
7. Jean Baudrillard cited by Malcolm Barnard (2014) "Fashion Theory" (3:30)
8. Interview for Vogue with Ryan Smith To Jessica Alba, cited by Malcolm Barnard (2014) "Fashion Theory" (7:90)
9. Clark J (2020) Por qué los tonos tierra nutritivos están ganando terreno. WGSN. https://www.wgsn.com/blogs/why-nourishing-earth-tones-are-gaining-ground/
10. Reiman J (2013) "The story of purpose"
11. Brene Brown cited by Joey Reiman (2013) "The story of purpose" (1:33)
12. Reiman J (2013) "The story of purpose" (1:39)
13. Business of Fashion, and McKinsey and Company (2017) The state of fashion 2018
14. Gómez C (2015) El desarrollo sostenible: conceptos básicos, alcance y criterios para su evaluación. https://www.unesco.org/new/fileadmin/MULTIMEDIA/FIELD/Havana/pdf/Cap3.pdf
15. Ulrich KT, Eppinger SD (2016) "Product design and development" (12:232)
16. Lewis and Gerlsakis cited by Ulrich KT, Eppinger SD (2016) "Product design and development" (12:235)
17. House of Commons, Environmental Audit Committee UK (2019) "Fixing Fashion: Clothing, consumption and Sustainability" (#:30)
18. The Sustainable Apparel Coalition—Higg Index https://apparelcoalition.org/the-higg-index/
19. Parker L, Dickson MA (2019) "Sustainable Fashion Handbook" (Slow Fashion:27)
20. The Sustainable Apparel Coalition. https://apparelcoalition.org/
21. Cradle to Cradle products innovation institute www.c2ccertified.org
22. Stewart Brand in his book "The clock of the long now", cited by Parker L, Dickson MA (2019) Sustainable fashion handbook. (Slow Fashion:27)
23. Henig RM (2020) National Geographic. https://www.nationalgeographic.com/science/2020/04/experts-warned-pandemic-decades-ago-why-not-ready-for-coronavirus/
24. Low Touch Economy (2020) Board of Innovation. https://info.boardofinnovation.com/hubfs/LTE-report-espan%CC%83ol.pdf
25. Sustainable Apparel Coalition y Higg Co.L (2020) Weaving a Better Future Rebuilding a More Sustainable Fashion After Covid-19 https://apparelcoalition.org/wp-content/uploads/2020/04/Weaving-a-Better-Future-Covid-19-BCG-SAC-Higg-Co-Report.pdf

Upcycling as a Tool for Participatory Critical Reflection

Lucia López Rodríguez

Abstract In recent years, there has been an increase in the integration of upcycling as a strategy to minimize the volume of textile solid waste, which has motivated alternative ways of producing clothing, as well as new business models. Although the common factor of this type of undertaking is the approach to upcycling as a technique of remanufacturing garments, many of these proposals take it as a mean of critical reflection towards the contemporary ways of making and consuming fashion. In this way, the simple production and sale of garments ceases to be the central activity and other actions such as co-design, open source platforms, user participation in production processes and critical reflection of the structures of fashion, become part of the main purpose. This chapter presents three cases of clothing upcycling ventures in Latin America (12NA in Chile, COMAS in Brazil and Estampa Crítica #TEXTOURGENTE in Uruguay), which works through the intersection of design, art, social projects and education. In all three cases, user participation in garment making is nodular, generating critical audiences that detach from the passive role of the postmodern consumer. In this type of projects, the integration of the local community becomes relevant and technical knowledge is not exclusive property of the designer or the brand. Instead, what is relevant, is the dissemination of tools that contribute to collective awareness and the reduction of the environmental impact associated with the fashion industry.

Keywords Upcycling · Textile Waste · Participatory Fashion Practices

1 Upcycling as a Waste Minimization Strategy

The dominant business model in the fashion system is characterized by the constant and massive production of garments, which leads to over-consumption and its consequent generation of waste [1]. To address the environmental impact related to the

L. López Rodríguez (✉)
Escuela Universitaria Centro de Diseño, Facultad de Arquitectura, Diseño y Urbanismo
Montevideo, Montevideo, Uruguay
e-mail: lucialopez@fadu.edu.uy

M. Á. Gardetti and R. P. Larios-Francia (eds.), *Sustainable Fashion and Textiles in Latin America*, Textile Science and Clothing Technology,
https://doi.org/10.1007/978-981-16-1850-5_16

volume of discarded clothing (pre or post- consumer), the industry has adopted different strategies based on the waste hierarchy, which was popularized in the late eighties and prioritizes waste prevention, minimization, reuse, recycling, and finally, as last resort, land filling [2]. Although it is globally recognized that preventive actions precede other alternatives such as reuse, remanufacturing or recycling, it is the latter that have had a greater implementation, especially recycling. Within the reuse strategies, upcyling has gained approval in recent years, especially in small-scale business and fashion projects, avoiding the consumption of new materials and the use of resources associated with textile finishes. Since the introduction of Upcycling as a concept in the nineties by Reiner Pliz, its use has been expanded and extended beyond the recycling of materials. At present, said term embraces different ways of transforming waste into new products of higher value. In the case of fashion upcycling, the textile material may come from cutting waste, end of rolls, other post-industrial textile waste, or discarded clothing, either pre or post-consumer. The cases that are presented in this chapter, show different examples of clothing upcycling, transforming fashion waste into new garments.

2 Upcycling as a Resistance Movement

Integrating reuse, remanufacturing or recycling strategies is crucial to minimize the global impact of fashion. However, because the technological processes and technical tools respond to the ideological system in which they are embedded, the mere incorporation of them is not enough to generate a paradigm shift. The search of a more sustainable future implies rethinking our modes of production and consumption, resisting and de-structuring the linear logic of "take, make and dispose" and incorporating new ways of making fashion based on a global systemic vision [17]. Among the many options that seek to tackle current socio-environmental problems are Slow Fashion, Circular Fashion, Green Fashion, Eco Fashion, Ethical Fashion, and others. If we look at the wide range of initiatives that pursue sustainability, we find that many of them share a holistic awareness of the environment, considering its ecological, social, cultural, political and economic dimensions. This multidimensional perception, allows different practices such as art, design, education, politics and science to merge, and, on the other hand, allows to emerge new concepts that were not previously attached to fashion, such as participation, community, activism and empowerment.

It is in these settings, in which the actors of the fashion universe begin to question: How could fashion practices define our living? From which platforms should we start to act for a deeper and rhizomatic transformation? The systemic vision of the world, spread by Lynn Margulis [3] and later by William McDonough [17] in the design field, allowed us to visualize ourselves as part of an interconnected whole, a global unit where the part affects the whole. This theory, in which organisms coexist in a balanced interdependence, has led many designers and other actors to question the environmental impacts caused by over-consumption and the disconnection of the fashion system. Therefore, the upcycling movement, which gains wider visibility each year, not only addresses the environmental problem of fashion from its ecological dimension, but also does so with a strong emphasis on the social, cultural and political dimensions.

This type of upcycling (which I will refer to as *participatory upcycling*), expands beyond its technical nature, embracing other practices of political, educational and reflective nature. As seen in many of the projects and business ventures in Latin America, although the reduction of the volume of textile waste through upcycling is part of its core activities, this is not the central one. There seems to be a clear interest in disseminating values associated with sustainability and conscious consumption, as well as in making the upcycling technique accessible. This dissemination is pertinent, considering the fact that the great impact of fashion is relative to its large scale, consequence of the massive consumption of products. Due to this, in order to achieve a positive impact, it would be necessary that as many actors (including consumers) as possible to become participants and become involved in the cause, transforming upcycling into a tool for participatory critical reflection. To accomplish this goal, different enterprises and fashion brands are implementing workshops, courses, educational projects, co-organizing activist actions, and generating synergies through collaboration with external actors such as suppliers, NGOs, foundations and other public and private institutions.

In this sense, taking as reference the ideas of the artist and educator Joseph Beuys, fashion could act as a catalyst for social change [4]. This would imply an important shift in the role of fashion brands, designers, manufacturers, consumers and fashion itself. It would imply taking an activist and formative position, a position of active exchange, where fashion values acquire a more humane character, defined by the contemplation of those who make it, as well as those who wear it. As the author Otto Von Busch [5] states "We can use fashion as a workshop for collective enablement where a community shares their methods and experiences. Liberating one part of fashion from the phenomenon of dictations and anxiety to become instead a collective experience of empowerment through engaged craft."

Having mentioned the above, we can say that *participatory upcycling* is characterized by using the teaching of upcycling as a means for collective critical reflection on contemporary ways of making and consuming fashion. This requires, on the one hand, reaffirming the participatory nature of fashion practices and, on the other hand, conceiving garments as a political object. Participatory instances are necessary to define the political sense of fashion, since it is not possible to let go of the social responsibility attached to our work [6]. The simple fact of rethinking and resignifying an object devoid of meaning (waste) through upcycling, allows us to reflect on the value system and the hidden externalities of conventional fashion. In regard to this point, the case examples addressed in this text recognize that clothing must act as a bearer of social values, and therefore each garment becomes the materialization of the human interactions and processes behind it. This means that the value of the garment will be, not in its materiality or in the status it represents, but in the care of the processes and the people who created it [7]. Hence, in *participatory upcycling* the value of clothing is widened; the global network that involves the creation of a garment and its perceived obsolescence becomes a topic of discussion. Aspects such as how it was made, who made it, where and why it would be necessary to remanufacture it, become relevant. Each garment is seen as a set of human decisions and recognizing that human dimension acknowledges "the agency of all people involved, doing justice to their active role of engagement and capacity to act" [8].

3 Participatory Processes in the Field of Fashion

In recent years, a greater involvement of users (or other actors) in production processes has been observed. Under a broad perspective, we can consider as *participatory practices* those that somehow involve external actors in the production, creation and/or design of a product life cycle. The type of participation can be specific, such as in mass customization where the user selects a characteristic of the product (color, texture, etc.), or it can involve more complex and profound actions such as co-creation or co-design workshops, in which the participants jointly develop products, services and experiences.

Although user participation in design processes has been widely used as a marketing strategy to strengthen ties between the consumer and the brand [9], this study focuses on cases where participation requires a deeper involvement and a critical reflection of contemporary production and consumption models. The authors López-Navarro and Lozano-Gómez [1], affirm that the most reflective participatory processes act as enhancers of value "favouring sustainability and meeting individual human needs rather than market demands. This approach seeks to integrate consumers into the value creation chain, in an effort to forge emotional involvement with the brand's philosophy of sustainability beyond the act of consumption. Thus, it stimulates users'personal involvement in generating satisfaction, breaking the cycle of passive acquisition of clothing and creating garments that are meaningful to the user."

3.1 Beyond Materiality

In the late twentieth century the practice of design began to change, going beyond the creation of things and expanding to design experiences, services and processes [10]. In the three cases of participatory upcycling present in this chapter, the results do not focus only on the transmission of technical knowledge or on the final product, but on the mobilizations that are generated through the human exchange and on the future ability of the participant to transform what has been learned and reproduce it in their daily lives. In this type of process, the quality of the experience depends on the impact generated through the interaction between the different agents, and on the stimulation of the individual, both cognitively and emotionally [9]. For this reason, it is important that the contents shared in the workshops could be able to juxtaposed with the participants' emotional, social, and environmental interests. In this way, what has been learned is stored in memory, creating a value that can be transmitted through other future experiences. Translating Borriaud's ideas into fashion, it could be said that in *participatory upcycling* the essence of fashion practice would lie in the relationships between subjects and their environment, and the work of each individual or collective would become, "a bundle of relations with the world, that in turn, would generate other relationships, and thus to infinity." [11].

3.2 New Role for Consumers, Designers and Producers

In participatory fashion and processes, the user performs an active role, breaking the "one-way information flow from designer to consumer" [6] of conventional fashion, in which the consumer has almost no intervention on what is consumed. The user is transformed into what Fletcher [6] calls the *user-maker*, a connected and engaged individual, capable of driving social change. Just as Joseph Beuys [4] stated that we are all artists, we could say that participatory design is based on the idea that we can all be designers, recognizing the potential of each individual and creating initiatives that involve a wide range of members of the community.

In addition to this, this exchange also generates a new role for the designer or producer, since to his/her creative role is added the role of facilitator in order to arrange the joints for a mutual exchange and collaborative learning space. The designer ceases to be the central figure that makes the decisions and becomes "... an orchestrator and facilitator, ...an agent of collaborative change. It is not the divine creator of the original and new, but a negotiator, questioning and developing design as a skill and practical production utility" [5]. By displacing the figure of the designer as a generator of objects, towards a role of producer of situations, we can once again find a parallelism between fashion and art, remarking the art displacements in the twentieth century when merging the artist work and social practice. As art critic Claire Bishop [12] said: "... the artist is conceived less as an individual producer of discrete objects than as a collaborator and producer of situations; the work of art as a finite, portable commodifiable product is reconceived as an ongoing or long-term project with an unclear beginning and end."

3.3 Education and Game-Like Tools

To improve participation, it is helpful to approach educational strategies. Numerous pedagogical tools are practical resources to exchange experiences and knowledge and facilitate collective creation. Whether in an upcycling workshop or in another similar activity, play is one of the best strategies to maintain a fluid involvement and strengthen the bond between the different actors. Trough tactic games, becomes easier to introduce technical design tools, and as a result, learning emerges [13, 14]. Game-like tools facilitate active participation, expanding the possibilities of creativity. It also diminishes the tensions that ignoring certain technique aspects could produce on individuals. Dissolving hierarchical structures in this type of process is important so that participants feel comfortable and creativity emerges without hindrance. For this, exploratory stages are necessary, because it is in these where each participant feels that their contributions are both free and relevant, and that the final products will be the result of the evolution of collective contribution. Free exploration, deconstruction, and role play are some of the tactics used in upcycling workshops so that through the game, participants surrender to the experience [13].

3.4 Sharing Knowledge

As we know, fashion brands and designers have always been very reserved about sharing the know-how of their production processes. However, participatory practices imply, to a certain extent, the liberation of the knowledge acquired over the years, since the very purpose of disseminating upcycling as a practice requires it. Although this may not be an easy task for many designers, sharing knowledge, as well as collaboration, allows for instances of exchange and mutual growth. In the three cases of *participatory upcycling* presented in this chapter, it is observed a critical position towards exclusive knowledge and exclusion as a mode of distinction in fashion. In many cases, information is shared through workshops and web platforms, opening up possibilities for expansion and globalizing audiences. In order to make learning easier, many designers have chosen to create open source pilots, a kind of step-by-step guide to remanufacture garments. One of the most noteworthy references, is the work of fashion designer and researcher Otto Von Busch, who created a web platform with free access to numerous remanufacturing methods. His project, "The ReForm manuals" is part of what he calls the "recyclopedia", a compilation of "cookbooks" for remanufacturing. These technical guides, called "manuals" by Von Busch or "recipes" by Agustina Comas, are also shared with users in face-to-face workshops. This instances allows technical knowledge to be easily shareable, but also intervened by the workshop participants. This is one of the most interesting points since it allows a round-trip exchange: the information is offered by an original source (ex: the designer), then it is processed by the users, who appropriate it and transform it, to later nurture the creativity of the original source. The design process is enriched and invigorated through this dialogue. Sharing knowledge also opens up a range of possibilities to generate synergies and collaborations with other actors and stakeholders, such as government entities, educational institutions, manufacturers, other public or private organizations, and even other fashion brands or projects.

4 Upcycling Aesthetics: An Aesthetic of Contents

Garment aesthetics is a necessary quality to seduce users as it acts as a "social attractor" [6]. At the same time, defining it carefully is key to intelligently materializing the message that we want to transmit. Therefore, the essence of the values contained in the textile object is reflected through the aesthetics that we build. But, what code determines the aesthetics of these type of projects? Is there a common factor? Possibly, to answer these questions a more exhaustive investigation needs to be made, finding clues in the texts of Nicolás Borriaud [11] in which he refers to *Relational Aesthetics*, such as that aesthetic that takes as its basis "the sphere of human interrelationships and their social context, more than the affirmation of an autonomous and private symbolic space". However, at first glance, it is observable that the aesthetics of some upcycling ventures, beyond the imprint of each designer, responds to the technical characteristics of the construction methods and the inherent values of the upcycling movement.

4.1 The Details of Discarded Garments as an Aesthetic Resource

There are several ways of remanufacturing or upcycling garments, but all of them face the same challenge: how to work with the material optimizing production scheduling. There are very few cases in which it is chosen to unstitch the garment pieces because the labor time it takes increases the costs. One way to solve this is to keep the original seams, enhancing and using them as visual components in the new garment. For the same reason, it is proposed to preserve the trimmings (buttons, zippers, brooches, etc.) and other textile finishes. This way of taking advantage of the material, while maintaining some of the characteristics (details) of the original garments, helps to reduce the environmental and economic impact of garment production (Fletcher 2014). But beyond this, another significant reason to keep these details is the symbolic value that they carry. In some way, by conserving them, the persons who made them are present in the new upcycled garment. By maintaining these constructive vestiges, the reminiscence of the work already done, stored in the memory of seams and finishes, is displayed. Maintaining these characteristics in upcycling processes becomes a way to value the work done by others, enhance it and appreciate it as a generator of value. It is to make a hidden treasure that was discarded visible again, rejecting the fast fashion dictum that conceives garments as disposable objects of no value. In this way, preserving the details of the garments used as material input engenders an aesthetic language that helps to identify upcycled garments per se.

4.2 Quality

Just as upcycling seeks to reduce the flow of textile waste through remanufacturing, another equally important objective is to generate high-quality garments so that they can be used as long as possible. For this to be possible, it is necessary to select clothing of good quality finishes and materials. However, while garments sold by upcycling brands usually have an excellent technical quality, those made in participatory workshops does not have to aim at the same level of demand. In the workshops, the quality of the garment is partially displaced in favor of educational and learning processes. The fact of making, being an active part of the production and reflecting on one's fashion practices are the main objective. In addition to this, the involvement of the user in manufacturing could make him/her aware of the time and technical knowledge that a garment requires to be made, valuing the labor and human time dedicated to its creation.

4.3 Less But Long-Lasting

The materials used in upcycling are very heterogeneous since the flaws of the clothing used as raw material may vary widely. To find the best way to take advantage of materials, designers spend a lot of time exploring different typologies. This extensive labor time is one of the main barriers to mass production of upcycled garments, so upcycling brands are characterized by having small collections, integrated by a few models and reduced quantities of each. Many times, the model can be repeated but the result may be still a unique piece since the original garments have different characteristics, such as color, or fabric patterns. In the cases presented below, aesthetic obsolescence and constant product variation is rejected, and therefore the garments do not lose value after one season. These business models respond to the paused times of slow fashion, in which the proposal is that consumers do not have several models of each garment, but few garments that are treasured. In these systems quality prevails over quantity.

5 Participatory Fashion Upcycling: Cases in Latin America

Below are three cases of entrepreneurial designers from Chile, Brazil and Uruguay that work with upcycling as a tool for participatory critical reflection. Although the three examples share in common the previously mentioned *participatory upcycling* characteristics, each one addresses them from their particularities. The information presented here was obtained in the last two years, through interviews, dialogues and the author's personal experience in the respective workshops of each case.

CASE I: 12NA Docena.

Headquarters: Valparaíso—Chile
Materials: Post-consumer garments
Upcycled garments: >10.000
Working since: 2004

Description: 12NA is a textile upcycling platform created by Mercedes Martínez (AR) and Mariano Breccia (AR). Their work covers different areas, like production of garments and accessories, co-creation educational workshops, production of artistic events and upcycling consulting services for other companies and institutions. Their work seeks to create interest in reflecting on ways of doing things. They also have been working for the last 5 years to make upcycling a movement, specially in Latin America. To achieve this, in addition to the activities they carry out, they have developed a community web and offer a residency program for artists and designers with an interest in textile recycling and sustainable development. 12NA's work is focused on the community, so most of their activities are open to the public, involving different audiences from the region. Local identity is a deeply worked concept that can be seen in products such as *Punchaw*, a remanufacturing method created from the basic structure of the *poncho*, a typical South American garment. Their apprehension for the local values is also seen in the vocabulary they use; although they sell their collections world widely, they are one of the few cases that use the word *upcycling* translated into Spanish (*supra-reciclaje*).

Why participatory processes and critical reflection: For Martínez y Breccia, the incorporation of co-creation workshops occurred naturally from the birth of the brand. They believed that it was essential to be able to share their knowledge with other people, so that together they could reflect on how the ways of dressing and consuming fashion affect our society. Martínez reaffirms the importance of "opening processes to new generations, because we are committed to expanding change. For us, education is a commitment and activism is a mission of social transformation". Today, the 12NA collective does not conceive design as an exercise that is practiced alone. Each of their designs is the sum of accumulated collective experiences, faced in the workshops, residences and other activities: "design is collaboration, you never work alone".

Shared knowledge: The exercise of sharing methods and work and "letting the ideas go" has become important for the growth of the brand. The designers believe that it is vital to return to collaborative work and re-learn how to co-create in order to challenge yourself as a designer and as a human being. For them, losing the fear of not being in control "is a difficult exercise, but once one you learn to get away from the ego and indulge in the process, the result is always surprising." In recent years, designs such as *Punchaw* and *Bolso Origami* have been reinterpreted and appropriated by several designers in the region, and these pieces have become successful garments in their collections. Another objective of the brand, is to recover other ancestral ways of production, in which authorship was not a notable component but rather that knowledge was shared among the community, and the competition did not exist as we know it today. These type of collaborations are promoted through Community events organized by 12NA (such as the Circular festival in Chile) and the suprarecycling.org, a web platform that aims to network all upcyclers in Latin America.

Game-like tools as a strategy: Games are always present in 12NA workshops, from dice games to dynamics that remind us of a dance round, or trying a thousand ways to put on a garment. For them, the game allows us to relax and let emotions flow, allowing us to "learn from emotion". According to Breccia, playful activities, are the most effective way of integrating knowledge, since it invites us to surprise and to see elements of our daily life from a different perspective. Since their inception, they have kept in mind that the enjoyment of activities minimizes prejudices and stimulates the creativity of the participants. One of their first projects was *Deconstructjoy*, a series of clothing remanufacturing videos in which the deconstruction process is an "eternal exploration game that offers infinite possibilities" (Fig. 1).

Upcycling aesthetics: The selection of products for the brand's collection implies a careful selection of vintage or other high-quality post-consumer clothing. The quality of the original materials, both of the fabrics and the clothing, is essential to be able to create products (or pieces, as they called them) of high value and long durability. In 12NA the details of the garments are kept in the produced pieces, cutting the garments in strategic places to preserve original visible seams, prints or trimmings. It is because of this, that they do not unstitch the garments, as Martínez mentions, their tool is the scissors and not the seam ripper. Their work process seeks to "rescue the *ki* of garments", since they consider that the original garments have an energy

Fig. 1 Workshop "Para que usarías una mascara". Festival Circular, Chile, 2019

and value that deserves to be enhanced and recovered. In this way, 12NA's garments are loaded with details, different patterns and multiple textures from the original materials. In its pieces, parts of sportswear coexist with pieces of suits, generating an urban and unique proposal that differentiates them. Since they work with limited stocks, although the designs can be repeated, each piece is unique and eclectic. Scale production is a challenge for the brand, and to be able to do it many times they must accumulate similar garments for long periods, months or even years, to be able to make only dozens of the same design (Figs. 2 and 3).

CASE II: COMAS.

Headquarters: São Paulo—Brazil
Materials: Pre-consumer garments
Upcycled garments: >3.000
Working since: 2008 (In-Use / Previous brand), 2015 (COMAS)

Fig. 2 *Punchaw* garment: Traditional local textile + leather jacket

Fig. 3 *Punchaw* garment: Traditional local textile + sportswear garment

Description: COMAS was created by the fashion designer Agustina Comas (UY), based on her concern about the enormous number of pre-consumer garments that were discarded annually in Brazil. The brand works mainly with men's shirts that are rejected by local factories due to defects such as flaws, stains, fitting problems, among others. For the designer, these garments are not waste but high-value resources, since the fabrics and clothing are of high quality. Although the remanufacturing of garments is one of the main objectives of the brand, the diffusion of upcycling as a movement and the creation of awareness are just as important. The brand, in addition to selling its collections, offers upcycling educational workshops and training and consulting services to companies that want to incorporate remanufacturing into their production systems.

Why participatory processes and critical reflection: Prior to the creation of her brand, Comas began to give workshops in order to bring participants closer to the environmental problems caused by deadstock clothing in Brazil, and to share different methods and approaches to upcycling. According to the designer, participatory processes are part of the essence of her entrepreneurship, since the workshops generate "a virtuous cycle of exchange, a source of creative energy that allows the revival of garments that were previously devoid of meaning." From collaborative work, new ideas are generated and that permit us to get out of the limitations of individual knowledge to generate something new, which will then be processed and transformed later by the participant, and also by the brand team. Hence, the fruit is the method, the exchange, and the things learned, and clothing are the materialization of all these processes. Beyond workshops designed for the general public, COMAS works collaboratively with other brands, textile factories and manufacturing workshops to facilitate the incorporation of upcycling on a large scale for other companies. It has also articulated production centers with seamstresses from different communities, with the aim of enhancing collective learning and the projection of micro-enterprises.

Shared knowledge: In order to systematize and accurately scale the upcycling of garments, the COMAS team created the *Upcycling de Raíz* (Upcycling of Root) method, a set of steps and tools that enable garments to be quickly and easily transformed into new products. In each workshop, along with this method, different "recipes" that guide the participants in the step-by-step process of making garments, are presented. But the most interesting aspect is that the recipes themselves act as a starting point for users to reconfigure and modify them, obtaining new results. Being able to open and release knowledge is an exercise that is also nutritious for Comas, and as she mentions: "it is crucial to learn how to release what you think is your own, because actually, it is not. We have been taught to be selfish and to believe that ideas should not be communicated because they could be copied, but we must understand that each designer was formed by taking and transforming things from others. The feedback that is collectively obtained from the workshops is the richest and we must continue in that direction".

Game-like tools as a strategy: In COMAS workshops, games are used as an instrument to unleash creativity. Game-like tools are part of the first two steps of the *Upcycling de Raíz* method, and those steps are: *Pensar con el Cuerpo* (Thinking with the Body) and *Gimnasia Cerebral* (Brain Gymnastics). This games consist of taking a garment and exploring all its possibilities to find new functions, that is, to question: what happens if the sleeves are used as pants? What happens if we put a shirt inside out? In this way, we forget about the original function of the garment and the participants begin to think with the body, to let the body look for new ways to occupy that garment. By doing the physical exercise of reformulating the limits of the garments, we are doing a brain gym, to "build the New from what has already been designed." It's about letting our imagination run wild through free exploration and losing our fear of being wrong (Figs. 4 and 5).

Upcycling aesthetics: COMAS aesthetic is minimalist, and is subject to the simplicity of the noble materials the designer works with. The classification of the male shirts focuses on materials such as linen, cotton and denim, and uses the palette of its menswear suppliers, covering colors like white, earth colors and the range of blues, representative of denim washing. The maximum conservation of the details of the original shirts, (collar, box plate buttons, pockets, cuffs, etc.) make its products highly recognizable. COMAS's garment construction method is based on the principles *Conservar la energía utilizada* (conserving the energy used) and *Preservar el*

Fig. 4 Workshop COMAS. Participants: Fashion Industry Workers

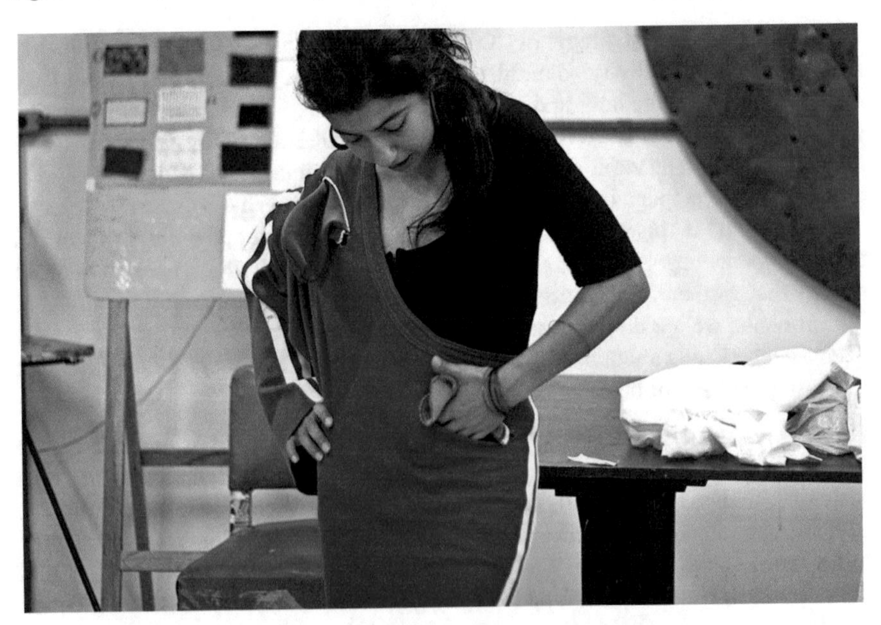

Fig. 5 Workshop COMAS. Exploration and testing of cutting sweatpants

conocimiento congelado (preserving frozen knowledge). Through these principles they seek to value the energy already used in the original clothing, that is, "if someone has already taken the trouble to sew, to make a stitching, to use resources and take time to dye the textile, why not respect it and preserve it?". Another fundamental principle of Comas' work is the consideration of *El defecto como efecto* (defect as an effect), which means that elements considered as "flaws" become an opportunity to give the garment a unique character. In this way, we could transform a stain into a print, or a patch into a graphic detail. This vision, which is developed by the designer in mending workshops, is related to the *visible mending movement*, which reflects on the obsolescence of fashion and overconsumption, making repairs of garments visible. What is intended is not to hide the mends, but to transform them into an aesthetic component and show the noble act of repairing a garment to extend its useful life (Fig. 6).

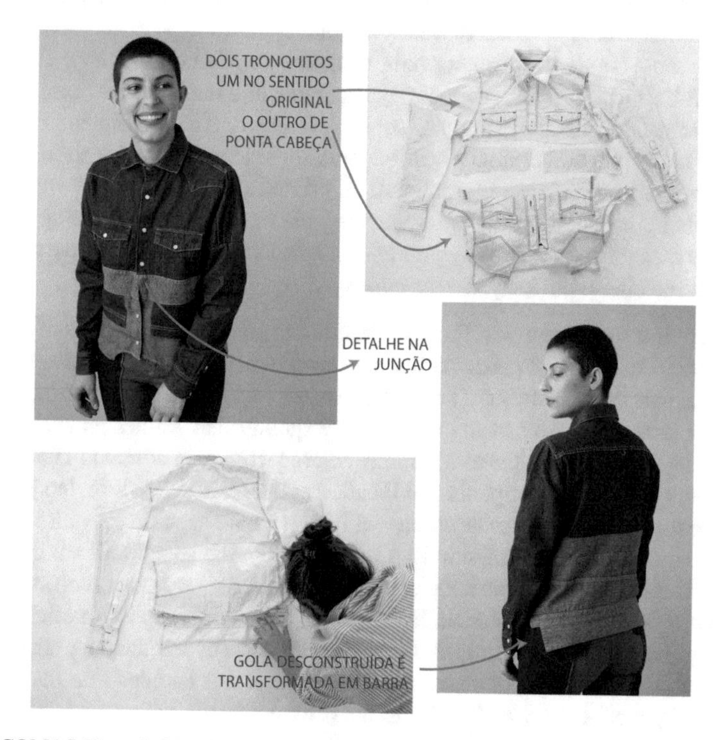

Fig. 6 COMAS Upcycled Denim *Escher* Shirt

CASE III: Estampa Crítica #TEXTOURGENTE

Headquarters: Montevideo—Uruguay
Materials: Pre and Post-consumer garments
Upcycled garments: >200
Working since: 2018

Description: In 2018 I created the proyect *Estampa Crítica* #TEXTOURGENTE (Critical Printing #TEXTOURGENTE), a nomadic upcycling workshop that seeks the exchange and collective reflection on urgent issues that affect our living, and pretends to answer the question: which is the urgent text that must be stated in our community? The project also seeks to reactivate our human ties to fashion, reinforcing the political nature of our garments, using them as a textual support, and conceiving clothing as an active agent and a communicating object. The workshops are organized as follows: first, a controversial discussion topic is chosen to be raised in a particular group of people (students, professionals, or members of a community). The chosen topic must be of high interest for the participants and needs to be communicated beyond the group in order to create social awareness. Then, from the reading of literature, narratives and dialogue, participants share ideas and write short texts that then are printed on a disused garment. The printed garments will be worn by the participants in their daily life, communicating the message and spreading it. In this way, a new meaning is given to the garment, and the user becomes an active consumer, or as Kate Fletcher would say, a *user-maker*.

Why participatory processes: From the beginning, upcycling was used in the workshop as a tool to promote participation and involvement, since collective reflection is the main axis of the project. The exchange with and between the participants has been a fundamental aspect of the workshop, a quality that defines and shapes it. In this way, the workshop adapts to what may emerge from each context (a new topic of discussion, variety of participants and locations, diverse textile waste, etc.), some of these aspects could remain indefinitely, while others are specific to each workshop. As an example, it is worth mentioning that, initially, the social and environmental impact of fashion was the topic node of the project, discussing issues such as overconsumption or fashion obsolescence. However, as different groups of participants got involved, it became evident that what was considered as urgent changes according to the particularities of each context. In consequence, it was decided that the theme of the workshop would vary, giving rise to the different needs of the participants. This enriched the range of possibilities of the workshop, addressing issues such as migration, inequality, or others of social or ecological nature, and exploring territories in Chile, Argentina and Uruguay.

Shared knowledge: A common characteristic in these *participatory upcycling* workshops is that prior knowledge is not necessary. As previously mentioned, the workshop focuses on two activities, on the one hand, the dialogue (critical reflection) and on the other, the upcycling of garments. Upcycling is a means to materialize the first, therefore it is important that the technical aspects associated with printing or garment making do not act as a limiting factor. To make sure of this, I opted for a much simpler and faster stamping method than screen printing: a roll stamping method using masking tape as a reserve. Once the text to be printed has been defined, each participant "writes" it on the garment by sticking the masking tape. This is a very simple thing that allows participants to feel safe without fear of "making mistakes", and even more importantly, allows each participant to reproduce or alter the method to be used or taught in other instances. Each stamp is unique and although they are similar graphically, the variations are infinite, enabling the groups to be encouraged in each instance by the variety of results. The other relevant aspect regarding the generation of knowledge, is that this does not derive only from the facilitators, but from the group reflection and its multiple perspectives.

Game-like tools as a strategy: Something I have learned in the first workshops is that in instances of collective reflection one of the most complex points is to ensure that a fluid and expressive dialogue is generated quickly in the group. Game-like tools are a key piece in this cases since it motivates participation and encourages the action of the participants, breaking the static and the possible distance between the actors. In this manner, different playful strategies, often adaptations of popular children's games, are used in order to generate trust, relax the dialogue and compose the texts (Figs. 7 and 8).

Fig. 7 Workshop #TEXTOURGENTE. Participants: Fashion Industry Workers

Fig. 8 Workshop #TEXTOURGENTE. Participants: Fashion designers and students

Upcycling aesthetics: In the case of *Estampa Crítica* #TEXTOURGENTE the strongest aesthetic component is given by the printing method used. The grayish or black background generated by the textile ink in contrast to the text (which maintains the original color of the fabric) predominates as a graphic element. In this case, a central aspect is that the text (the message) is the protagonist and for this reason many times the garments are not intervened at a formal level, but are only printed, hiding flaws or stains of the garments with tapes or decorative ribbons. As in other cases, the aesthetics change depending on the material with which we work. For example, making upcycling from medical uniforms will generate an aesthetic different from that generated from garments donated by a sportswear brand. The details of the original garments are also kept "alive", showing the origin and valuing the previous existing work. In some cases, the print carefully seeks to enhance the details, like seams and buttons, by increasing their contrast and making them even more visible (Fig. 9).

6 Final Thoughts

The current fashion system is based on a linear production model, characterized by a high consumption of resources and a high generation of waste. In recent years, garment upcycling has gained popularity as a strategy for minimizing the volume of textile solid waste. Despite the fact that upcycling is recognized as a technical

Fig. 9 Exhibition of upcycled garments made after workshop in Centro de Exposiciones SUBTE, Uruguay, 2019

method, it is observed that many of the fashion initiatives that address it respond to a socio-environmental motivation, using the teaching of upcycling as a tool for participatory critical reflection, in order to generate social change.

This generates a very important transformation since it implies that the field of action of fashion magnifies, expanding the universe of concepts that define it as such. Some of the most notorious shifts in fashion are: the incorporation of education into the company's core activities, involvement in activist actions, the inclusion of users (or other actors) in the production process, and the fusion of the role of the designer and the role of the facilitator [5]. This chapter shows three cases of clothing upcycling entrepreneurial designers in Latin America that incorporate these shifts, showing a strong interest in participatory design and in the activation of the critical capacity of individuals.

Expanding the practice of fashion towards collaborative and participatory actions, allows generating new synergies and exchanges between the different agents that make up its universe, and also enable it to be intertwined with other disciplines and areas. We can ask ourselves how these interactions will derive in the coming years, perhaps giving rise to a more equitable and human-centered fashion. A kind of fashion that does not focus on materiality but on the values it represents. According to Fletcher [6] "participatory design and individual and social action will probably define an important component of sustainable fashion and textiles activity into the future", it is our decision to activate this vision in our daily actions. Today, participatory community-based art is already a recognized dimension of contemporary art [12]. Will fashion follow a similar path and become an agent of social change in the near future?

References

1. López-Navarro M, Lozano-Gómez C (2013) Co-creation Experiences as the Basis for Value Creation in the Sustainable Fashion Industry. In: Kandampully J (ed) Customer Experience Management: Enhancing Experience and Value through Service Management, Kendall Hunt, Dubuque, p 133–152
2. Hultman J, Corvellec H (2012) The european waste hierarchy: From the sociomateriality of waste to a politics of consumption. Environ Plan A 44(10):2413–2427
3. Margulis L (1998) Symbiotic Planet. Basic Books, New York, A New Look on Evolution
4. Bodenmann-Ritter C (ed) (1975) Joseph Beuys. Verlag Ullstein GmbH, Berlín, Jeder Mensch ein Künstler
5. Von Busch O (2008) Fashion-able. Hacktivism and Engaged Fashion Design. Dissertation, University of Gothenburg.
6. Fletcher K (2008) Sustainable fashion and textiles. design journeys. Earthscan, London
7. Lopez L (2019) The intangible expansion of the textile object. Available via Lablaco J https://journal.lablaco.com/sustainability/the-intangible-expansion-of-the-textile-object/
8. Bruggeman D (2018) Dissolving The Ego Of Fashion. Publisher Artez Press, Arnhem, Engaging With Human Matters
9. Pardo D, Prado V (2018) Involving customers trough co-creation: An approach from the fashion industry. In: Information Resources Management Association USA. Fashion and Textiles: Breakthroughs in Research and Practice, IGI Global, USA, p 69–98
10. Suri FJ (1990) IDEO: The experience evolution: developments in design practice. The design journal 6(2):39–48
11. Borriaud N (2006) Estética Relacional. Adriana Hidalgo, Buenos Aires
12. Bishop C (2012) Artificial hells: participatory art and the politics of spectatorship. Verso Books, London
13. Brown T (2008) Serious play (video file). Available via TED Ideas Worth Spreading. https://www.ted.com/talks/tim_brown_on_creativity_and_play?language=es#t1679423 Accessed 06 May 2020
14. Lee Y (2008) Design participation tactics: the challenges and new roles for designers in the co-design process. Co-Design 4(1):31–50
15. Cumming D (2017) A case study engaging design for textile upcycling. J Textile Des Res Practice 4:113–128
16. Lopez L (2020) Desplazamientos del Sistema de la Moda. In: Alonso S (ed) Plaza Mario. Proyecto CasaMario Reside en Centro de Exposiciones Subte: Programa Público. CasaMario Ediciones, Montevideo, pp 284–301. Accessed 15 Dec 2019
17. Mc Donough W, Braungart M (2002) Cradle to Cradle: Remaking the way we make things. North Point Press, New York

Printed in the United States
by Baker & Taylor Publisher Services